Daniel Brunner, Miguel C. Soriano, and Guy Van der Sande (Eds.)
Photonic Reservoir Computing

Also of Interest

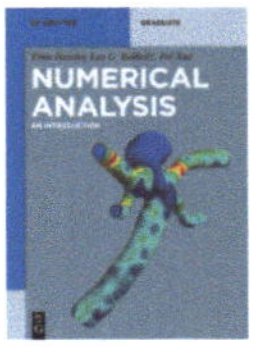

Numerical Analysis. An Introduction
Timo Heister, Leo. G. Rebholz, Fei Xue, 2019
ISBN 978-3-11-057330-5, e-ISBN (PDF) 978-3-11-057332-9,
e-ISBN (EPUB) 978-3-11-057333-6

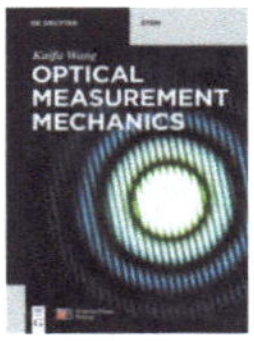

Optical Measurement Mechanics
Kaifu Wang, 2018
ISBN 978-3-11-057304-6, e-ISBN 978-3-11-057305-3

Plasma and Plasmonics
Kushai Shah, 2018
ISBN 978-3-11-056994-0, e-ISBN: 978-3-11-057003-8

Multiphoton Microscopy and Fluorescence Lifetime Imaging. Applications in Biology and Medicine
Karsten König, 2017
ISBN 978-3-11-043898-7, e-ISBN 978-3-11-042998-5

Scientific Computing. For Scientists and Engineers
Timo Heister, Leo. G. Rebholz, 2015
ISBN 978-3-11-030823-5, e-ISBN (PDF) 978-3-11-035942-8,
e-ISBN (EPUB) 978-3-11-038680-6

Photonic Reservoir Computing

Optical Recurrent Neural Networks

Edited by
Daniel Brunner, Miguel C. Soriano, and
Guy Van der Sande

DE GRUYTER

Physics and Astronomy Classification Scheme 2010
Primary: 42.79.Ta, 07.05.Mh, 05.45.-a; Secondary: 89.75.-k, 42.30.-d, 89.20.-a

Editors

Dr. Daniel Brunner
Institut FEMTO-ST
15B avenue des Montboucons
25030 Besancon Cedex
France
daniel.brunner@femto-st.fr

Prof. Dr. Guy Van der Sande
Vrije Universiteit Brussel
Applied Physics Research Group
Pleinlaan 2 1050 Brussels
Belgium
guy.van.der.sande@vub.be

Dr. Miguel C. Soriano
IFISC (CSIC-UIB)
Universitat de les Illes Balears
Carr. de Valldemossa
07122 Palma
Spain
miguel@ifisc-uib-csic.es

ISBN 978-3-11-058200-0
e-ISBN (PDF) 978-3-11-058349-6
e-ISBN (EPUB) 978-3-11-058211-6

Bibliographic information published by the Deutsche Nationalbibliothek
The Deutsche Nationalbibliothek lists this publication in the Deutsche Nationalbibliografie; detailed bibliographic data are available on the Internet at http://dnb.dnb.de.

Cover image: from2015 / iStock / Getty Images
Typesetting: VTeX UAB, Lithuania
Printing and binding: CPI books GmbH, Leck

www.degruyter.com

Preface

This book is devoted to a comprehensive compilation of the first hardware platforms employed for photonic reservoir computing. Reservoir computing is a machine learning technique that is susceptible by design to be implemented in hardware. We here show how it has been successfully ported to a number of photonic platforms, paving the way to all-optical information processing. We dedicate this book to Jan Van Campenhout, who first launched the idea of photonic reservoir computing in 2008.

The first chapter starts with a historical overview, as well as a future-minded motivation for photonic computing. Starting with the oldest known mechanical computer, the reader is guided from the first implementation of optical memory to the controversial topic of all-optical computing logic. This walk through the archives of optical computing naturally arrives at an abstract comparison between electrons and photons in terms of their usefulness within the framework of logic computing. While photonic hardware may have some deficiencies in the perspective of logic computing, it can support the transition from the Von Neumann architecture toward analogue and neuro-morphic computing schemes with parallel computations. In this context, this chapter also includes an introduction to the concept of neural networks. In a series of small but detailed examples, the reader becomes familiar with the basic concepts and relevant parameters which constitute either a feed-forward or a recurrent artificial neural network. The end of this chapter firmly grounds the newly introduced concepts of neural networks and Hopfield networks by discussing their physical counterparts, implemented using photonic hardware.

Chapter 2 starts by contrasting digital computing with analogue computing schemes. A very mature industry exists around digital computing, leading to highly optimized digital computers with wide availability and creating a high entry barrier for any alternative computing scheme. Physical Reservoir Computing stands out as an analogue computing system that uses the natural dynamics of a hardware substrate to perform calculations. This is in stark contrast with most other computing schemes where a computational model is forced onto a substrate that will support it. Reservoir computers are often considered a branch of neural networks. However, it is important to note that they do not necessarily share the same discrete network topology with localized neurons and interconnections. In the last section, the concept of linear and nonlinear memory capacity is thoroughly discussed. These figures of merit represent an important step in the direction of creating generic design methodologies for physical reservoir computers, characterizing them in a task-independent manner.

Chapter 3 introduces the state of the art in reservoir integration, mostly focusing on the realization of passive architectures. In those systems, the chip provides the complex projection of the input data, while the nonlinear transformation is typically created in the optical-electrical signal conversion step via a photo detector. First, the authors introduce the implementation in networks of discrete mixers connected

https://doi.org/10.1515/9783110583496-201

in a swirl topology, followed by concepts based on spatially continuous propagation. These concepts are more abstract in terms of their mapping on the original reservoir architecture consisting of discrete nodes. Based on the introduced systems, the authors demonstrate and discuss a large variety of applications, including high-bandwidth solutions for challenges in the field of telecommunication as well as applications in the field of bio-medicine.

Chapter 4 elaborates on the huge potential of large scale photonic reservoirs. The authors discuss several optical configurations that can generate complex network couplings. Taking advantage of their inherent parallelism, these optical networks can easily be made of thousands of nonlinear nodes. The results in Chapter 4 demonstrate the first large scale photonic reservoir computer that has learning capabilities, with up to 2000 network nodes. In addition, the authors sketch the first implementation of a network of coupled semiconductor lasers. Altogether, this chapter contains a state-of-the-art overview of the properties of photonic reservoirs with diffractive coupling.

Chapter 5 gives a general introduction of reservoir computing based on delay systems. Delay-system RC are among the first successful RC's hardware implementations, lending particular importance to this chapter. Multiple chapters to follow within this book will build upon the foundation laid here, and crucially the authors succeed in preparing this solid base by complementing the information that is to follow in later chapters. The authors start out from fundamental delay-system properties and their relevance to RC computing. From there, they describe the more basic RC-delay implementations and introduce nonlinear system-inspired tools for the analysis of such RC. Finally, they treat fundamental properties of hardware implementations and their consequence for computation.

Chapter 6 elaborates on the relevance of Ikeda delay dynamics for the control and development of photonic reservoir computers. This chapter contains first a historical overview on Ikeda-like dynamics as a paradigmatic toy example of complexity in optics. It then jumps into a detailed description of several hardware implementations of this representative system, namely based on the modulation of the optical intensity, wavelength, or phase. The author focuses on the importance of the different time-scales involved in Ikeda-like hardware implementations for information processing using a space-time analogy. Altogether, this chapter is a must read for any reader interested in the physics of complex systems and how they can be operated as photonic reservoir computers.

Chapter 7 deals with the implementation of photonic RC using semiconductor lasers as the physical substrate. External optical feedback provides the recurrent loop in this approach, following the conceptual basis of delay-based RC presented in Chapter 5. As highlighted by the authors, experimental realizations have been often carried out with single-mode semiconductor lasers. Thus, the authors provide a detailed description of these experiments together with their corresponding numerical modeling using the Lang–Kobayashi rate equations. This chapter also covers other substrates for photonic RC, such as semiconductor ring lasers, erbium-doped microchip lasers, and semiconductor optical amplifiers.

Chapter 8 focuses on how to build advanced photonic reservoir computers, using an optoelectronic system as their workhorse. First, the authors present novel designs to implement fully analog reservoir computers, including analog input and output layers, that are able to function without the support of digital components. The authors then move on to lay out several strategies to train photonic reservoir computers on the fly, such that they can adapt to changing environments. The final section of the chapter focuses on the challenging task of implementing photonic reservoir computers that can operate autonomously. This becomes possible when the output of the system is fed back to its own input, serving as a generator of arbitrarily complex waveforms. Taken as a whole, this chapter represents the spearhead of the development of fully functional photonic reservoir computers.

Contents

List of Contributing Authors

Daniel Brunner
FEMTO-ST, CNRS
15B Avenue des Montboucons
25030 Besançon
France
daniel.brunner@femto-st.fr

Piotr Antonik
LMOPS EA 4423
CentraleSupélec
Université Paris-Saclay
2 Rue Edouard Belin
57070 Metz
France
piotr.antonik@centralesupelec.fr

Xavier Porte
FEMTO-ST, CNRS
15B Avenue des Montboucons
25030 Besançon
France
javier.porte@femto-st.fr

Joni Dambre
Ghent University/imec
Electronics and Information Systems Department
Technologiepark-Zwijnaarde 126
B-9052 Gent
Belgium
Joni.Dambre@UGent.be

Andrew Katumba
Ghent University/imec
Dept. of Information Technology
Technologiepark-Zwijnaarde 126
B-9052 Gent
Belgium
Andrew.Katumba@UGent.be

Matthias Freiberger
Ghent University/imec
Electronics and Information Systems Department
Technologiepark-Zwijnaarde 126
B-9052 Gent
Belgium
Matthias.Freiberger@UGent.be

Floris Laporte
Ghent University/imec
Dept. of Information Technology
Technologiepark-Zwijnaarde 126
B-9052 Gent
Belgium
Floris.Laporte@UGent.be

Alessio Lugnan
Ghent University/imec
Dept. of Information Technology
Technologiepark-Zwijnaarde 126
B-9052 Gent
Belgium
Alessio.Lugnan@UGent.be

Stijn Sackesyn
Ghent University/imec
Dept. of Information Technology
Technologiepark-Zwijnaarde 126
B-9052 Gent
Belgium
Stijn.Sackesyn@UGent.be

Chonghuai Ma
Ghent University/imec
Dept. of Information Technology
Technologiepark-Zwijnaarde 126
B-9052 Gent
Belgium
Chonghuai.Ma@UGent.be

Peter Bienstman
Ghent University/imec
Dept. of Information Technology
Technologiepark-Zwijnaarde 126
B-9052 Gent
Belgium
Peter.Bienstman@UGent.be

Julian Bueno
IFISC, University of the Balearic Islands
E-07122 Palma de Mallorca
Spain
julian@ifisc.uib-csic.es

Sheler Maktoobi
FEMTO-ST, CNRS
15B Avenue des Montboucons
25030 Besançon
France
sheler.maktoobi@femto-st.fr

Louis Andreoli
FEMTO-ST, CNRS
15B Avenue des Montboucons
25030 Besançon
France
louis.andreoli@femto-st.fr

Silvia Ortín
IFISC, University of the Balearic Islands
E-07122 Palma de Mallorca
Spain
silvia@ifisc.uib-csic.es

Luis Pesquera
CSIC-Universidad de Cantabria
Instituto de Fisica de Cantabria
Avd. de los Castros s/n.
E-39005 Santander
Spain
pesquerl@ifca.unican.es

Guy Van der Sande
Vrije Universiteit Brussel
Applied Physics Research Group
Pleinlaan 2
1050 Brussels
Belgium
guy.van.der.sande@vub.be

Miguel C. Soriano
IFISC, University of the Balearic Islands
E-07122 Palma de Mallorca
Spain
miguel@ifisc.uib-csic.es

Laurent Larger
FEMTO-ST, CNRS
15B Avenue des Montboucons
25030 Besançon
France
laurent.larger@femto-st.fr

Serge Massar
Université libre de Bruxelles
Laboratoire d'Information Quantique CP224
Av. F. D. Roosevelt 50
B-1050 Bruxelles
Belgium
smassar@ulb.ac.be

François Duport
III-V Lab
1 avenue Augustin Fresnel
91767 Palaiseau Cedex
France
javier.porte@femto-st.fr

Daniel Brunner, Piotr Antonik, and Xavier Porte

1 Introduction to novel photonic computing

1.1 Photonics for computing

Rather than a feature of modern times, *computing* captivated the interest of humans long before any of the technological revolutions we typically associate to computers. One of the earliest historic realizations of a computer is the Antikythera-mechanism; see Figure 1.1(a), a fascinating example indeed. This intricate device was found in a highly corroded state around a century ago, and was dated back to 200 BC. Due to its corrosion and complexity, deciphering purpose and functionality posed a superb challenge which still continues today. By now, we know that the Antikythera-mechanism calculates moon phases and planetary positions within the lunisolar calendar [1]. Most impressively, the prediction accounts for corrections made necessary by the elliptical orbits of planets and the moon. The *algorithm* executed by the Antikythera-hardware appears to be based on a theory about lunar and planetary motion developed by Hipparchos of Rhodes [1]. The Antikythera-mechanism therefore is the physical implementation of a computing algorithm producing nontrivial and important information on the basis of a simple external input—a date to be set by turning a handle. Transformation of input information is therefore established by a mechanical automatism, which can be re-executed.

Manufacturing this mechanical computer with technology available at the time must have been a formidable challenge. The mechanism consists of at least 37 bronze gears, and such a multistage system most certainly demands high manufacturing accuracy for a reliable operation. We can therefore assume that its creation required the investment of substantial resources in terms of expertise and dedication. Due to agricultural and religious reasons, lunar cycles and the positions of planets were information of high value during that historic period, apparently justifying the associated efforts. This illustrates that already two millennia ago humans highly valued the automatized creation of important information by computers.

Upon close inspection, progress such as the acceleration of computing by a factor of one billion can be identified during the last two millennia. It was understood that computers strongly benefit from practical re-programmability, a functionality Conrad Zuse realized in 1941 with his digital-electronic computer Z3. Interestingly, the Z3 was already applied to complex technological challenges related to mechanical properties of airplane wings, an operation which demands practical *re-programming*. Many more conceptual and technological developments have finally led us to the extremely powerful computers of today. We however intentionally avoid an exhaustive discussion of computing's history. Focusing on our mechanic and electronic computing substrate examples serves the illustration of a general observation: the physical nature of the

https://doi.org/10.1515/9783110583496-001

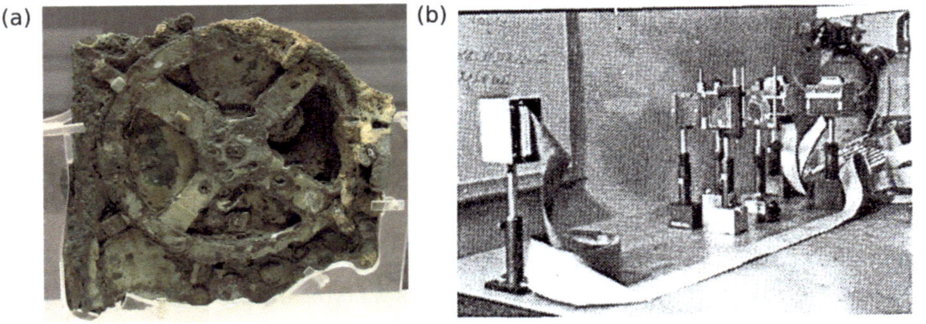

Figure 1.1: (a) Shows a photo of the Antikythera-mechanism, a historic example of a computer created in the second century BC. (b) Shows an optical setup by Farhat et al. [4], creating an optical content addressable memory.

substrate's interactions providing the basis for computing an algorithm changed over time.

One might assume that a transition—or not—between substrates was always in service of further advancing computational performance. However, performance is most likely not the only motive. At least equally important is the technological readiness of a substrate and the presence of a manufacturing platform for its implementation. The last part of this argument leads directly to economic considerations, which today are ever present. These certainly should and cannot be ignored; simultaneously, they should not shroud our vision in the search of next generation solutions and substrates. As the transition from mechanics to electronics set the basis for the current, Turing complete, integrated and powerful computing processors of today, one is equally obliged to seriously consider the possibilities arising from a transition from electronic to photonic computing—or from a merger of both. Arguments of the style *it did not work before, lets discard photonics* ignore (i) its undeniably existing fundamental advantages, recent progress in (ii) novel photonic substrates and (iii) computational schemes. We will show that in particular the advent of artificial neural networks (ANNs) demands a decisively open-minded approach to the question of next-generation computing substrate.

For our purpose, substrates might be grouped according to which particle acts as carrier of information. Limiting our selection to substrates of current technological relevance, we are left with electrons and photons. Nowadays, everybody is familiar with some representative of an electronic computer. Optical computers remain more exotic, yet they are pursued since over five decades [2, 3] due to the potentially high prospects. Multiple concepts have been demonstrated, e. g., Figure 1.1(b) shows an optical content addressable memory by Frahat *et al.* [4] together with Psaltis *et al.* [5]. Such a system is capable to provide the content of stored memory, even if the original query is noise-corrupted or inaccurate.

1.1.1 Benefits of optics for computing

Numerous electronic, electro-optical and all-optical computing concepts have been demonstrated in proof-of-concept experiments. The result is a wide selection of concepts and hardware substrates, which brings us to the question about the fundamental advantages and disadvantages of particular substrates. In 1990, Lohmann approached the problem by asking what in principle is required from a substrate for realizing computation, establishing a connection between particular substrates and their suitability to implement fundamental computing operations [6]. This approach places our generic discussion on a solid foundation. On a most basic level, Lohmann identified *logic / interactions* and *transport / communication* as essential ingredients, certainly a view biased toward computing based on binary logic operations. Paying tribute to the recent impact of ANNs and simultaneously to our book's title, we will modify the first category to *nonlinear transformations*, which include logical operations. One can quickly enter the realm of philosophical discussions about what is computing and what is not. While these discussions can be enlightening, they can also obscure the practical implications of scientific arguments. An interesting and rigorous contribution to discussions of this kind can be found in Horsman *et al.* [7]. Here, we will restrict computing to operations which include nonlinearity at some stage; other operations involved in information processing we will simply refer to as transformations.

Electrons and photons each have unique physical properties, and hence favor the physical realization of different processes. In 2013, Shamir [8] linked these fundamental differences to the doom of *digital* optical computing. However, we would like to elaborate the subject from a slightly different angle. We will show that novel-material systems might open new avenues to pursue. Most importantly, we emphasize that neural networks (NNs) do *not* rely on digital logic and that fundamental differences to the architecture of logical computers will demand different properties and scaling from a substrate.

Due to their fermionic nature and electric charge, electrons strongly interact. Electrons are therefore excellently suited to induce nonlinear responses, and consequently to perform nonlinear transformations. At the same time, these strong interactions reduce the suitability of electrons for analog information transport: originally encoded information will be tainted by electron-electron interactions, e. g. induction between adjacent interconnects and with the substrate. Furthermore, the ever-present electromagnetic permittivity unavoidably links an electronic communication line to a capacitance. Combined with the information carriers' charge, this limits the modulation bandwidth of electronic transmission lines. In Table 1.1, electrons therefore are listed as excellently suited to implement nonlinear transformations, but only as mediocre for information transmission. This has direct implications for the implementations of ANNs and their ultra large-scale connectivity. More details on the limits and their practical implications of modern electronic computation can be found here [9, 10].

Table 1.1: Suitability of electronic, photonic and exotic/hybrid substrates to physically implement fundamental operations of computing. Table adapted from [6].

	Electrons	**Photons**	**Exotic/hybrid substrates**
Nonlinear transformations	√√√	×	√√√
Information transport	√√	√√√	√√

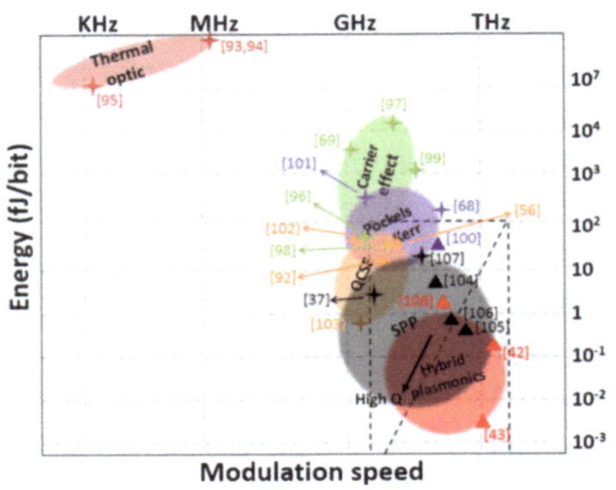

Figure 1.2: Switching energies of different nonlinear optical effects. The figure was taken from [13]. Today's lowest switching energies reach the atto-Joule regime, making them compatible with current semiconductor technology [14].

Photons, on the other hand are charge-less bosons. Consequently, they do not interact directly for optical intensities relevant to our discussion. Only a physical substrate acting as mediator can introduce such action. This implies that the electric field (magnetic fields can mostly be ignored) of one photon first needs to modify the mediating medium's properties, which in turn modifies the properties of other photons. Unfortunately, interaction coefficients of traditional materials are rather small, and when compared to electronics the result is a bad rating for nonlinear transformation in Table 1.1. On the other hand, signal transmission strongly benefits from the very same aspects: as long as physical properties allow the separation of signals, i. e., using photon polarization or wavelength, a large number of communication channels can be transmitted along a single line without significant interaction. Furthermore, in contrast to the electronic case, the length of a transmission line does not limit the modulation bandwidth. Photons are therefore excellently suited for communication, which we experience on a daily basis: modern communication and the Internet largely rely on optical information transport. The rating in Table 1.1 for optical information transport is therefore excellent.

Motivated by the frustratingly linear behavior of photons, countless strategies were developed in order to bring nonlinearity to photonics. A highly active field of research is the use of advanced materials and novel waveguide geometries. Figure 1.2 gives a generic overview of different nonlinear effects, their required switching energies and their maximum modulation bandwidths. Recently, this diagram was successively extended to faster and more energy efficient processes. Originally, optical nonlinearities were mostly realized based on the Pockels and Kerr effect. In waveguides of 1 mm length, modern structures based on these processes require ~ mW to create one period of a $\sin^2$ nonlinearity all-optically [11, 12]. However, nonlinearity does not need to be produced all-optically. Exploiting plasmonic effects, switching energies for an electro-optical modulation are now approaching the atto-Joule regime at potentially THz bandwidths [13]. Such switching energies are even competitive with current semiconductor technology [14], and with simple incorporation of a photosensitive element one could create efficient all-optical nonlinearities. This shows the great potential of novel photonic technology to implement the nonlinear transformations for optical computing.

Judging from the past developments and research in computing, photonics and electronics will remain locked somewhere between competition and synergy. Optical computing is a field with a surprisingly long history; the first books were already published in the early 1970s [15]. For now, electronics clearly wins the day by providing cheap high-performance circuitry and decades of exponential scaling in number of transistors per chip and per dollar. Equally, integrated photonic components for computing were already considered in the 1990 by Laval *et al.* [16], yet for now they cannot match the energy efficiency and area-footprint of electronics. However, this conversation is strongly biased toward the field a majority identifies (maybe wrongly so) synonymous with computing: Boolean logic in a mostly serial architecture. Already today, a standard CPU's performance limit is not imposed by the size or speed of individual transistors, but by the energy it consumes inside the small volume of its substrate. By now, a significant fraction of dynamic energy dissipation originates from interconnections [17, 18], and in that light the advent of neural networks creates a fundamental shift in demands placed upon a computing core. Providing some numbers: currently the switching energy of a CMOS transistor gate lies somewhere in the range of 40 aJ–3 fJ; switching an interconnect wire of 10 mm length consumes 600 fJ ([19] and references therein). Connections in a classical von Neumann computer are mostly short ranged, yet they already limit scaling of these chips. The impact interconnects exert upon the energy budget of current architectures will be dwarfed by neural networks (NN), whose fundamental feature *is* large scale connectivity. Combined with recent advances in integrated photonics, exploiting the fundamental advantage of optics over electronics for signal transmission could prove essential for next generation NN hardware. These considerations are confirmed by a modification in the electronic-chip industry's hardware development. Fueled by the growth of NN applications, Google even developed its own hardware platform, the Tensor Processing Unit (TPU) [20]. At its heart lies a systolic-array circuit, which is better suited for calculations of large-scale vector-matrix products. Yet, the device is not capable of carrying out such operations fully in parallel—something demonstrated in optics since decades [21]. We therefore conclude that the transformative impact of NN-concepts extends deeply into fundamental properties of a processor's architecture, and will therefore drive future computing hardware development.

All these—unfortunately quickly—approaching ANN-hardware roadblocks should make us seriously reconsider photonics as an alternative or complimentary technology. Photonics offers crucial advantages addressing many of the discussed key issues. Today we cannot seriously contemplate mechanical computers. The same might be true in the future for ANN-processors with their almost countless interconnects. An open-minded approach still striking a balance between real-world potential as well as long term *possibilities* is therefore paramount.

1.1.2 Logical optical computers

Owing the success of computing based on Boolean logic operations, optical logic for computing received substantial attention from an early stage. Among others, a large European consortium within the *European Joint optical Bistability Project*, consisting of 18 universities, explored the field in detail [22]. Using photonics instead of electronics was mainly motivated by three factors. First, avoid the opto-electronic conversion. Communication is carried out optically, today even more than at the time of first interest in optical logic. Interfacing with an electronic computer therefore fundamentally requires opto-electronic conversions, and vice versa. Among others, these conversions are associated to an increased power-consumption. The second factor is the potentially high bandwidth of optical processes, readily achieving sub-pico second switching times. One has to acknowledge that today's transistors can also reach such switching speed [14]. Yet, for full systems, one again needs to consider the bandwidth limitations induced by electronic interconnects, and today this limit lies in the GHz regime. Finally, already at the field's foundation, a clear focus was placed on the parallel and spatially distributed nature of optical signal transmission [23]. Highly parallel circuitry was therefore an important objective of optical logic computers.

Implementing Boolean logic based on a latching optical bi-stability was already proposed in 1969 by Szöke *et al.* [2]. In their publication, they discussed how a saturable cavity can stably be switched between high or low transmission states using the signal of an external laser source(s). Generally, systems supporting such bistable switching behavior are based on nonlinearity combined with some feedback or coupling mechanism. Such an interaction term can either be a dedicated feedback channel, or as is the case for the saturable absorber cavity, the constant and mutual interaction between light and material [24]. Firth [24] introduces a generic model for such bistability in the chapter on *Theory of optical Bistability and optical Memory* in [25]:

$$\frac{dI}{dz} = -\alpha(\varphi)I \tag{1}$$

$$\frac{d\varphi}{dt} = I_0 f(\varphi) - \Gamma(\varphi - \varphi_0) = A(\varphi) - B(\varphi). \tag{2}$$

Here, I is optical intensity, φ (φ_0) is the device (ambient) temperature, α a temperature-dependent absorption coefficient, $f(\varphi)$ is the nonlinear light-matter interaction. This

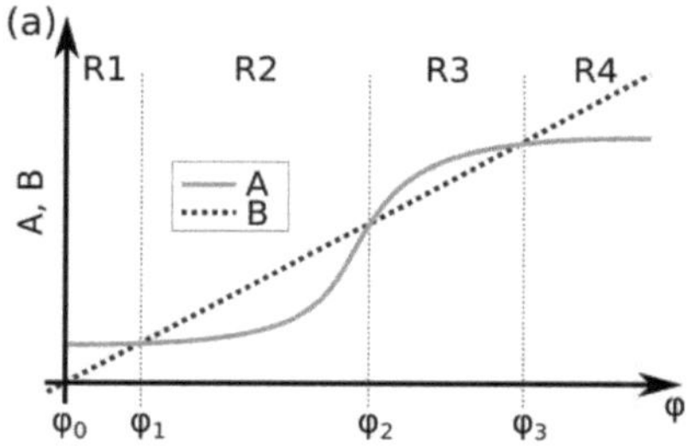

(b)

	φ	$d\varphi/dt$	φ_s
R1	$\varphi<\varphi_1$	>0	φ_1
R2	$\varphi_1<\varphi<\varphi_2$	<0	φ_1
R3	$\varphi_2<\varphi<\varphi_3$	>0	φ_2
R4	$\varphi_3<\varphi$	<0	φ_2

Figure 1.3: Schematic illustration of optical bistability. Panel (a) shows the system's nonlinear ($A(\varphi)$) and linear ($B(\varphi)$) response to changes in φ in equation (2). Fixed points are located where linear and nonlinear terms are balanced. Given by the sign of $\dot{\varphi}$ in their vicinity, these are stable or unstable fixed points. A summary of the relevant properties are given in panel (b).

example is based on a nonlinearity induced by temperature changes due to optical absorption, but this model can simply be extended for, e. g., saturable absorbers. In Figure 1.3, we schematically illustrate the underlying mechanism. In panel (a), $A(\varphi)$ is the nonlinear and $B(\varphi)$ the system's linear response, respectively. Panel (b) identifies the characteristic regions of the system. According to equation (2), if initialized in region R1 (R3), the gradient $d\varphi/dt$ is positive and the system is attracted towards φ_1 (φ_3). For region R2 (R4), the gradient is negative, again resulting in attraction towards φ_1 (φ_3). As the gradient at φ_1 and φ_3 is zero, both values are attracting fixed points or stable steady states of the system if $A(\varphi)$ and $B(\varphi)$ agree with the general properties highlighted in Figure 1.3(a). For the case of using two external input signals, this configuration can implement the fundamental logic *AND* or *OR* operations. When realizing an optical *OR* logic-gate, each external input would have to provide a power, setting the system into $\varphi > \varphi_2$, while in the *AND* system only the power of both inputs combined would set the system to that point. For a saturable-cavity system, the result is the implementation of two fixed points corresponding to low or high optical transmission. Such optical bistability was reported based on a Fabry–Perot cavity filled with sodium vapor [26] or an InSb semiconductor [27]. Smith *et al.* [23] provides an overview of the demonstrated systems and possible configurations. Finally, the cascading of multiple such elements was also realized [28], demonstrating restoring digital logic.

Today, the prospects, challenges and feasibility of computing with optical logic are still being discussed intensely and controversially. An illustrating example is a commentary article by Caulfield *et al.* [29], followed by a series of comments by Tucker *et al.* [30] and Miller *et al.* [31] on optical transistors. Realizing an optical transistor, the principle unit for all-optical computing logic, is highly nontrivial and a device relevant for real-world applications, Miller highlights multiple criteria which must be fulfilled [32]. Owing this complexity, only one large-scale and multistage all-optical logic computer has been reported [33]. The system consists of six cascaded stages, each stage home to a 32×32 array or bistable optical elements; see Figure 1.4. While it is certainly

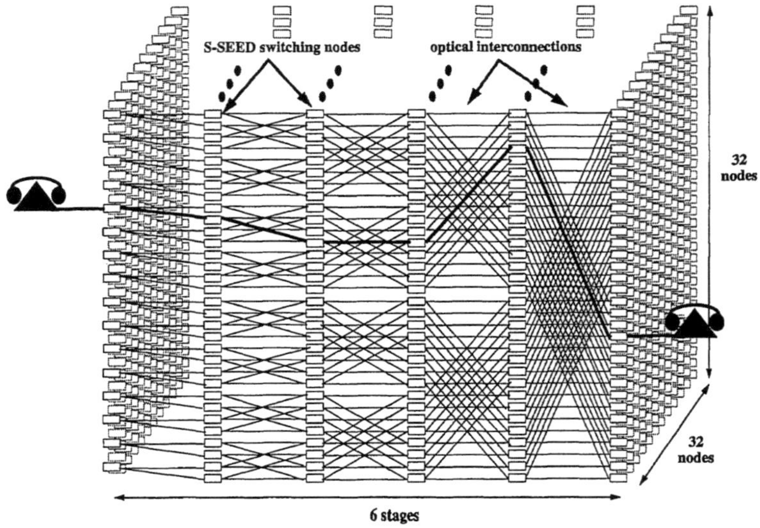

Figure 1.4: Schematic illustration of a 6-stage 32 × 32 array of bi-stable optical elements. Figure taken from [33].

true that such a system-dimensionality is insufficient for a computer based on logic operations, 6 layers of 1024 nonlinear nodes are already reaching a sufficient scale for interesting machine learning applications. Finally, in support of ongoing interest into optical computing, Miller precisely highlights the arguments of interconnects that we previously leveraged by pointing out their increased importance for future NN computing substrates.

1.1.3 Optical computing with spatial transformations

A different approach to computing is fundamentally based on the spatial nature of optical imaging. A single optical lens transforms an arbitrarily complex optical waveform simultaneously. All information contained within a lens's field of view and resolvable by its impulse response function is therefore transformed fully in parallel, potentially resulting in a very large space-bandwidth product. Other than with serial optical binary logic, transformation of spatially distributed optical information therefore leverages fundamental advantages of optics by exploiting its massive spatial parallelism. Of fundamental importance in schemes following this approach is often the Fourier transformation property of the simple optical lens [34, 35]. If an object is placed in front of a lens at focal distance f, hence in its focal plane, then the object's spatial Fourier transformation is created at f behind the lens. An important limitation is that the Fourier transformation of a lens is only exact in amplitude and phase within the area where the paraxial approximation is satisfied [35].

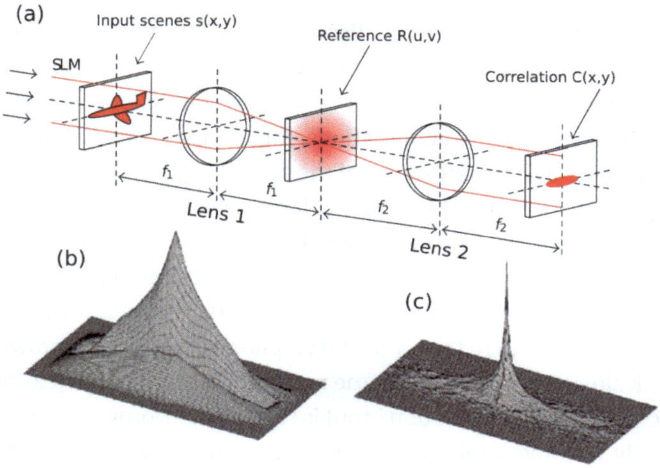

Figure 1.5: (a) An optical correlator based on free-space optics. A scene $s(x,u)$ is placed in the focal plane of Lens 1 and illuminated by a plane wave. The object's Fourier transformation is then present in the back-focal plane of the lens. If an object of interest's complex conjugate Fourier transformation ($R(u,v)$) is placed in the same plane, then Lens 2 carries out a convolution between the Fourier spectrum of $s(x,y)$ and $R(u,v)$. In case that $s(x,y)$ corresponds to the object of interest, then the convolution results in a large correlation signal. Panel (b) shows the autocorrelation peak for amplitude and phase, (c) only for phase matched reference $R(u,v)$. Figure taken from [3].

A popular application of this optical transformation is realizing optical convolution, which was demonstrated by Weaver and Goodman [36] and later even applied to support numerical computation by Psaltis *et al.* [37]. In Figure 1.5, we schematically illustrate how this concept can be extended to realize object recognition via an optical correlator. The object to be analyzed $s(x,y)$ is placed at position $z = 0$, corresponding to one focal distance f in front of the first lens, and is coherently illuminated via a plane wave. At position $z = 2f$, the resulting Fourier spectrum is multiplied with a filter or reference $R(u,v)$. Here, $R(u,v)$ is the complex conjugate Fourier spectrum of the identification target with u and v the spatial frequencies in x and y, respectively. This arrangement is followed by a second lens placed at $z = 3f$, which transforms the multiplication at position $z = 2f$ into a convolution at $z = 4f$. The resulting correlation between the object and the classification target is then provided at a distance $z = 4f$ away from the object, hence the setup illustrated in Figure 1.5 is typically referred to a $4f$-correlator. Figure 1.5(b) and (c) show the resulting spatial correlation for the case reference $R(u,v)$ provides amplitude and phase or only phase information, respectively. The optical correlator was applied to relevant tasks such as road-sign recognition [38]. One severe limitation of this approach is its sensitivity on tampering with the input data. The 4f-optical correlator is only invariant against spatial translation of the optical input, in which case only the position of the correlation peak would be shifted at position $z = 4f$. Rotation and stretching, on the other hand, will reduce the correlator's performance. Beyond the discussed system, multiple different information processing applications realized by various arrangements of lenses, filters, etc. are derived and demonstrated by Cutrona *et al.* [39]. While most of such spatial-transformation concepts utilize linear operations, they can be extended by incorporating nonlinear responses at different stages of the setup [40].

Other more fundamental mathematical operations have been demonstrated via optical techniques, too. One example is the subtraction of two images based on Stokes reversibility, which was realized via phase-conjugate Michelson interferometry [41]. Another classical operation is the optical implementation of matrix products [21], also based on a systolic array approach [42]. In the initial development of the field, a nowadays surprising bottleneck slowed down progress: the nonavailability of high-quality and practical input devices [3]. Since information input is typically a two-dimensional image, this requires readily reconfigurable screens. Only after the first spatial light modulator (SLM) was demonstrated by Labrunie [43], this hurdle was mostly overcome.

1.2 Neural networks

Artificial neural networks (ANNs) are composed of nonlinear computational elements, operating in parallel and arranged in a manner similar to biological neural interconnections [44]. These models have been extensively studied with the aim of achieving human-like performance in the field of pattern recognition. Nowadays, neural networks are mainly considered from two perspectives: cognitive science—an interdisciplinary study of the mind, and connectionism—a theory of information processing [45]. In the present book, neural networks are considered for designing photonic systems. Therefore, the questions relating their design and purpose with how the brain might work are considered out of scope.

The pioneers of the field—McCulloch and Pitts—studied, in the early 1940s, the potential of the interconnection of a model-neuron [46]. Later in 1949, Donald Hebb proposed a learning rule for adapting the connections of artificial neurons [47]. The name "perceptron" was coined by Rosenblatt in 1958 [48], who developed the theory of statistical separability. In 1969, Minsky and Papert (1969) provided a rigorous analysis of the perceptron [49] and temporarily slowed down the development of the field by demonstrating that perceptrons were incapable of processing the basic exclusive-or (XOR) circuit. The field was revived by Werbos, who, in 1971, developed the backpropagation learning algorithm, published in his doctoral thesis [50, 51].

Neural networks are composed of elementary computation units, i. e., the neurons. A biological neuron is a cell capable of producing a rapid train of electric spikes. Simulating its complex internal dynamics (using, e. g., the Hodgkin–Huxley model

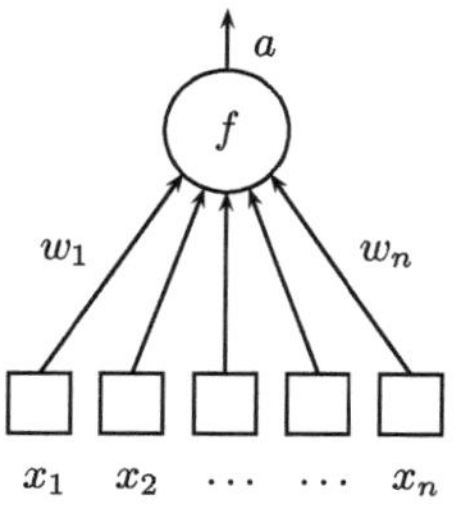

Figure 1.6: Scheme of an analog neuron, represented by a nonlinear function $a = f(x_i, w_i)$ of inputs from other neurons x_i and weights (or parameters) w_i. Adapted from [57].

[52]) is impractical for real-world applications of ANNs. For this reason, artificial neurons have been introduced, keeping the spiking behavior but greatly simplifying the internal dynamics (see, e. g., [53–56]). Further simplification of the artificial neuron model is possible by ignoring individual spikes and introducing an average spiking rate a. Such neurons are called *analog neurons* and their behavior is described by

$$a = f\left(\sum_i w_i x_i\right), \tag{3}$$

where a is the output of the neuron (also referred to as the current state of the neuron, or the activation), x_i are the inputs from other neurons in the network, w_i are the weights of these connections, and f is the activation function. A common choice for the latter is a sigmoid function (i. e., a s-shaped function), such as the tanh function or the inverse tangent function [44].

In summary, a neuron, schematized in Figure 1.6, is a nonlinear function, parameterized by the coefficients w_i, often called *weights* or, stemming from the biological origins, *synaptic weights*. One should be careful with the term "connection" and take it metaphorically. In most hardware applications of artificial neural networks, the neurons are not physical objects. Instead, they could be implemented electronically, for instance, in silicon, or time-multiplexed in an analog signal. Therefore, connections between neurons rarely have any actual existence, and merely indicate how individual neurons, i. e., the hardware blocks that implement them, are connected and interact with each other.

Neural networks come in two classes: feedforward and recurrent (or feedback) networks, which are the topics of Section 1.2.2 and Section 1.2.3, respectively. But first, we introduce their main predecessor—the perceptron.

1.2.1 Perceptron

Figure 1.7 shows a toy example of a binary classification task with two classes separable by a hyperplane. Included in the figure are two instances of infinitely many possible hyperplanes, or in this case, straight lines. Classifiers that take a linear combination of the input features and produce a binary output were called *perceptrons*

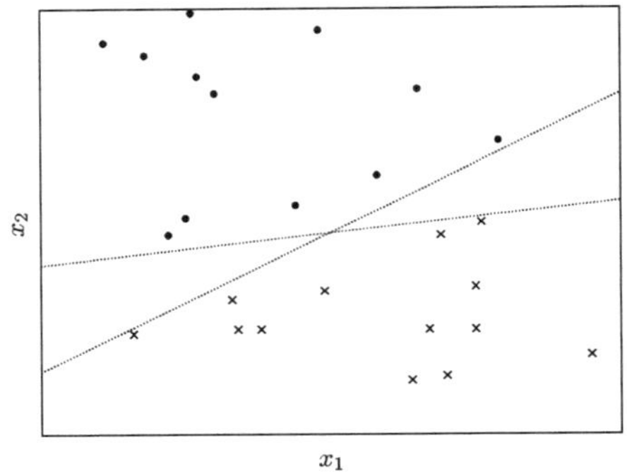

Figure 1.7: A toy example of a linearly separable classification problem, with two of the infinitely many possible separation hyperplanes.

[48, 58]. Perceptrons set the foundations for the neural network models of the 1980s and 1990s [59].

The *perceptron learning algorithm* is a binary classifier that looks for a separation hyperplane by minimizing the distance of misclassified points to the decision boundary. The real-value input vector $x \in \mathbb{R}^n$ is mapped to a binary value $y(x)$ following:

$$y(x) = \begin{cases} 1 & \text{if } w \cdot x + w_0 > 0, \\ -1 & \text{otherwise,} \end{cases} \tag{4}$$

with

$$w \cdot x = \sum_{i=1}^{n} w_i x_i,$$

where $w_i, \ldots, w_0$ are the parameters of the separating hyperplane, also called a decision boundary. In the context of neural networks, a perceptron is an artificial neuron with the Heaviside activation function. The present version of the algorithm (equation (4)) is commonly termed a *single-layer perceptron*, to distinguish it from a multilayer perceptron, which corresponds to a more complicated neural network. As a linear classifier, the single-layer perceptron is a simple feedforward neural network (that will be presented in Section 1.2.2).

The search for a decision boundary can be carried out through the stochastic gradient descent algorithm, starting from a random guess [59]. If a response $y(x) = 1$ is misclassified, then $w \cdot x + w_0 < 0$, and vice versa for a misclassified response $y(x) = -1$. The error to minimize is defined by

$$E(w_i) = -\sum_{j \in \mathcal{E}} y_j (w \cdot x + w_0), \tag{5}$$

where $\mathcal{E}$ is the set of misclassified points. The gradients are given by

$$\frac{\partial E}{\partial w_0} = -\sum_{i \in \mathcal{E}} y_i, \tag{6}$$

$$\frac{\partial E}{\partial w_{i\neq 0}} = -\sum_{j\in\mathcal{E}} y_j x_i, \tag{7}$$

and the parameters w_i are updated recursively by visiting each misclassified input in $\mathcal{E}$ and applying the gradients as follows:

$$\begin{pmatrix} w_0 \\ w_1 \\ \cdots \\ w_n \end{pmatrix} \leftarrow \begin{pmatrix} w_0 \\ w_1 \\ \cdots \\ w_n \end{pmatrix} + \lambda \begin{pmatrix} y_i \\ y_i x_1 \\ \cdots \\ y_i x_n \end{pmatrix}, \tag{8}$$

where λ is the learning rate, which in this case can be taken to be 1 without loss of generality [59].

If the problem is linearly separable, the algorithm converges to a separating hyperplane in a finite number of steps [59]. Two solutions, obtained from different random guesses, are shown in Figure 1.7.

The perceptron learning algorithm presents a series of downsides. First, a single-layer perceptron can only solve a linearly separable problem. If the inputs are not linearly separable, the algorithm will not converge and develop cycles. The cycles can be long and, therefore, hard to detect. The most famous example is the perceptron's inability to solve the Boolean XOR problem. Second, for a linearly separable problem, there are many solutions, and the result of the algorithm strongly depends on the starting values, that is, the initial random guess. Finally, the proof of convergence in a finite number of steps does not guarantee a reasonable convergence time—the smaller the gap, the longer the time to find it [59].

Perceptrons found many applications in speech or image recognition in the 1980s, but have been superseded by much simpler support vector machines [60–63].

1.2.2 Feedforward neural networks

A feedforward neural network is a function of its inputs, and can be seen as a composition of the functions of its neurons [57]. In most cases, the former function is nonlinear, as are the individual functions of the neurons. However, in particular cases, a linear function can be chosen for the neurons, which would result in a linear feedforward network.

Graph representation of a neural network is an intuitive and effective way of visualising the structure of the system. In such a graph, the neurons are vertices and the edges correspond to the connections. The graph of a simple feedforward neural network is shown in Figure 1.8. By definition, the graph is acyclic, i. e., no path in the graph forms a closed loop. The neurons that perform the final computations and produce the outputs of the network are called *output neurons*. The other neurons, which perform intermediate computations, are called *hidden neurons*. The inputs (squares

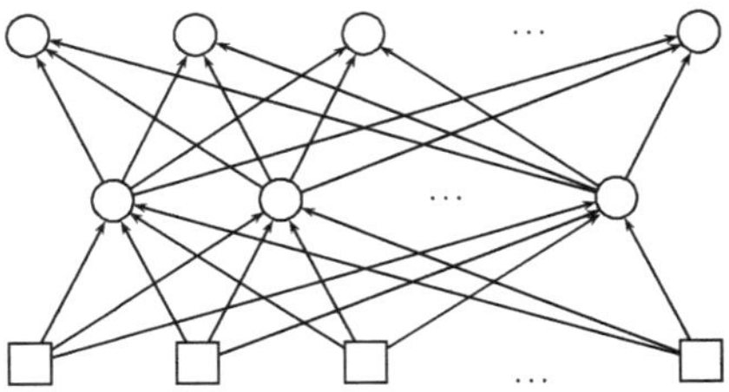

Figure 1.8: Graph representation of a feedforward neural network with n inputs, one layer of N_h hidden neurons, and N_o output neurons. Adapted from [57].

in Figure 1.8) and input neurons (first layer of circles, connected to the input squares) are often confounded in the literature. This may seem confusing because, technically, the inputs are not neurons: they do not perform any computation, but simply deliver the input variables to the neurons.

A large number of different network topologies can be imagined, under the sole constraint of an acyclic graph. However, most of neural network applications implement multilayer networks, as illustrated in Figure 1.8 [57].

A feedforward network with n inputs, N_h hidden neurons and N_o output neurons computes N_o nonlinear functions of its n input variables. These are compositions of the N_h functions computed by the hidden neurons. Owing to the acyclic graphs, feedforward networks are static. That is, if the inputs are constant, so are the outputs. For this reason, feedforward neural networks are often termed static networks, as opposed to recurrent or dynamic networks, which will be described Section 1.2.3. Feedforward multilayer networks with sigmoid nonlinearities are often termed *multilayer perceptrons*, or MLPs.

To illustrate the notions above, consider an example of a feedforward neural network with a single layer of nonlinear hidden neurons (with a sigmoid activation function) and a single linear output neuron. This example corresponds to a class of feedforward neural networks that is particularly important in practice [57].

The output of this network is given by

$$\begin{aligned} g(\boldsymbol{x}, \boldsymbol{w}) &= \sum_{i=1}^{N_h} \left[w_{N_h+1,i} \tanh\left(\sum_{j=1}^{N_k} w_{ij} x_j + w_{i0} \right) \right] + w_{N_h+1,0} \\ &= \sum_{i=1}^{N_h} \left[w_{N_h+1,i} \tanh\left(\sum_{j=0}^{N_k} w_{ij} x_j \right) \right] + w_{N_h+1,0}, \end{aligned} \tag{9}$$

where $\boldsymbol{x}$ is the input vector of $N_k + 1$ input variables, and $\boldsymbol{w}$ is the weights vector of $(N_k+1)N_h+(N_h+1)$ parameters. As introduced above, the hidden neurons are numbered from 1 to N_h, and the output neuron is labeled $N_h + 1$. Conventionally, the weight w_{ij} is assigned to the connection that links the neuron j (or the network input j) to neuron i. The input x_0 is usually set to a constant value $x_0 = 1$ and used as a *bias* term for all the neurons in the network, with the corresponding weights w_{i0}.

From equation (9), one notes that the output $g(\boldsymbol{x}, \boldsymbol{w})$ of the network is a *linear* function of the parameters of the last connection layer (from the N_h hidden neurons to the output neuron $N_h + 1$), and a *nonlinear* function of the parameters of the first layer of connections (from the $N_k + 1$ inputs of the network to the N_h hidden neurons). Therefore, the output of a multilayer perceptron is a nonlinear function of its inputs and of its parameters.

1.2.3 Recurrent neural networks

Recurrent neural networks represent the most general neural network architecture [57]. Their connection graph contains at least one path that forms a closed loop, i. e., following the connections, one returns back to the starting neuron. Such a path is called a *cycle*. Since the output of a neuron cannot be a function of itself, time must be explicitly taken into account for such architectures. In other words, the output of a neuron cannot be a function of itself at the same moment of time, but it can be a function of its past values.

Given the predominance of digital systems for hardware applications of (e. g., standard computers or dedicated digital circuits), discrete-time systems are the natural framework for studying recurrent neural networks. Therefore, recurrent neural networks are often described by recurrent equations (hence the name), which are discrete-time equivalents of continuous-time differential equations.

In this discrete-time framework, each connection is assigned two parameters (one more than for feedforward neural networks): a weight and a delay (possibly equal to zero). Delays are integer multiples of the elementary time unit. For causality reasons, the sum of the delays in a cycle in the graph of a causal recurrent neural network must be nonzero.

An example of a recurrent neural network is shown in Figure 1.9. The delays assigned to the connections, expressed as integer multiples of a time unit T, are marked in the diamonds. The network is casual, since the only cycle from neuron 1 back to itself through neuron 2 features a nonzero sum of delays.

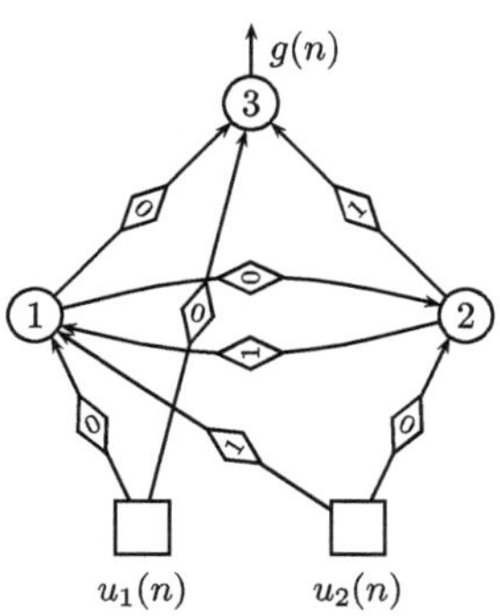

Figure 1.9: Graph representation of a recurrent neural network with two inputs. Digits in diamonds indicate the delays assigned to each connection, expressed in terms of the unit time or sampling period T. The network holds one cycle: from 1 to 1 through 2. Adapted from [57].

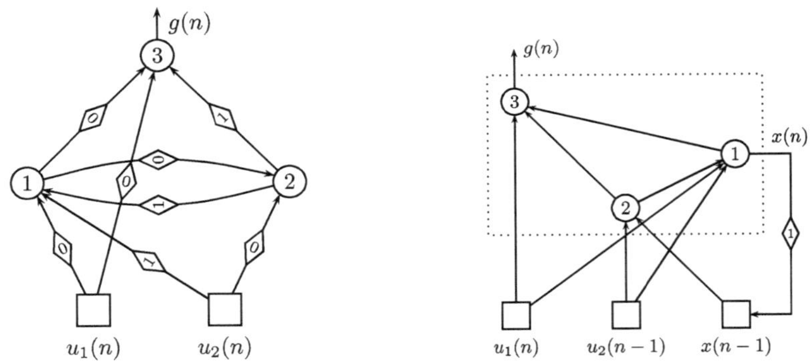

Figure 1.10: The canonical form (right-hand side) of the network shown on Figure 1.9 (left-hand side). The dotted line highlights the feedforward neural network within the canonical form. Adapted from [57].

Similar to feedforward networks, we present the general mathematical description of recurrent neural networks. The general equations of a linear system are

$$\boldsymbol{x}(n) = A\boldsymbol{x}(n-1) + B\boldsymbol{u}(n-1), \tag{10a}$$

$$\boldsymbol{g}(n) = C\boldsymbol{x}(n-1) + D\boldsymbol{u}(n-1), \tag{10b}$$

where $n \in \mathbb{Z}$ is the discrete time, expressed in terms of the unit time or sampling period T, so that $t = nT$, $\boldsymbol{x}(n)$ is the state vector at time nT, $\boldsymbol{u}(n)$ is the input vector, $\boldsymbol{g}(n)$ is the output vector, and A, B, C, D are matrices. Similarly, the canonical form of a nonlinear system is defined as

$$\boldsymbol{x}(n) = \phi[\boldsymbol{x}(n-1), \boldsymbol{u}(n-1)], \tag{11a}$$

$$\boldsymbol{g}(n) = \Psi[\boldsymbol{x}(n-1), \boldsymbol{u}(n-1)], \tag{11b}$$

where ϕ and Ψ are nonlinear vector functions (e. g., neural functions). Nerrand *et al.* [64] have demonstrated that any recurrent neural network, however complex, can be expressed in a canonical form, made of a feedforward neural network, some outputs of which (termed state outputs) are fed back to the inputs through unit delays [57].

For instance, the neural network in Figure 1.9 can be transformed into the canonical form, shown in Figure 1.10. That network has a single state variable: the output of neuron 1. In that example, neuron 1 is a hidden neuron, but in general, a state neuron can also be an output neuron. Further details on recurrent neural networks and their canonical forms can be found in [57].

1.2.4 Deep neural networks

Until recently, most artificial neural network methods had exploited shallow-structured architectures, that typically contain a few (one or two, at most) layers of nonlinear neurons [65]. Shallow networks have been shown effective in solving many simple problems. However, their limited modeling and representational power can cause difficulties when tackling real-world applications involving complex data such as human speech, language, and natural visual scenes. Biological information processing mechanisms in the human brain, such as vision and audition, which are clearly equipped with layered hierarchical structures [66], suggest the need of deep architectures for building accurate representations from complex and rich data.

Historically, deep learning originated from the field of artificial neural networks. That is, multilayer perceptrons (or feedforward neural networks, introduced in Section 1.2.2) with many layers of hidden neurons are often referred to as deep neural networks (DNNs). At the moment of writing these lines, no clear boundary between shallow and deep architectures, in terms of the number of layers, could be found in the literature.

The back-propagation (BP) algorithm, mentioned in Section 1.2, was used for learning the optimal parameters of these deep networks. However, BP alone only did well in practice for networks with a few hidden layers [67, 68]. The main problem was the high number of local optima in the nonconvex cost function of the deep networks. The BP algorithm, based on the gradient descent method, starting from some random initial point, often gets trapped in a poor local optimum. The severity increases significantly with the depth of the network. This issue is partially responsible for steering the machine learning research away from deep architectures toward shallow models with convex loss functions with an easily accessible global optimum.

The above issue was empirically solved when unsupervised learning algorithms were introduced in [69, 70]. A new class of deep generative models, called deep belief network (DBN), was introduced. A DBN is a deep neural network with connections between the layers but not between the units within each layer [69, 71]. It is trained in two phases. At the first *unsupervised* phase, it learns to reconstruct its inputs, so that the layers could act as feature detectors. At the second *supervised* stage, the system learns to perform classification.

Initializing the weights of a deep neural network or MLPs with a correspondingly configured DBN often produces much better results than that with the random weights [65]. Therefore, deep MLPs can be pretrained with unsupervised DBN and then fine-tuned by back-propagation. Other training techniques have been developed, but they lie outside of the general scope of this introduction.

Nowadays, deep learning has various similar definitions. LeCun *et al.* [72] proposed the following one as the first line of their field-defining paper: "Deep learning allows computational models that are composed of multiple processing layers to learn representations of data with multiple levels of abstraction."

That is, the first key aspect of deep learning is that it is based on models consisting of multiple layers of nonlinear information processing. The second key aspect is that, similarly to the human information processing systems, it develops multiple representations of the data at different layers of abstraction. Deep learning would not meet such a huge success if it had not been assisted by several advances in technology, such as the drastically increased chip processing abilities (general-purpose GPUs) and the significantly increased size of data used for training. As pointed out by Geoffrey Hinton—a pioneer in the field of ANNs and coauthor of the first paper on the backpropagation algorithm for training MLPs [73]—at his Royal Society talk, the deep learning was already known back in 1986, but it could not take off at that time for the following four reasons:

1. Our labeled data sets were thousands of times too small.
2. Our computers were millions of times too slow.
3. We initialized the weights in a stupid way.
4. We used the wrong type of nonlinearity.

To conclude this brief introduction to deep learning, we mention three popular state-of-the-art deep learning techniques:

Multilayer perceptrons. Deep feedforward neural networks, discussed in Section 1.2.2. At present, the most popular nonlinear activation function is the rectified linear unit (ReLU), or the half-wave rectifier $f(z) = \max(z, 0)$ [72].

Convolutional neural networks. Introduced by LeCun *et al.* [74] for analysis of visual imagery, these networks preserve the spatial structure of the data. They can be considered as a variation of multilayer perceptrons and are inspired by biological processes, since the connectivity pattern between the neurons resembles the organisation of the visual cortex [75].

Long short-term memory networks The LSTMs are recurrent neural networks, trained using BP and composed of memory blocks that are connected into layers. Introduced by Jurgen Schmidhuber [76], they can be used to create large (stacked) recurrent networks. These systems are well suited for classification, processing, and prediction of time series.

1.2.5 Hopfield networks

In his seminal paper published in 1982 [77], Hopfield asked the question if a computational task can be performed by a large number of simple neurons. A particular emphasis was placed on the *collective* nature of the system, and that complex processing functionality can arise as an emergent property of neural systems. Based on what we today would call an ANN, the system would be *delocalized* and make use of *extensive asynchronous processing*. Highlighting these technical aspects suggests Hopfield was interested in an implementation on dedicated hardware. The specific task to be solved by such a *Hopfield Network* is content addressable memory. Other than in classical computers, in Hopfield Networks content addressable memory should be robust against noise and accessible even if only parts of the information is available. The

Hopfield Network therefore is to incorporate robust error correction into its architecture.

As pointed out in [77], multiple physical systems inherently possess such error correction properties. In terms of physical dynamical systems, a stable content addressable memory corresponds to a locally stable fixed point. In their vicinity, the system's flow field—describing its equations of motion—points toward the locally stable positions; see Figure 1.3. Assuming damped motion and an initialization in such a vicinity, the system will therefore relax toward these fixed points and the correct memory is recalled. For a system of N neurons and $\mathbf{x}^m = \{x^1, x^2, \ldots, x^N\}$ as the system's mth fixed point, an initial system state of $\mathbf{x}^{m'} = \mathbf{x}^m + \mathbf{\Delta}$ converges to $\mathbf{x}^{m'} \to \mathbf{x}^m$ if noise or lack of information $\mathbf{\Delta}$ is sufficiently small. A specific memory value $\mathbf{x}^m$ therefore is addressable by non-ideal initialization vectors.

The possible state of neuron $i = \{1, 2, \ldots, N\}$ is defined on the bases of rate coding: $x_i = 1$ ("firing at maximum rate") or $x_i = 0$ ("not firing"). Hence, state $\mathbf{x}$ corresponds to a binary word vector. Evolution along time n is governed by the network's connectivity matrix $\mathbf{W}$ and the neuron nonlinearities $f(\cdot)$:

$$x_i(n+1) = f\Big(\sum_j W_{i,j} x_j(n)\Big) \tag{12}$$

$$f(x_i) = \begin{cases} x_i \to 1 & x > D \\ x_i \to 0 & \text{else,} \end{cases} \tag{13}$$

where D is a discriminator value typically set to zero. Hopfield goes on and defines the coupling matrix based on the $m = \{1, 2, \ldots, M\}$ memory values $\mathbf{x}^m$ according to

$$W_{i,j} = \sum_m (2x_i^m - 1)(2x_j^m - 1) \tag{14}$$

$$W_{i,i} = 0. \tag{15}$$

In Figure 1.11(a), the resulting network is schematically illustrated. An injection vector $\mathbf{x}^{m'}$ queries a specific memory stored in $\mathbf{W}$. The query word vector sets the network into its initial state, from which it then evolves only according to internal interactions as described by equation (12):

$$x_i^{m'}(n+1) = f\Big(\sum_m (2x_i^m - 1)\Big[\sum_j x_j^{m'}(n)(2x_j^m - 1)\Big]\Big) \tag{16}$$

$$= f(H_i^{m'}(n)). \tag{17}$$

For $\mathbf{x}^m$ uniform statistically distributed Booleans, one can make the following observations regarding the system's temporal evolution. The term in square-brackets of equation (16) averages to 0 unless $m = m'$, for which it is $\frac{N}{2}$. As a result, the internal state of neuron i becomes $H_i^{m'}(n) \approx (2x_i^m - 1)N/2$, which is positive (negative) for $2x_i^m = 2$

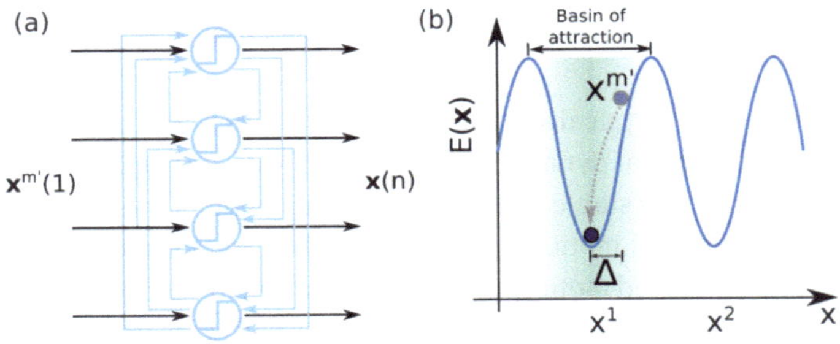

Figure 1.11: Schematic illustration of a Hopfield network (a). (b) Basin of attraction and Energy associated to a state **x**.

($2x_i^m = 0$), respectively. Due to the transformation by the threshold nonlinearity $f(\cdot)$, a stored state is stable except for noise originating from the $m' \neq m$ terms. Moreover, $f(\cdot)$ forces inputs toward one of the saved binary word vectors. Generally, the binary word vector provided in response to a query converges to the stored $\mathbf{x}^m$ with the shortest Hamming distance to initial input $\mathbf{x}^{m'}$ [77].

What makes Hopfield networks particularly fascinating systems is their conceptual simplicity and a deep connection to physical systems. This simplicity allows deriving general properties which would be difficult to obtain for more ubiquitous neural networks. For example, one can show [5] that deviations induced by $m \neq m'$ terms in equation (17) to an input vector result in a signal-to-noise ratio of SNR $= [(M - 1)n'/2]^{1/2}$, where n' is the number of correct bits in $x^{m'}$. One therefore obtains a limit for how many word vectors M can be stored in **W** for achieving memory of a specific storage-finesse. In practice, storage should be limited to $M \leq 0.1N$, beyond which the probability of association to a wrong entry grows significantly.

Finally, one can associate storage matrix **W** to an energy function, a common approach for physical systems:

$$E(\mathbf{x}) = -\frac{1}{2}\sum_{j\neq i} x_i x_j. \tag{18}$$

As schematically illustrated in Figure 1.11(b), saved word-vectors $\mathbf{x}^m$ correspond to local minima in the energy function of equation (18). Whenever the system is initialized, its state evolves along the negative gradient of equation (18), creating basins of attraction around $\mathbf{x}^m$. As long as the query-vector's distance $\boldsymbol{\Delta}$ to the desired memory value is smaller than the attraction basin's size, the Hopfield network will provide the correct response.

Several additional aspects of Hopfield networks favor its potential implementation in networks consisting of physical connections and neurons. Hopfield investigated the possibility to *clip* connection matrix **W** to values $W_{i,j} = \text{sgn}(W_{i,j})$, where sgn is the sign function [77]. For physical systems, such *clipping* strongly reduces system complexity. Network connections would be Boolean and either excitatory or inhibitory. And again, it is possible to analytically derive the impact of such drastic action: the storage-finesse is reduced through a reduction of the *SNR* by a factor of $(2/\pi)^{1/2}$. If one therefore wants to operate a simplified hardware Hopfield network, it is possible to estimate an adequate memory size M based on experience obtained from full-resolution model systems. Of particular interest for physical networks are measures identified by Denker et al. [78]. In optics, subtraction of signals, as required by negative entries in $\mathbf{W}$, can be of formidable complication as it typically requires phase-stability. However, one can leverage an adaptable threshold in function $f(\cdot)$ as defined in equation (15). This additional degree of freedom enables adding an offset to $\mathbf{W}$ until all its entries are exclusively positive. Motivated by such forgiving criteria, the Hopfield network was received enthusiastically by the physics and engineering community, and numerous implementations in physical hardware networks were created. A selection of the most relevant realizations is given in Section 1.3.1, the state of the art at the time can be appreciated in a special issue [79].

1.3 Early photonic implementation

It was quickly realized that neural networks feature a fundamentally different architecture than that of classical von Neumann computing. Updating a neural network's state typically involves vector or matrix products, followed by accumulations and transformations with simple nonlinear local functions creating neuron outputs. The concept simultaneously profits and facilitates optical realizations. As discussed at the beginning of this chapter, the principle carrier of optical information, the photon, possesses a property fundamentally different to its electronic counterpart, the electron. Photons are bosons, and as such they do not directly interact when occupying the same space. Linear multiplications and accumulations can therefore be executed in parallel, even for signals partially occupying the same physical space. Neural networks can maximally capitalize from this advantage. At the same time, the neural network approach to computation reduces the complexity of local nonlinear transformations to a minimum. Simple thresholding, saturating or linear-rectifying functions are highly popular and equally powerful neuron nonlinearities.

Yet, optically creating complex link architectures is not straightforward. And since photons do not directly interact, it is a continuing struggle to cap the energy consumption for simple optical nonlinear transformations. A careful evaluation of an optical neural network in terms of its global performance is therefore mandatory. Aspects as energy efficiency, input-output isolation, cascadability, and scaling properties need to be considered. The detail in which such considerations need to be taken into account is excellently illustrated by an article from Caulfield et al. [29], in which they argue that future supercomputing requires optics. Swift responses by Tucker [30] and Miller [31] highlighted multiple points of controversy. Despite this controversy, creating generally fosters understanding, and prospects of optical neural networks are of such scale that the struggle towards a competitive, practical, and functioning system should be considered mandatory.

1.3.1 Optical perceptron

As discussed in Section 1.2.1, in a perceptron the network or input state $\mathbf{x}$ is weighted according to connections $\mathbf{W}$, followed by a thresholding nonlinearity:

$$y_i = f\Big(\sum_j W_{i,j} x_j\Big), \quad f(z) = \begin{cases} 1, & z > 0 \\ 0, & \text{else.} \end{cases} \tag{19}$$

Depending on the dimensionality of output $\mathbf{y}$, connections either correspond to a vector or a matrix. At the heart of perceptron-learning, we therefore find either vector or matrix products. Consequence of the parallelism of optics, it was realized that such operations can strongly benefit from an optical implementation. Identical considerations are at the heart of optical interconnects, where optical signals are routed between input and output channels in parallel. As routing corresponds to such vector and matrix products, interconnect-techniques are transferable to an application as optical perceptron learning, simply by adding nonlinear thresholding by function $f(\cdot)$. Finally, one requires a rule with which entries of $\mathbf{W}$ can be calculated or obtained from interactive optimization by supervised learning. In general, weights are updated according to $W_{i,j}(n+1) = W_{i,j}(n) + \Delta W_{i,j}(n)$ and $\Delta W_{i,j}(n) = g(\mathbf{x}(n), \mathbf{W}(n), \mathbf{y}(n))$, where $g(\cdot)$ is some function [80].

Research into optical perceptrons was a highly active field in the late 1980s. Multiple approaches to create and optimize connections $\mathbf{W}$ were investigated. Among others, holographic realizations appeared particularly interesting. Psaltis et al. [80] introduced the possibility of creating as well as iteratively optimizing $\mathbf{W}$ based on photorefractive crystals. As illustrated in Figure 1.12(a), a hologram is created using interference between pattern $\mathbf{W}$ loaded onto a spatial light modulator (SLM) and a reference wave. Subsequent illumination of the recorded hologram with state $\mathbf{x}$ then creates an optical signal corresponding to the product $\mathbf{W} \cdot \mathbf{x}$ at positions determined by the previously used reference sources. As the adjustment of weights requires that too strong entries in $\mathbf{W}$ must be reduced, Psaltis et al. [80] superimposed the pattern with noncoherent illumination, causing a reduction of the coupling strength.

Creating and modifying holographic patterns for multiple sources is not independent: actions dedicated to a specific holographic pattern will simultaneously modify other patterns stored in the same crystal. To avoid this undesired effect, Psaltis et al. suggested to arrange spatial positions in a fractal configuration [82]. Under these and other conditions arising from geometric considerations, a holographic medium with

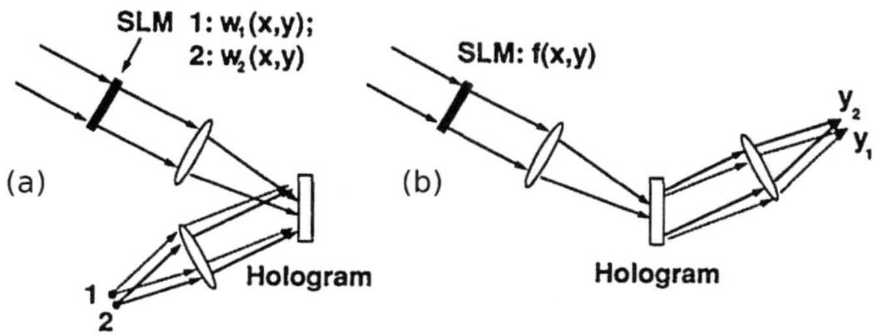

Figure 1.12: Schematic illustration of an optical perceptron learning. (a) Writing of weights W_1 and W_2 into the holographic medium, associating the current SLM states to class 1 or 2 via illumination with sources 1 and 2, respectively. (b) Upon reappearance of states 1 (2), the outputs $\mathbf{y}_1$ ($\mathbf{y}_2$) are generated. Figure taken from [81].

volume $V = 1\,\text{cm}^3$ can store approximately 10^{10} connection parameters at an optical wavelength of $\lambda = 1\,\mu\text{m}$ [80]. Using an optical setup following these principles, they experimentally demonstrated perceptron learning, separating binary input patterns via thresholding at the output [80]. Furthermore, Wagner and Psaltis showed that based on a slightly more advanced optical functionality, NNs consisting of more than one layer could be trained, including the today heavily exploited scheme of error back-propagation [80, 83]. Yet, these elaborations remained limited to theoretical considerations.

Conceptually comparable, Hong et al. [81] implemented a different approach to create positive and negative modifications to entries in **W** according to perceptron learning. Using Stoke's principle of reversibility, based on a double Mach–Zehnder interferometer geometry, they created two write beams with a relative phase-shift of π. As a consequence, the hologram-write patterns from one or the other beam are equally out of phase, and hence one can counteract the impact of the other. They experimentally demonstrated the coherent writing and erasure of target patterns, and the dichotomization of 12 bit binary patterns via numerical simulations of their holographic perceptron learning.

McAulay et al. [84] chose a different approach. Instead of a holographic medium, they used a spatial light repeater to implement perceptron learning. In experiments, they optically dichotomized 4 Chinese characters. These characters were of significantly higher complexity than the previously used binary patterns. Further augmenting the computational challenge, they chose highly similar symbols to be separated into two different classes. Based on these similarities, the authors investigated the system performance for exclusively positive or bipolar weights in **W**. They demonstrated significantly better separation between the two classes for the bipolar weight system.

In general, it can be said that the early optical perceptron learning experiments are of considerable elegance in their implementation. Perceptron learning and error back propagation remain at the heart of todays NNs concepts. Exploiting material or architecture inherent physical effects for their implementation certainly illustrates the large potential of what can be gained with dedicated hardware computing substrates. Yet, these early concepts also suffer significant drawbacks. It is hard to ignore, e. g., the inefficiency of such implementations. Holographic coupling is fundamentally limited by the diffraction efficiency, which mostly lies below 10 % [85]. For McAulay et al. [84], we estimate that less than 1 % of the optical input arrives at the output. Furthermore, all experiments are based on bulk optics, and it is not apparent how the holographic concept could be transferred into integrated devices.

1.3.2 Optical Hopfield networks

Content addressable memory realized via a Hopfield Network (see Section 1.2.5) simply requires iterative update of the RNN according to connection weights **W** and a thresholding function; see equations (16) and (17). In complexity, this goes a small step beyond the simple perceptron, as the system's state vector **x** is not just simply externally provided input, but takes center stage in the system's evolution. Yet, in terms of optical architecture concepts, this corresponds to a simple extension of the system by a feedback path. Figure 1.13 shows two different approaches to close this feedback loop, [5]. An electrical feedback is created in panel (a), while panel (b) illustrates how such recurrent neural networks could be realized using optical feedback. Crucially, both systems remain electro-optical, as they rely on electronic detection and thresholding. The reasoning behind it is that in a network of N neurons, electronics are restricted to the system section scaling with $O(N)$, while optics takes care of the crucial part where scaling is according to $O(N^2)$. However, the authors point out that these electronic components could be replaced by bistable optical amplifiers, requiring multiple pinhole and one microlens arrays. The resulting system would be all-optical and coherent [5].

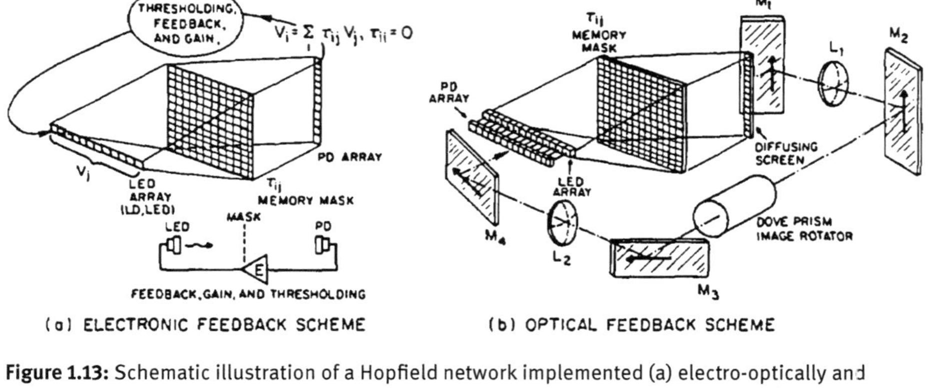

Figure 1.13: Schematic illustration of a Hopfield network implemented (a) electro-optically and (b) all-optically. Figure reproduced from [4].

In their seminal publication, Farhat et al. [4] implement the electro-optical version of Figure 1.13(a) in an experimental setup. As in Section 1.3.1, they addressed the problem of bipolar entries in the connection matrix **W** by multiplexing the implementation of the optical weights: one section was dedicated to positive, the other to negative weights. Their system consisted of a linear array of 32 light emitting diodes creating state **x**. Positive and negative values of bipolar matrix **W** were stored in an optical mask each, each followed by a linear array of 32 photo detectors, measuring the bipolar multiplication's results. The system's electronic high-frequency cut-off was at $(30\,\text{ms})^{-1}$, yet the experiment was performed at lower speeds.

Weights of **W** were obtained based on the analytical procedure of equations (14) and (15), [77]. Psaltis et al. stored four memory vectors $\mathbf{x}^m$ and studied the reliability with which the system could address this content using faulty addressing vectors. They showed that the system was robust against errors on up to 12 out of the 32 bits.

In [86], the scheme is extended to a two-dimensional optical implementation of state vector **x**. Jang et al. created a 4×4 state vector using a liquid crystal spatial light modulator and holographically implemented the connections according to **W**. However, for a setup with a two-dimensional state vector, the connection matrix becomes a tensor with rank four [87]. The authors solve this problem by creating 4×4 separate sub-hologram with 4×4 matrix entries each. In an unfocused setup, bipolar levels in **W** are encoded in positive values levering a constant offset and a corresponding adjustment of threshold-level in equation (13) [78]. The authors saved two 16-valued word vectors and their content was addressable using vectors with fewer than 3 bit errors. More detailed considerations regarding the fabrication of the required holographic element have been treated by Keller et al. [88]. Finally, Yeh et al. [85] followed a similar approach with a network consisting of 8×8 neurons.

Ito et al. [89] extended the general concept and used optical fiber couplers to literally create an optical network consisting of connection lines. With their approach, the authors implemented a Hopfield network for a 5×5 network state, hence consisting of up to $25^2 = 625$ fiber connections. The state matrix **x** was encoded in a two-dimensional laser array ($\lambda = 0.87\,\mu\text{m}$), detection was realized via a two-dimensional array of photo diodes. Bipolar weights were again implemented with a paired set of connections, and the fiber network was optimized with regards to the required connection for the three particular binary word vectors to be saved. Furthermore, they amended the original concept by an additional bit-significance vector. Experimental performance was evaluated against the address bit error and compared with numerical simulations, identifying noise in the coupling ratios as the overall limiting factor. As the authors used standard macroscopic fibers to implement their system, the resulting experiment was rather bulky. However, one can possibly consider this publication as the first computer based on optical waveguides, and once implemented in integrated optics the approach is of great potential.

1.4 Conclusion

During the past decades, ANNs have established themselves as one of the leading information processing concepts. Their originally neuro-inspired architecture makes them fundamentally different from the Turing–von Neumann computing concept. This architectural difference has a direct and profound impact on the performance when ANNs are emulated on classical computing infrastructure. Among others, the main reason is that current computing substrates do not provide the massive parallelism required to efficiently compute the state of an ANN. Already in the 1980s, the potential of optics in addressing this bottleneck was intensively discussed. Early demonstrations from that time already show the impressive potential of these early days proof-of-concept experiments. Full-scale technology transfer was then hindered by a fading interest into ANN; partially by limited performance of the concepts, partially by lack of pressing applications. The recent surge in ANN activity has removed these two stumbling-blocks, and today photonic ANNs are a highly attractive and active field of research with enormous potential for next-generation computing substrates. In the remainder of the book, we introduce and photonically implement Reservoir Computing, an ANN-concept particularly attractive for implementation in dedicated physical substrates.

Bibliography

[1] T. Freeth et al. "Decoding the ancient Greek astronomical calculator known as the Antikythera Mechanism". In: *Nature* 444.7119 (Nov. 2006), pp. 587–591. ISSN: 0028-0836. DOI: 10.1038/nature05357. URL: http://www.nature.com/articles/nature05357.

[2] A. Szoke et al. "Bistable optical element and its applications". In: *Applied Physics Letters* 15.11 (Dec. 1969), pp. 376–379. ISSN: 0003-6951. DOI: 10.1063/1.1652866. URL: http://aip.scitation.org/doi/10.1063Z1.1652866.

[3] P. Ambs. "Optical computing: a 60-year adventure". In: *Advances in Optical Technologies* 2010 (May 2010), pp. 1–15. ISSN: 1687-6393. DOI: 10.1155/2010/372652. URL: http://www.hindawi.com/journals/aot/2010/372652/.

[4] N. H. Farhat et al. "Optical implementation of the Hopfield model". In: *Applied Optics* 24.10 (May 1985), p. 1469. ISSN: 0003-6935. DOI: 10.1364/AO.24.001469. URL: https://www.osapublishing.org/abstract.cfm?URI=ao-24-10-1469.

[5] D. Psaltis and N. Farhat. "Optical information processing based on an associative-memory model of neural nets with thresholding and feedback". In: *Optics Letters* 10.2 (Feb. 1985), p. 98. ISSN: 0146-9592. DOI: 10.1364/OL.10.000098. URL: https://www.osapublishing.org/abstract.cfm?URI=ol-10-2-98.

[6] A. W. Lohmann. "Principles of optical computing". In: *Nonlinear optics and optical computing*. Boston, MA: Springer US, 1990, pp. 151–157. DOI: 10.1007/978-1-4613-0629-0_10. URL: http://link.springer.com/10.1007/978-1-4613-0629-0_10.

[7] C. Horsman et al. "When does a physical system compute?" In: *Proceedings. Mathematical, physical, and engineering sciences/the Royal Society* 470.2169 (2014), p. 20140182. ISSN:

1364-5021. DOI: 10.1098/rspa.2014.0182. URL: http://rspa.royalsocietypublishing.org/content/470/2169/20140182.

[8] J. Shamir. "Half a century of optics in computing—a personal perspective". In: *Applied Optics* 52.4 (Feb. 2013), pp. 600–612. ISSN: 1539-4522. URL: http://www.ncbi.nlm.nih.gov/pubmed/23385897.

[9] H. Esmaeilzadeh et al. "Dark silicon and the end of multicore scaling". In: *IEEE MICRO* 32.3 (2012), pp. 122–134. ISSN: 0272-1732. DOI: 10.1109/MM.2012.17. URL: http://portal.acm.org/citation.cfm?doid=2000064.2000108.

[10] I. L. Markov. "Limits on fundamental limits to computation". en. In: *Nature* 512.7513 (Aug. 2014), pp. 147–154. ISSN: 1476-4687. DOI: 10.1038/nature13570. URL: http://www.nature.com/nature/journal/v512/n7513/full/nature13570.html?WT.ec_id=NATURE-20140814.

[11] D.-I. Yeom et al. "Low-threshold supercontinuum generation in highly nonlinear chalcogenide nanowires". In: *Optics Letters* 33.7 (Apr. 2008), p. 660. ISSN: 0146-9592. DOI: 10.1364/OL.33.000660. URL: https://www.osapublishing.org/abstract.cfm?URI=ol-33-7-660.

[12] B. J. Eggleton, B. Luther-Davies, and K. Richardson. "Chalcogenide photonics". In: *Nature Photonics* 5.3 (Mar. 2011), pp. 141–148. ISSN: 1749-4885. DOI: 10.1038/nphoton.2011.309. URL: http://www.nature.com/articles/nphoton.2011.309.

[13] K. Liu et al. "Review and perspective on ultrafast wavelength-size electro-optic modulators". In: *Laser & Photonics Reviews* 9.2 (2015), pp. 172–194. ISSN: 1863-8899. DOI: 10.1002/lpor.201400219.

[14] M. Moore. "International roadmap for devices and systems". In: *IEE* (2016), p. 23. URL: https://irds.ieee.org/images/files/pdf/2016_MM.pdf.

[15] K. Jr. Preston. "Use of the Fourier transformable properties of lenses for signal spectrum analysis". In: *Optical and Electro-Optical Information Processing*. Cambridge, MA: MIT Press, 1965.

[16] S. Laval and N. Paraire. "Dynamic operation of nonlinear waveguide devices for fast optical switching". In: *Nonlinear optics and optical computing*. Boston, MA: Springer US, 1990, pp. 21–35. DOI: 10.1007/978-1-4613-0629-0_2. URL: http://link.springer.com/10.1007/978-1-4613-0629-0_2.

[17] S. W. Keckler et al. "GPUs and the future of parallel computing". In: *IEEE MICRO* 31.5 (Sept. 2011), pp. 7–17. ISSN: 0272-1732. DOI: 10.1109/MM.2011.89. URL: http://ieeexplore.ieee.org/document/6045685/.

[18] V. Adhinarayanan et al. "Measuring and modeling on-chip interconnect power on real hardware". In: *2016 IEEE international symposium on workload characterization (IISWC)*. IEEE, Sept. 2016, pp. 1–11. ISBN: 978-1-5090-3896-1. DOI: 10.1109/IISWC.2016.7581263. URL: http://ieeexplore.ieee.org/document/7581263/.

[19] D. A. B. Miller. "Attojoule optoelectronics for low-energy information processing and communications". In: *Journal of Lightwave Technology* 35.3 (2017), pp. 346396. ISSN: 0733-8724. DOI: 10.1109/JLT.2017.2647779.

[20] K. Sato, C. Young, and D. Patterson. *An in-depth look at Google's first tensor processing unit*. 2017. URL: https://cloud.google.com/blog/big-data/2017/05/an-in-depth-look-at-googles-first-tensor-processing-unit-tpu.

[21] R. A. Athale, W. C. Collins, and P. D. Stilwell. "High accuracy matrix multiplication with outer product optical processor". In: *Applied Optics* 22.3 (Dec. 1986), pp. 368–370. ISSN: 0003-6935. URL: http://www.ncbi.nlm.nih.gov/pubmed/22777773.

[22] P. Mandel, S. D. Smith, and B. S. Wherrett. *From optical bistability towards optical computing*. 1987. URL: http://www.inoa.it/home/arecchi/SezA/fis141.pdf.

[23] S. D. Smith. "Lasers, nonlinear optics and optical computers". In: *Nature* 316.6026 (July 1985), pp. 319–324. ISSN: 0028-0836. DOI: 10.1038/316319a0. URL: http://www.nature.com/doifinder/10.1038/316319a0.

[24] W. J. Firth. “Theory of optical bistability and optical memory”. In: *Nonlinear optics and optical computing*. Boston, MA: Springer US, 1990, pp. 3–20. DOI: 10.1007/978-1-4613-0629-0_1. URL: http://link.springer.com/10.1007/978-1-4613-0629-0_1.

[25] S. Martellucci and A. N. Chester. *Nonlinear optics and optical computing*. Springer US, 1990, p. 285. ISBN: 978-1-4612-7900-6.

[26] H. M. Gibbs, S. L. McCall, and T. N. C. Venkatesan. “Differential gain and bistability using a sodium-filled Fabry–Perot interferometer”. In: *Physical Review Letters* 36.19 (May 1976), pp. 1135–1138. ISSN: 0031-9007. DOI: 10.1103/PhysRevLett.36.1135. URL: https://link.aps.org/doi/10.1103/PhysRevLett.36.1135.

[27] D. A. B. Miller, S. D. Smith, and A. Johnston. “Optical bistability and signal amplification in a semiconductor crystal: applications of new low-power nonlinear effects in InSb”. In: *Applied Physics Letters* 35.9 (1979), pp. 658–660. ISSN: 0003-6951. DOI: 10.1063/1.91245. URL: https://doi.org/10.1063/1.91245.

[28] S. D. Smith et al. “The demonstration of restoring digital optical logic”. In: *Nature* 325.6099 (Jan. 1987), pp. 27–31. ISSN: 0028-0836. DOI: 10.1038/325027a0. URL: http://www.nature.com/articles/325027a0.

[29] H. J. Caulfield and S. Dolev. “Why future supercomputing requires optics”. In: *Nature Photonics* 4.5 (May 2010), pp. 261–263. ISSN: 1749-4885. DOI: 10.1038/nphoton.2010.94. URL: http://www.nature.com/doifinder/10.1038/nphoton.2010.94.

[30] R. S. Tucker. “The role of optics in computing”. In: *Nature Photonics* 4.7 (July 2010), p. 405. ISSN: 1749-4885. DOI: 10.1038/nphoton.2010.162. URL: http://www.nature.com/articles/nphoton.2010.162.

[31] D. A. B. Miller. “The role of optics in computing”. In: *Nature Photonics* 4.1 (2010), p. 406. ISSN: 0147-8389. DOI: 10.1111/j.1540-8159.2009.02627.x. URL: http://doi.wiley.com/10.1111/j.1540-8159.2009.02627.x.

[32] D. A. B. Miller. “Are optical transistors the logical next step?” In: *Nature Photonics* 4.1 (Jan. 2010), pp. 3–5. ISSN: 1749-4885. DOI: 10.1038/nphoton.2009.240. URL: http://dx.doi.org/10.1038/nphoton.2009.240.

[33] F. B. McCormick et al. “Six-stage digital free-space optical switching network using symmetric self-electro-optic-effect devices”. In: *Applied Optics* 32.26 (Sept. 1993), p. 5153. ISSN: 0003-6935. DOI: 10.1364/AO.32.005153. URL: https://www.osapublishing.org/abstract.cfm?URI=ao-32-26-5153.

[34] P. M. Duffieux. *L'integrale de Fourier et ses applications a l'optique*. Ed. by Masson. Besanôn: Faculte des Sciences Besanôn, 1946.

[35] J. W. Goodman. *Introduction to Fourier optics*. Roberts & Co, 2005, p. 491. ISBN: 9780974707723.

[36] C. S. Weaver and J. W. Goodman. “A technique for optically convolving two functions.” In: *Applied Optics* 5.7 (July 1966), pp. 1248–1249. ISSN: 0003-6935. URL: http://www.ncbi.nlm.nih.gov/pubmed/20049063.

[37] D. Psaltis et al. “Accurate numerical computation by optical convolution”. In: *SPIE 1980 international opitca computing conference*. 1980, pp. 151–156.

[38] G. Keryer et al. “On-board optical joint transform correlator for real-time road sign recognition”. In: *Optical Engineering* 34.1 (Jan. 1995), p. 135. ISSN: 0091-3286. DOI: 10.1117/12.183999. URL: http://opticalengineering.spiedigitallibrary.org/article.aspx?doi=10.1117/12.183999.

[39] L. Cutrona et al. “Optical data processing and filtering systems”. In: *IEEE Transactions on Information Theory* 6.3 (June 1960), pp. 386–400. ISSN: 0018-9448. DOI: 10.1109/TIT.1960.1057566. URL: http://ieeexplore.ieee.org/document/1057566/.

[40] S. H. Lee (ed.). *Optical information processing: fundamentals*. Topics in applied physics 48. Springer, Berlin Heidelberg, 1981. ISBN: 978-3-540-10522-0, 978-3-540-38521-9. URL: http://gen.lib.rus.ec/book/index.php?md5=6fb89e58aad3e2b5bd8f57d3c83b6aef.

[41] A. E. Chiou and P. Yeh. "Parallel image subtraction using a phase-conjugate Michelson interferometer". In: *Optics Letters* 11.5 (May 1986), p. 306. ISSN: 0146-9592. DOI: 10.1364/OL.11.000306. URL: https://www.osapublishing.org/abstract.cfm?URI=ol-11-5-306.

[42] H. J. Caulfield et al. "Optical implementation of systolic array processing". In: *Optics Communications* 40.2 (1981), pp. 86–90. URL: http://www.sciencedirect.com/science/article/pii/0030401881903333.

[43] G. Labrunie, J. Robert, and J. Borel. "A 128 × 128 electro-optical interface for real time data processing". In: *Revue de Physique Appliquee* 10.3 (1975), pp. 143–146.

[44] D. P. Mandic, J. A. Chambers, et al. *Recurrent neural networks for prediction: learning algorithms, architectures and stability*. Wiley Online Library, 2001.

[45] D. A. Medler. "A brief history of connectionism". In: *Neural Computing Surveys* 1 (1998), pp. 18–72.

[46] W. S. McCulloch and W. Pitts. "A logical calculus of the ideas immanent in nervous activity". In: *The Bulletin of Mathematical Biophysics* 5.4 (1943), pp. 115–133.

[47] D. O. Hebb. *The organization of behavior: a neurophysiological approach*. 1949.

[48] F. Rosenblatt. "The perceptron: a probabilistic model for information storage and organization in the brain." In: *Psychological Review* 65.6 (1958), p. 386.

[49] M. Minsky and S. Papert. *Perceptrons: an introduction to computational geometry*. MIT Press, Cambridge, Mass, 1969.

[50] P. Werbos. *Beyond regression: new tools for prediction and analysis in the behavioral sciences*. 1974.

[51] P. J. Werbos. "Backpropagation through time: what it does and how to do it". In: *Proceedings of the IEEE* 78.10 (1990), pp. 1550–1560.

[52] A. L. Hodgkin and A. F. Huxley. "A quantitative description of membrane current and its application to conduction and excitation in nerve". In: *The Journal of Physiology* 117.4 (1952), p. 500.

[53] R. FitzHugh. "Mathematical models of threshold phenomena in the nerve membrane". In: *The Bulletin of Mathematical Biophysics* 17 (1955), pp. 257–278.

[54] W. Gerstner. "A framework for spiking neuron models: the spike response model". In: *Handbook of Biological Physics* 4 (2001), pp. 469–516.

[55] W. Gerstner and W. M. Kistler. *Spiking neuron models: single neurons, populations, plasticity*. Cambridge university press, 2002.

[56] E. M. Izhikevich. "Which model to use for cortical spiking neurons?" In: *IEEE Transactions on Neural Networks* 15.5 (2004), pp. 1063–1070.

[57] G. Dreyfus et al. *Neural networks: methodology and applications*. Springer Berlin, 2005.

[58] F. Rosenblatt. *Principles of neurodynamics. Perceptrons and the theory of brain mechanisms*. Tech. rep., Cornell Aeronautical Lab Inc Buffalo NY, 1961.

[59] T. Hastie, R. Tibshirani, and J. Friedman. *The elements of statistical learning: data mining, inference, and prediction*. Springer, New York, 2013.

[60] N. Cristianini and J. Shawe-Taylor. *An introduction to support vector machines and other kernel-based learning methods*. Cambridge university press, 2000.

[61] V. Kecman. *Learning and soft computing: support vector machines, neural networks, and fuzzy logic models*. MIT press, 2001.

[62] I. Steinwart and A. Christmann. *Support vector machines*. Springer Science & Business Media, 2008.

[63] S. Salcedo-Sanz et al. "Support vector machines in engineering: an overview". In: *Wiley Interdisciplinary Reviews: Data Mining and Knowledge Discovery* 4.3 (2014), pp. 234–267.
[64] O. Nerrand et al. "Neural networks and nonlinear adaptive filtering: unifying concepts and new algorithms". In: *Neural Computation* 5.2 (1993), pp. 165–199.
[65] L. Deng, D. Yu, et al. "Deep learning: methods and applications". In: *Foundations and Trends in Signal Processing* 7.3–4 (2014), pp. 197–387.
[66] C. F. Cadieu et al. "Deep neural networks rival the representation of primate IT cortex for core visual object recognition". In: *PLoS Computational Biology* 10.12 (Dec. 2014). Ed. by Matthias Bethge, e1003963. ISSN: 1553-7358. DOI: 10.1371/journal.pcbi.1003963. URL: https://dx.plos.org/10.1371/journal.pcbi.1003963.
[67] Y. Bengio et al. "Learning deep architectures for AI". In: *Foundations and Trends in Signal Processing* 2.1 (2009), pp. 1–127.
[68] X. Glorot and Y. Bengio. "Understanding the difficulty of training deep feedforward neural networks". In: *Proceedings of the thirteenth international conference on artificial intelligence and statistics*. 2010, pp. 249–256.
[69] G. E. Hinton, S. Osindero, and Y.-W. Teh. "A fast learning algorithm for deep belief nets". In: *Neural Computation* 18.7 (2006), pp. 1527–1554.
[70] G. E. Hinton and R. R. Salakhutdinov. "Reducing the dimensionality of data with neural networks". In: *Science* 313.5786 (2006), pp. 504–507.
[71] G. E. Hinton. "Deep belief networks". In: *Scholarpedia* 4.5 (2009), p. 5947. DOI: 10.4249/scholarpedia.5947.
[72] Y. LeCun, Y. Bengio, and G. Hinton. "Deep learning". In: *Nature* 521.7553 (2015), pp. 436–444.
[73] D. E. Rumelhart, G. E. Hinton, and R. J. Williams. "Learning representations by back-propagating errors". In: *Nature* 323 (1986), pp. 533–536.
[74] Y. LeCun et al. "Handwritten digit recognition with a back-propagation network". In: *Advances in neural information processing systems*. 1990, pp. 396–404.
[75] M. Matsugu et al. "Subject independent facial expression recognition with robust face detection using a convolutional neural network". In: *Neural Networks* 16.5–6 (2003), pp. 555–559.
[76] S. Hochreiter and J. Schmidhuber. "Long short-term memory". In: *Neural Computation* 9.8 (1997), pp. 1735–1780.
[77] J. J. J. Hopfield. "Neural networks and physical systems with emergent collective computational abilities". In: *Proceedings of the national academy of... 79*. April 1982, pp. 2554–2558. URL: http://www.pnas.org/content/79/8Z2554.short.
[78] J. S. Denker. "Neural network refinements and extensions". In: *AIP Conference Proceedings* 151.1 (AIP, June 1986), pp. 121–128. DOI: 10.1063/1.36267. URL: http://aip.scitation.org/doi/abs/10.1063/1.36267.
[79] K. Wagner and D. Psaltis. "Optical neural networks: an introduction by the feature editors". In: *Applied Optics* 32.8 (Mar. 1993), p. 1261. ISSN: 0003-6935. DOI: 10.1364/AO.32.001261. URL: https://www.osapublishing.org/abstract.cfm?URI=ao-32-8-1261.
[80] D. Psaltis, D. Brady, and K. Wagner. "Adaptive optical networks using photorefractive crystals". In: *Applied Optics* 27.9 (May 1988), p. 1752. ISSN: 0003-6935. DOI: 10.1364/AO.27.001752. URL: https://www.osapublishing.org/abstract.cfm?URI=ao-27-9-1752.
[81] J. H. Hong, S. Campbell, and P. Yeh. "Optical pattern classifier with perceptron learning". In: *Applied Optics* 29.20 (July 1990), p. 3019. ISSN: 0003-6935. DOI: 10.1364/AO.29.003019. URL: https://www.osapublishing.org/abstract.cfm?URI=ao-29-20-3019.
[82] D. Psaltis et al. "Holography in artificial neural networks". In: *Nature* 343.6256 (Jan. 1990), pp. 325–330. ISSN: 0028-0836. DOI: 10.1038/343325a0. URL: http://www.nature.com/doifinder/10.1038/343325a0.

[83] K. Wagner and D. Psaltis. "Multilayer optical learning networks". In: *Applied Optics* 26.23 (Dec. 1987), p. 5061. ISSN: 0003-6935. DOI: 10.1364/A0.26.005061. URL: https://www.osapublishing.org/abstract.cfm?URI=ao-26-23-5061.
[84] A. D. McAulay, J. Wang, and X. Xu. "Optical perceptron learning for binary classification with spatial light rebroadcasters". In: *Applied Optics* 32.8 (Mar. 1993), p. 1346. ISSN: 0003-6935. DOI: 10.1364/AO.32.001346. URL: https://www.osapublishing.org/abstract.cfm?URI=ao-32-8-1346.
[85] S. L. Yeh, R. C. Lo, and C. Y. Shi. "Optical implementation of the Hopfield neural network with matrix gratings". In: *Applied Optics* 43.4 (Feb. 2004), p. 858. ISSN: 0003-6935. DOI: 10.1364/AO.43.000858. URL: https://www.osapublishing.org/abstract.cfm?URI=ao-43-4-858.
[86] J.-S. Jang et al. "Optical implementation of the Hopfield model for twodimensional associative memory". In: *Optics Letters* 13.3 (Mar. 1988), p. 248. ISSN: 0146-9592. DOI: 10.1364/OL.13.000248. URL: https://www.osapublishing.org/ol/abstract.cfm?uri=ol-13-3-248.
[87] H. J. Caulfield. *Parallel A/ 4 weighted optical interconnections*. In: URL: https://www.osapublishing.org/ao/viewmedia.cfm?uri=ao-26-19-4039&seq=0.
[88] P. E. Keller and A. F. Gmitro. "Computer-generated holograms for optical neural networks: on-axis versus off-axis geometry". In: *Applied Optics* 32.8 (Mar. 1993), p. 1304. ISSN: 0003-6935. DOI: 10.1364/AO.32.001304. URL: https://www.osapublishing.org/abstract.cfm?URI=ao-32-8-1304.
[89] F. Ito and K.-i. Kitayama. "Optical implementation of the Hopfield neural network using multiple fiber nets". In: *Applied Optics* 28.19 (Oct. 1989), p. 4176. ISSN: 0003-6935. DOI: 10.1364/AO.28.004176. URL: https://www.osapublishing.org/abstract.cfm?URI=ao-28-19-4176.

Joni Dambre

2 Information processing and computation with photonic reservoir systems

2.1 Introduction

2.1.1 The boundaries of digital computing

The recent surge of research into alternatives for conventional (mostly digital) CMOS-based computing hardware is pushed forward by the limits of scalability. The fundamental assumptions, upon which the success of digital computing is based are mainly its robustness, resulting in extremely low error probabilities, and its inherently low static power consumption, both of which break down for very small devices. The digital computing paradigm and the associated design methodology rely on (almost) error-free operation and cannot deal with *inaccurate* devices.

Already in 1956, Von Neumann discussed how to approximate high-precision digital computing with unreliable components [1] by introducing redundancy and *majority voting*. However, the bounds he derives to prove that with enough redundancy, error probabilities can be pushed below any desired threshold, are based on the assumption that errors occur independently. When the errors are correlated, as is usually the case in real life, an ensemble rarely performs worse than the individual models, but the convergence of the accuracy is no longer guaranteed. In practice, taking an *ensemble* of unreliable models is now common practice in *machine learning*.

The second most important property to break down is the fact that, using CMOS, we can build implementations of digital gates that consume very little static power. As feature dimensions and isolation layer thickness get thinner, MOSFET transistors and CMOS gates start leaking current in all directions. Whereas the power consumption of a computer was long considered an unimportant issue, it has now become more important than the speed of computation. A few powerful GPUs in the room are an excellent replacement for other heating devices, but our hunger for ever-increasing and ubiquitous computing does not fade, especially in the presence of the huge leaps that are being made with AI and deep learning.

Many early attempts of computing with alternative devices stay within the digital model of computation. Clearly, this has its benefits: if you can build gates and flip-flops, the whole design methodology can stay in place and the chances of industrial uptake of your new technology improve dramatically. However, thus far, no mapping between inherently analog devices and the basic digital building blocks (binary gates, binary gated memory cells) has been found that can sufficiently outperform transistors on at least one dimension of performance (size, power consumption per operation or power density, speed) without overly compromising the others and at the same time

https://doi.org/10.1515/9783110583496-002

is commercially viable in a mass production setting. Until now, the proposed devices were either too large, too slow, too power hungry, too inaccurate or too difficult to produce, or did not offer sufficient benefits to trigger a change of course in an entire industry built around integrated CMOS circuits.

If we take a step backwards, the digital computing model is maybe no longer the most efficient. This is especially true for the huge amount of information processing that handles signals, e. g., image processing and sensor processing, where, essentially real numbers are translated into real numbers and small errors are usually acceptable. From an energy perspective, it is no longer the most efficient to represent analog values as bit sequences and to use analog devices to mimic theoretical mathematical models known as digital gates and memory cells. The field of research that addresses ways to optimally exploit relaxed precision requirements using more or less traditional computing devices is generally called *approximate* computation [2, 3]. It mostly proposes incremental changes to traditional design flows at different design levels (algorithms, compilers, synthesis, and devices). Different authors often target different performance goals, although the general focus is on power efficiency.

2.1.2 Analogue computing

The alternative to digital computing is to compute directly in the analogue domain. Unfortunately, for general-purpose (or broad-purpose) analogue computing, useful combinations of a clear-cut computational model and an automated hardware design methodology are rare. One approach is to move to *neural networks* as a computational model. Their design is based on *learning from examples* and tuning parameters to minimize a given cost function. They are usually used as machine learning *software models* that are run on digital computers to approximate input-output relationships for which no perfect solution is known or does even exist.

Many variants exist within the broad category of neural networks, and for some, their implementation can also be realized with analogue building blocks. From the artificial ones used in machine learning to the biologically plausible ones studied in computational neuroscience, all are computational models which come with ways to tune their parameters (weights). Essentially, they are extremely powerful *function approximators*. Some types of neural networks can be proven to be universal approximators [4, 5], which means that they can approximate a very broad class of input-output relationships with any desired precision if they are allowed to grow large enough. In practice, they are inherently imperfect, i. e., even the most powerful neural networks make mistakes. These can be frequent but small, or very infrequent but very large. Depending on the application, either one can be worse than the other. Since neural network training and evaluation is usually based on the average quality of the output across a large number of examples, only detailed analysis of the errors made can make the distinction.

In general, it is useful to make a distinction between the computational model that is used and the physical medium in which this model is implemented [6]. The mapping between both should result in a set of computational elements or building blocks that can be realized in the chosen medium (with guarantees on all aspects of their performance) and a methodology for combining them into a system that realizes the desired behavior. From this perspective, it is very helpful if the requirements for these building blocks are not very rigid. It is this rigidity in digital computation that makes it so difficult to realize digital building blocks from analogue devices. In comparison, in neural networks, we need to be able to compute a weighted sum of signals at the neuron inputs, but the exact behavior of the neurons turns out not to be very important for their performance. In artificial neural networks, several different functions are already being used, most notably: sigmoidal, exponential, stepwise linear, and Gaussian. Exactly and efficiently matching those with analogue devices is also not trivial, and the same is true for the building blocks used in the models from computational neuroscience. However, it turns out that most input-output relationships that resembles them should give rise to good performance. In fact, this robustness of the neural network computational model is one of the historical reasons for the emergence of *physical reservoir computing* and *photonic reservoir computing*, the topic of this book, from computational models that are closely related to neural networks.

A crucial aspect for the success of analogue computing is the availability of a design methodology that can be automated to provide the mapping between a desired behavior to (a hierarchical composition of) the computational elements of the model and from those to the physical medium. Clearly, for digital computing, this is in place. Therefore, in order for any new hardware approach to become competitive, this mapping should outperform the existing solutions according to at least one quality measure (e. g., speed, power consumption, size, accuracy, noise-robustness) without compromising too much on the others.

2.2 Reservoir computing

2.2.1 A more relaxed model of computation

In this book, we describe recent work that uses *physical reservoir computing* [7] to design photonic computational devices. This is an altogether different approach to computation, in which no abstract computational model is enforced onto the implementation substrate. Instead, the natural dynamics of the substrate are exploited and combined to approximate the desired input-output relationships of a given computational task.

Instead of in the computational model, the constraints are now in the optimization approach, which is based on the principles of *reservoir computing*. In its broadest

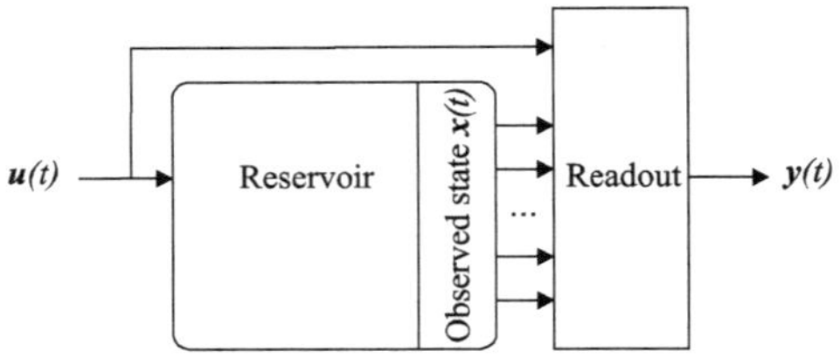

Figure 2.1: Schematic representation of a reservoir computing system. The input signal $u(t)$ is fed into the reservoir and the resulting reservoir states $x(t)$ are used to learn a linear readout that is then used to generate the output signal $y(t)$.

interpretation, reservoir computing consists of two parts: a *reservoir* and a *readout* as depicted in Figure 2.1. The reservoir is a dynamical system that is perturbed (*driven*) by the input signal $u(t)$, which affects its current internal state $x(t)$ as well as the future evolution of that state. In the readout, the reservoir's internal state is observed and a linear combination of the observations is optimized to optimally approximate a desired output signal. In contrast to the readout, the reservoir itself is not changed in this stage, although a few global system parameters are usually tuned during design to make the overall dynamics of the reservoir more suitable for the task.

The reservoir computing approach emerged almost simultaneously from the fields of neuroscience and artificial neural networks around the end of the previous and the beginning of this century. The two most frequently cited foundational works are those of Herbert Jaeger [8, 9] and Wolfgang Maass [10]. Jaeger and Maass used *echo state networks* (ESN) and *liquid state machines* (LSM), respectively, as their reservoirs. Both are simulated recurrent neural networks, operating in discrete or discretized time. ESNs consist of discrete-time analogue sigmoidal or hyperbolic tangent neurons. The network they form is usually fully connected, and the input signals are also connected to all nodes. The connection weights are randomly initialized. LSMs consist of (simple models of) biological spiking neurons. In both cases, the network's internal state space is finite and consists of all neuron outputs. Overviews of the progress in reservoir computing can be found in [11–14].

2.2.2 How to train a reservoir computer

In summary, a reservoir computing system consists of a reservoir, which transforms the input signal(s) into features that depend on the input history, and a readout layer. In most work on reservoir computing, this is a simple linear regression layer, which is trained by minimizing the mean squared error (equation (1))

$$\mathrm{MSE}(\mathbf{y}, \hat{\mathbf{y}}) = E\left[(\mathbf{y} - \hat{\mathbf{y}})^2\right] \tag{1}$$

between the generated output sequences and the desired output sequences for a set of examples (where E indicates the expected value).

In order to minimize the MSE, we minimize its approximation based on one or multiple input sequences of finite length. For this purpose, the reservoir is driven with the input sequence(s).[1] The N observed reservoir state signals are sampled in time (leading to S samples for each of the signals) and recorded into the $S \times (N + 1)$ *augmented state matrix* $\tilde{\mathbf{X}}$, in which the first N columns contain the N state signals, and the last column is an all-ones column. Using the notation $\tilde{\mathbf{w}}_{\text{out}} = (\mathbf{w}_{\text{out}}, w_0)$, the optimal output weights are then found by the normal form:

$$\tilde{\mathbf{w}}_{\text{out}} = (\tilde{\mathbf{X}}^T\tilde{\mathbf{X}})^{-1}\tilde{\mathbf{X}}^T\mathbf{y}, \tag{2}$$

yielding a closed-form expression for the approximated target signal:

$$\hat{\mathbf{y}} = \tilde{\mathbf{X}}\tilde{\mathbf{w}}_{\text{out}} = \tilde{\mathbf{X}}(\tilde{\mathbf{X}}^T\tilde{\mathbf{X}})^{-1}\tilde{\mathbf{X}}^T\mathbf{y}. \tag{3}$$

In machine learning, it is generally advised to use some form of regularization to avoid overfitting to the training data. The most common approach is to use *ridge regression*. This minimizes the augmented cost function:

$$\text{MSE}_{\text{ridge}} = \text{MSE} + \lambda\mathbf{w}_{\text{out}}^T\mathbf{w}_{\text{out}}. \tag{4}$$

The intuition behind this is the fact that keeping the weights small constrains the complexity of the model. Note that w_0 should not be included in the regularisation term as it is needed to set the average value of the approximated target signal to its correct value.

When using ridge regression, it is generally advisable to normalize the state signals before applying linear regression, to avoid the regularization from overly penalizing signals with small magnitude. In this case, we use

$$\mathbf{x}'[n] = \frac{\mathbf{x}[n] - \text{avg}(\mathbf{x}[n])}{\text{stdev}(\mathbf{x}[n])}$$

instead (and similarly construct the corresponding state matrix $\mathbf{X}'$). Augmenting the state matrix with an all-ones column is no longer necessary in that case, since any linear combination of the recorded state signals will have an average value of 0 and the optimal value of w_0 is simply given by

$$w_0 = \frac{1}{n_{\text{max}}} \sum_{n=1}^{n_{\text{max}}} \mathbf{y}[n].$$

The optimal readout weights $\mathbf{w}_{\text{out}}$ are then given by

$$\mathbf{w}_{\text{out}} = (\mathbf{X}'^T\mathbf{X}' + \lambda\mathbf{I})^{-1}\mathbf{X}'^T\mathbf{y} \tag{5}$$

1 Usually, in order to reduce the impact of the initial state, the first part of each sequence is ignored for both training and evaluation.

where $\mathbf{I}$ is the identity matrix and $\mathbf{X}'^T\mathbf{X}'$ is the normalized covariance matrix of the state signals. The regularization parameter λ is usually small (order 10^{-5}–10^{-3}), but its optimal value can vary a lot from case to case. Extremely small optimal values of λ indicate that the combination of task and training data was not prone to overfitting, whereas large values of λ suggest the opposite. For this reason, λ needs to be tuned using an appropriate form of validation. Usually, k-fold cross-validation is used. Once the optimal value of λ is found, all of the training data is used to train the final model and a sufficiently large and previously unused test data set serves to assess the final performance of the model.

Alternative training approaches for the data-driven optimization of linear models exist and can also be used, e. g., *recursive least squares* (RLS) for the on-line optimization in regression tasks or logistic regression[2] for classification tasks, but probably mainly for historical reasons, they are less common in the literature on reservoir computing.

2.2.3 Measures for reservoir performance

Since the readout of a reservoir is traditionally trained to minimize MSE, this is the natural measure of performance. Note that for classification tasks, better performance measures exist. These will be discussed at the end of this section.

As a baseline for MSE, consider the poorest possible reservoir: one for which the observed states do not contain any useful information at all. Instead of a reservoir, one could even imagine using N independent noise sources that have no relation to the input signal. If no overfitting occurs, the readout should produce a constant output equal to $\hat{\mathbf{y}} = \mu_y$, the expected value of the target signal. In this case, the MSE equals

$$\mathrm{MSE}[\mathbf{y}, \hat{\mathbf{y}}] = E[\mathbf{y} - \mu_y] = \sigma_y^2, \tag{6}$$

i. e., the variance of the target signal. Any reservoir with states that are remotely useful should lead to a MSE *on the training data* that is smaller (formal proofs exist but are omitted here). This bound only applies to the MSE measured on the training data, but the MSE on the test data for a properly trained and regularized model should not be very different.

One disadvantage to MSE is the fact that it is sensitive to the variance of the target signal, which is why other measures than MSE are often used. For example, *normalized mean squared error* (NMSE)

$$\mathrm{NMSE}[\mathbf{y}, \hat{\mathbf{y}}] = \frac{\mathrm{MSE}}{\sigma_y^2} \tag{7}$$

2 In fact, from the perspective of machine learning, logistic regression is greatly preferable over MSE-based linear regression for classification tasks.

has the advantage that it no longer depends on the signal variance and *normalized root mean squared error* (NRMSE = $\sqrt{\text{NMSE}}$) has the additional advantage that its size can be interpreted as a fraction of the target signal's standard deviation and that its value is $\in [0, 1]$ when evaluated on the training data. However, one of the most frequently used measures, especially in statistics, is the *coefficient of determination*, or R^2:

$$R^2[\mathbf{y}, \hat{\mathbf{y}}] = 1 - \frac{\text{MSE}}{\sigma^2(\mathbf{y})}. \tag{8}$$

Like NMSE, R^2 is normalized. In addition, given the fact that (on the training data) MSE $\leq \sigma^2(\mathbf{y})$, $0 \leq R^2 \leq 1$, where a value of 0 represents the useless reservoir mentioned above and a value of 1 corresponds to perfectly reproducing the target signal.

Using the orthogonality principle, the MSE of the resulting estimator can also be written as

$$\text{MSE}(\mathbf{y}, \hat{\mathbf{y}}) = \sigma^2(\mathbf{y}) - \text{cov}(\mathbf{y}, \hat{\mathbf{y}}) \tag{9}$$

and as a consequence, R^2 equals:

$$R^2[\mathbf{y}, \hat{\mathbf{y}}] = \frac{\text{cov}(\mathbf{y}, \hat{\mathbf{y}})}{\sigma^2(\mathbf{y})}. \tag{10}$$

Note that, in the context of linear regression (with minimization of the MSE) and when evaluated on the training data, it can be proven that R^2 equals the square of the correlation coefficient R which is defined as

$$R[\mathbf{y}, \hat{\mathbf{y}}] = \frac{\text{cov}[\mathbf{y}, \hat{\mathbf{y}}]}{\sigma(\mathbf{y}, \hat{\mathbf{y}})}. \tag{11}$$

A comprehensive explanation for this can be found in, e. g., [15].

When using a reservoir for classification tasks, it would often be better to use a linear classification model, such as logistic regression, to train the readout. This is the case whenever the outputs are thresholded to determine the correct class. Unfortunately, this is rarely done in practice, but after training, it is common to use typical classification performance errors, such as *accuracy* (the fraction of samples for which the predicted class was the correct one), *misclassification loss* (the fraction of wrongly classified samples). When addressing bitwise tasks in telecommunication, the reservoir outputs are typically sampled once per bit and the *bit error rate* (BER) is reported.

2.2.4 Echo state networks as a model system

In what follows, we briefly summarize echo state networks (ESN). In the early days of reservoir computing, these were the most frequently used types of reservoirs. As researchers have extensively studied them, their properties are well understood. In addition, their simplicity makes it easier to understand the interaction between several important system parameters that are also relevant for most physical reservoir implementations. Understanding the interaction between these parameters and the suitability of a reservoir for a given task is crucial for designing good physical reservoirs. For these reasons, ESNs will also be used as a model system further on in this chapter.

As stated above, ESNs are discrete time analogue recurrent neural networks with hyperbolic tangent neuron activation functions. Mathematically, using bold lowercase letters for column vectors, bold uppercase letters for matrices and $n = 1 \dots n_{\max}$ to indicate discrete time, they can be described by the following equations:

$$\mathbf{a}[n] = \mathbf{W}_{\text{res}}\mathbf{x}[n-1] + \mathbf{W}_{\text{in}}\mathbf{u}[n-1] + \mathbf{w}_{\text{bias}} \tag{12}$$

$$\mathbf{x}[n] = \tanh(\mathbf{a}[n]). \tag{13}$$

Here, $\mathbf{a}$ are the neuron *activations*, the input signals to the neuron non-linearities. Note that some sources use $\mathbf{u}[n]$ instead of $\mathbf{u}[n-1]$ to calculate the activations, which results in shifting the reservoir states, the readout signals and the desired output signal one step forward in time. In ESNs, this difference can be interpreted as assigning the discrete unit delays entirely to the interneuron connections (but not to the input connections) or entirely to the neurons. Neither view maps to the reality of physical implementations, where all connections and nonlinear elements have their own physical delay and their relative sizes depend on the specific implementation. The notation chosen here seems more natural from that perspective.

The $N \times k$ input weights $\mathbf{W}_{\text{in}}$ feed the inputs to the reservoir nodes, the $N \times N$ reservoir weights $\mathbf{W}_{\text{res}}$ provide the internal feedback in the reservoir and the N internal bias weights $\mathbf{w}_{\text{bias}}$ set the operating point of the reservoir nodes. ESNs can be fully or sparsely connected. Early studies indicate that for networks with real-valued signals, this decision does not impact performance a lot [16]. The input or inputs are also connected to all neurons or a subset thereof and all neurons also receive a constant input (a bias). A typical reservoir creation recipe is as follows: the weights for all connections are sampled from some probability distribution (e. g., the standard normal distribution or a uniform distribution between –1 and 1). In the case of sparse connectivity, a randomly sampled fraction of them are set to zero. After initialization, each set of weights (the input weights, the bias weights, and the reservoir weights) is rescaled with its own scaling factor: the *input scaling* (IS), the *bias scaling* (BS), and the *spectral radius scaling* (SR), all of which impact the overall dynamical regime of the reservoir and the computations it is suitable for.

The diagonal elements of $\mathbf{W}_{\text{res}}$ are usually nonzero. Since they couple each node's previous state back to itself, the neurons act as nonlinear low-pass filters. Since these values affect the overall bandwidth and dynamics of the reservoir, it is often desirable to control their relative magnitude compared to the other weights explicitly. This led to several descriptions of *leaky ESNs* [17], one of which is

$$\mathbf{a}[n] = \mathbf{W}_{\text{res}}\mathbf{x}[n-1] + \mathbf{W}_{\text{in}}\mathbf{u}[n-1] + \mathbf{w}_{\text{bias}} \tag{14}$$

$$\mathbf{x}[n] = \tanh((1-\alpha)\mathbf{a}[n-1] + \alpha\mathbf{a}[n]) \tag{15}$$

where α is the *leak rate*, $0 \leq \alpha \leq 1$. It is also possible to randomize the leak rates, such that they are different for each neuron. Since the self-feedback of each neuron is now explicitly regulated by the leak rate, the diagonal elements of $\mathbf{W}_{\text{res}}$ are often set to zero. The *leak rate* α is yet another tuning parameter for the reservoir.

Finally, ESNs can also be used with *output feedback*. In this case, the output of the readout projected back into the reservoir nodes, again with random weights and a tuneable rescaling factor. Since this approach is mostly used for signal generation tasks, we will not consider it further here. More information can be found in [8, 18, 19].

For an ESN, as for most other incarnations of reservoir computing, the target signal $\mathbf{y}[n]$ is approximated by optimising the $N+1$ readout weights $\{\mathbf{w}_{\text{out}}, w_0\}$ in a linear combination of the state signals at each time t:

$$\hat{\mathbf{y}}[n] = \mathbf{w}_{\text{out}}^T\mathbf{x}[n] + w_0 \tag{16}$$

where $\mathbf{u}[n]$ is the k-dimensional input vector at time n, N is the number of nodes (*neurons*) in the network, $\mathbf{x}[n]$ is the N-dimensional state vector, and $\mathbf{y}[n]$ is the output of the reservoir system. Note that in early works on reservoir computing, the input $\mathbf{u}[n]$ is also used in the linear combination, yielding additional input weights in equation (16).

In many ways, echo state networks are very different from most physical systems. For one, they operate in discrete time. In addition, they are usually fully connected, whereas this is not feasible for most physical systems. However, most physical systems have tuning parameters that have a very similar impact on computing to the scaling parameters and leak rate in an ESN. In addition, they can often be simplified in such a way that they resemble an ESN with a very specific connection pattern an setting of the scaling parameters. The knowledge of how these parameters affect computation in ESNs has often proved invaluable in tuning physical reservoir parameters. Studying ESN-like simplified models of a new physical system often helps to gain additional insights.

2.2.5 General requirements to reservoirs

In Section 2.2.2, we have described how to exploit the natural dynamics of any driven dynamical system for computation. Obviously, this in itself is not a recipe for success, since it only works well if there is a good match between the system's dynamics and what we need for a given task. In this section, we briefly review the most important theoretical concepts that have been developed to reason about the way reservoirs process their inputs and which kind of processing is required for a task. The focus here will be on an intuitive understanding of what these concepts mean. For an in-depth treatment, we refer to the original papers.

The fading memory property. Both ESNs and LSMs are nonlinear dynamical systems, which can be tuned to display a variety of dynamical regimes, including oscillatory and chaotic ones, by increasing their internal connection strengths. However, for the purpose of reservoir computing, the connection weights need to be tuned such that the network displays the *fading memory* property. Intuitively, this means that when the network is undriven, any transient dynamics eventually die out and the network evolves toward a unique stable state. The more slowly the transients die out, the longer information about past inputs *echoes* in the internal state. This *memory* about past states and inputs is largest when the system dynamics are close to a bifurcation that pushes them away from stability. Such a regime is often termed *edge of chaos*, but *edge of stability* seems more appropriate in this case, as it stresses that the driven system should operate at the stable side of the bifurcation. Tuning the system to be closer to or further away from the edge of stability is one of the ways to tune its memory to match task requirements.

Fading memory also means that, if you start the same reservoir from any two different initial states and drive both versions with the same input sequences, their internal states will evolve toward identical trajectories. When operating close to the edge of stability the effects of the initial state or past inputs remain present in the system for a long time as (nonlinearly transformed) *echos* of the past. Fading memory also implies that *similar* input patterns lead to similar states. In other words: reservoir computing extracts information from time-varying signals by extracting information that is spread in time and projecting it onto present time in such a way that a linear function is able to extract it. This implies that, for most tasks, the reservoir not only needs to remember the past, but also to nonlinearly transform it on the way.

In machine learning terms, the ESN reservoir acts as a high-dimensional spatiotemporal transformation (a filter bank) of the input signal(s). Obviously, the quality of the result largely depends on whether or not the internal state variables (the *features* for the linear regression) are suitable for the task. First, they should be tuned such that the reservoir indeed has the fading memory property. This can be ensured by analysing the Jacobian of the reservoir and finding the maximal local Lyapunov exponent (LLE). If this is < 1, the reservoir is at the stable side of the edge of stability. However, this analysis is rather complex. For tanh-ESNs, an approximation is usually made by linearizing the ESN at the point where its neurons have maximal gain (equal to $+1$), i. e., when their inputs are zero. Ensuring stability in this linearized reservoir boils down to setting the largest eigenvalue of the reservoir weight matrix W_{res} (its *spectral radius*) to 1. In most ESNs, this will result in stability [7].

Universal approximation properties. One of the cornerstones of the theory of (nonrecurrent) neural networks are the early universal approximation theorems, e. g., [4, 5], which basically state that a neural network with two layers of sigmoidal neurons (one hidden layer and one output layer) can approximate any desired input-output relationship with any desired precision *if the number of neurons on the hidden layer is made large enough*. This relies on one very important principle: by making the network larger, you will eventually be able to realize all functions that you were not able to realize with smaller networks.

Although the proofs are less strict, similar theorems exist in reservoir computing, stating that some types of reservoirs with a linear readout can be universal approximators for filters with fading memory [20]. A first requirement for this is that the reservoir itself must fulfill the fading memory property. A second requirement is that it has the *separation property*, which can be interpreted as the fact that, if the reservoir is made large enough, any difference in the input sequence must eventually result in linearly separable differences in the state space.

A reservoir is a universal approximator *only* if, by making the reservoir larger, a linear combination of the observed states can eventually cover all possible input transformations. The state signals of a reservoir are transformations of their common input history, and because of the recurrent nature of reservoirs, these signals are coupled. For this reason, the universal approximation property should be treated with some care. It is not sufficient to have a scalable nonlinear dynamical system in which some parameters are sampled from some random distribution. In fact, it is quite easy to construct a class of reservoirs that does *not* fulfil the universality. As a simple example, consider ESNs for which all input bias weights $\mathbf{w}_{\text{bias}}$ are set to zero. Since the tanh function is an odd function around its origin ($\tanh(-x) = -\tanh(x)$), such a reservoir can never output an even function of its inputs ($f(-x) = -f(x)$), however large we make it.

Although they are part of the theoretical foundations of and motivation for neural networks in general and reservoir computing in particular, universal approximation theorems are often not very useful in practice. Merely making a reservoir larger is often not the best way to improve its approximation quality. If this were the case, it should be possible to approximate any input-output relationship with any desired precision simply by making the reservoir larger, *while using the same approach for constructing the reservoir*. Unfortunately, this is not the case, as in practice, reservoir performance is known to saturate when increasing reservoir size.

In general, for physical reservoir computing, there is no systematic methodology to even check whether a given physical system could serve as a universal approximator and from which distributions its parameters should be sampled to achieve this. In addition, such theorems state nothing about how fast the approximation quality converges.

2.2.6 Physical reservoir computing

Very soon after the almost simultaneous introduction of the principles of reservoir computing from different research backgrounds, the common basis of these early works was identified and unified under the term *reservoir computing* [12, 21]. Also, researchers quickly started applying the approach to other (mostly non-linear) dynamical systems, and in particular to physical systems. In fact, the term reservoir computing was inspired by one of the first physical realizations [22], in which sound vibrations were transferred to a basin of water and the ripple patterns were used for spoken digit classification. The term *physical reservoir computing* was launched in [7], where a mechanical (robot) body was used as a reservoir to generate its own closed loop control. Soon after that, the subfields of *photonic* reservoir computing and *mechanical* reservoir computing (in the context of *morphological computing*) were launched. This book is dedicated to the first field. In the second, the focus is mostly on facilitating robotic control by using its body as a reservoir [23–25], responding to the motor actuations and the interaction with the outside world. In order to develop a theoretical foundation in this field, model systems, consisting of masses, passive springs, and dampers, were also studied [26–28]. Besides mechanical and photonic systems, physical reservoir computing is now being used with a range of different physical systems, such as memristive networks [29–34], carbon nanotubes [35], or molecular computing [36].

In general, the dynamical systems that serve as physical reservoirs operate in continuous time and are driven by continuous time input signals $\mathbf{u}(t)$. They are described by the temporal evolution of their internal state space $\mathbf{x}(t)$, which can be finite- or infinite-dimensional.

However, for the purpose of reservoir computing, they are driven in discrete time $t = nT$, $n \in \mathbb{Z}$, where T is the *sampling period*. This means that their externally applied input signals $\mathbf{u}(t)$ are generated from a discrete time sequence of values $\mathbf{u}[n]$ which are presented to the system at a given input rate $1/T$. An input signal generator then transforms this sequence into a continuous time signal using an appropriately chosen encoding scheme. The simplest scheme is to generate a piecewise constant signal, changing the input only at discrete moments $t = kT$.

Physical reservoirs are also observed in discrete time, i. e., their observed state variables are sampled with the sampling rate $1/T$. In practice, a *time shift* is sometimes used between the discrete moments $t = kT$ at which the input signals are applied and the corresponding moments of observation $t' = kT + \tau$. In most physical situations, the measured signals $x_i[n]$ will not be the state values themselves, but signals that have been transformed by the measurement process. For example, optical signals are converted to the electrical domain by measuring their intensity.

When studying a physical dynamical system, we often do not have access to the entire internal state space $\mathbf{x}(t)$. This corresponds to experimental situations where many variables may not be measurable, and is also the case if the dimensionality of the state space is infinite. We therefore assume that we can only access the instantaneous state of the system through a finite number N of the internal variables: $x_i(t)$, $i = 1, \ldots, N$. From here onward, the *dimension* of the reservoir will refer to the number of *observed* state space signals N, independent of the actual dimensionality of the state space of the physical system.

2.3 Information processing in reservoirs

In this section, we analyze how reservoirs process their input signals and how the most important high-level reservoir parameters interact to determine the suitability of a reservoir for a given task. We also discuss a few measures that allow to quantify this.

2.3.1 Reproduction memory

One crucial property of a reservoir is how long information about past inputs remains in the system. This is the *memory* that can be exploited by the readout to approximate the desired output. One of the very first ways in which this memory was analyzed is by quantifying how well a linear readout can reproduce past inputs. Since these inputs must be untransformed, this type of memory is also termed *linear memory*.

A way to quantify this, the *linear memory capacity*, was first proposed by Herbert Jaeger in one of his seminal works about echo state networks [37]. His proposed measure is the squared correlation coefficient between the target signal and its approximation $\hat{\mathbf{y}}$ as defined in equation (16):

$$C[\mathbf{X}, \mathbf{y}] = \frac{\mathrm{cov}^2(\hat{\mathbf{y}}, \mathbf{y})}{\sigma^2(\hat{\mathbf{y}})\sigma^2(\mathbf{y})}. \tag{17}$$

By definition, its values lie in $[0, 1]$. In the context of reservoir computing, a value of 1 implies that the readout can achieve a perfect reproduction of the target signal, whereas a value of 0 indicates that the state signals carry no information whatsoever that can be used to approximate the target signal with linear regression.

Using this measure, Jaeger proposed to quantify a reservoir's memory by assessing its capacity for reconstructing the input signal $\mathbf{u}_{-d} = \mathbf{u}[n-d]$ of d time steps in the past, for $d = 1 \ldots d_{\max}$:

$$CM_d[\mathbf{X}, \mathbf{u}_{-d}] = \max_{(\mathbf{w}_{\mathrm{out}})} C[\hat{\mathbf{y}}, \mathbf{u}_{-d}] \tag{18}$$

in which the maximum corresponds to the best possible MSE, i. e., the MSE for a theoretical readout that is trained on sequences of infinite length. Using these, the total linear memory capacity can be defined as

$$CM[\mathbf{X}, \mathbf{u}] = \sum_{d=1}^{\infty} CM_d[\mathbf{X}, \mathbf{u}_{-d}]. \tag{19}$$

Note that the capacity $CM_d[\mathbf{u}, \mathbf{u}_{-d}]$ for reconstructing a signal from its own past equals its squared autocorrelation function. However, the original aim of linear memory capacity was to characterize the processing that occurs in the dynamical system itself. By taking the inputs to be sequences of i.i.d. values, any measured memory will be due to the dynamical system and not due to the input's autocorrelation or *self-memory*. The distribution $p(u)$ from which the input values are sampled can be chosen according to what is most relevant for the dynamical system under study, but it is most common to use the uniform distribution in, e. g., $[-1, 1]$.

It should be noted that in many, if not all, real world situations the inputs are not i.i.d. In this case, it is useful to quantify how a reservoir processes the actual input signals. In order to quantify this, one can still calculate linear memory capacities using equation (18) with the actual input signal, but the upper bound from equation (20) is no longer guaranteed because the target signals of LMC, $\mathbf{u}_{-d}$, are no longer uncorrelated. An alternative would be to decorrelate the target signals in order of increasing values of d. This can in principle be achieved using a Gramm–Schmidt orthogonalization procedure, in which case the bound of equation (20) would again hold. However, as numerical errors quickly build up in the orthogonalization procedure, this is usually only feasible up to small maximal values of d.

An additional reason for measuring capacities with i.i.d. input sequences is that under these conditions the proposition holds that the *total linear memory capacity is bounded by the number of state signals that is used in the readout*:

$$CM_{\text{lin}}[\mathbf{X}, \mathbf{u}] \leq N. \tag{20}$$

In principle, this bound can be achieved, e. g., for linear ESNs for which W_{res} is an orthogonal matrix [38]. As soon as a reservoir operates in a nonlinear regime, CM_{lin} quickly deteriorates. The theoretical upper bound for linear memory can be approximated in nonlinear reservoirs, by operating them in their small-signal regime and assuming an unbounded signal-to-noise ratio. For example, in ESN reservoirs this can be achieved by setting the input scaling and bias scaling parameters so small that each tanh neuron's input is close to zero. However, as soon as some noise is introduced into the system or the readout, this theoretical linear memory quickly disappears. For a given value of the total linear memory capacity, different memory profiles can be achieved by tuning the reservoir's hyperparameters.

2.3.2 Nonlinear processing capacity

Clearly, a reservoir's linear memory does not tell the whole story, since most signal processing or classification tasks require much more than linear memory: they require a nonlinear transformation of the input history. This is typically what occurs in nonlinear dynamical systems such as ESNs. This is also the basis of nonlinear photonic reservoirs such as the original SOA-based integrated photonic reservoirs [39] (simulated only) or delayed feedback reservoir computing architectures that use a nonlinearity in the feedback loop (see Chapter 5 for an introduction). However, nonlinear transformations can also be achieved by combining a linear reservoir (providing memory) with a nonlinear readout. If sufficient memory can be provided by the reservoir *and* if a nonlinear (instead of a linear) readout is used that is in itself universal for memoryless computations, then such a reservoir can also be a universal approximator of nonlinear filters with fading memory [26]. In this case, sufficient memory means that *all* past inputs that are necessary for solving the task can be perfectly reconstructed from the observed reservoir state signals. This approach was discussed in [26] for mechanical systems. In practice, it turns out that a rather simple but nonlinear readout already offers a lot of the computations required for many real-world tasks. This principle is used in, e. g., the integrated passive photonic reservoirs described in [40–42].

Early on, researchers working with ESN-based reservoir computing noticed that for a fixed reservoir size with a linear readout, there is a trade-off between memory and nonlinearity. As soon as a reservoir is tuned into a slightly non-linear regime, the total linear memory capacity quickly drops. In a first attempt to quantify this, [43], the spectral transformation of reservoirs was used as a measure of their nonlinearity. This was measured by driving the reservoirs with a sinusoidal input signal and measuring the fraction of the energy in the state signals that remained at that frequency.

In [44], an extension of linear memory capacity was proposed which allows to more accurately quantify the transformations a reservoir performs. Intuitively, the approach can be explained as follows. In principle, any fading memory filter in discrete time can be approximated with any desired precision by a polynomial expansion on a potentially very large but finite number of values in the input history. This means that, for any given joint probability density of the relevant input values, a Hilbert space of fading memory functions can be constructed with an orthogonal polynomial basis (see [44] for rigorous definitions). The construction of such a basis is very similar to the construction of a polynomial basis for real functions of multiple real variables, where these variables are the past input values $\mathbf{u}_{-d}$. If we assume, as we did for linear memory capacity, that the input sequences are generated by an i.i.d. uniform process with values in $[-1,1]$, a polynomial basis can be constructed from finite products of normalized Legendre polynomials for each time step:

$$\mathbf{y}_\mathbf{k} = \prod_i \mathcal{L}_{k_i}(\mathbf{u}_{-i}) \tag{21}$$

where $\mathcal{L}_{k_i}(\cdot)$, $k_i \geq 0$, is the normalized Legendre polynomial of degree k_i. The normalized Legendre polynomial of degree 0 is a constant. The normalized Legendre polynomials of degree 1 are the u_{-d} that were used for computing the linear memory capacity. The set $\mathbf{k}$ lists the polynomial degrees for each past input, where k_i is the degree corresponding to the input of i time steps in the past. For each basis function, we can compute its total degree as the sum of the individual degrees for each delay: $\mathcal{D}_{\mathbf{k}} = \sum_i k_i$ and its memory depth as the largest index in k for which k_i is nonzero.

Expanding on the theory for linear memory capacity, we can again define the capacity for approximating these basis functions as the expected value for infinitely long sequences of the squared correlation coefficient between the target signal and its best approximation by the reservoir. We can also define the *total information processing capacity* as the sum of these capacities across all basis functions and prove that this sum, too, always lies in $[0, N]$.

In practice, capacities have to be estimated with limited precision from finite input sequences. In [44], a procedure is described to iterate through them. It uses relevance thresholds to decide how large a capacity estimate has to be in order to be sufficiently accurate. Constraining the set of possible basis functions to a finite set can be done either by setting upper bounds on the total degree and memory depth, or by assuming that a physical system's approximation capacity decreases monotonically for increasing degree and/or memory depth. This was found to be a reasonable assumption for many physical systems.

2.3.3 Memory, nonlinearity, and noise-sensitivity

As stated in the previous section, the functions used to compute linear memory capacity form a subset of those used to compute total memory capacity. The fact that the total capacity of any reservoir system is bounded by the total number of state signals used in the readout necessarily implies that there must be a trade-off between linear and nonlinear processing. By adding all capacities with the same memory depth or the same total degree, we can visualize the capacity profile of a given reservoir, as shown in Figure 2.2. This accords with our previous observations that an ESN's capacity for exactly reproducing past inputs rapidly decreases as the neurons are operated in a more nonlinear regime. In Figure 2.3, this is made even more clear by plotting the fraction of total capacity that is due to linear memory and nonlinear memory, respectively.

In most cases when a system is made more nonlinear, a rapidly increasing number of nonlinear terms emerges in the polynomial approximation of the reservoir states. In particular, when the spectral radius is high enough for the inputs to reverberate in the system's dynamics for a while, each nonlinearity it passes performs a nonlinear mixture of its inputs, spreading the input information across more and higher degree

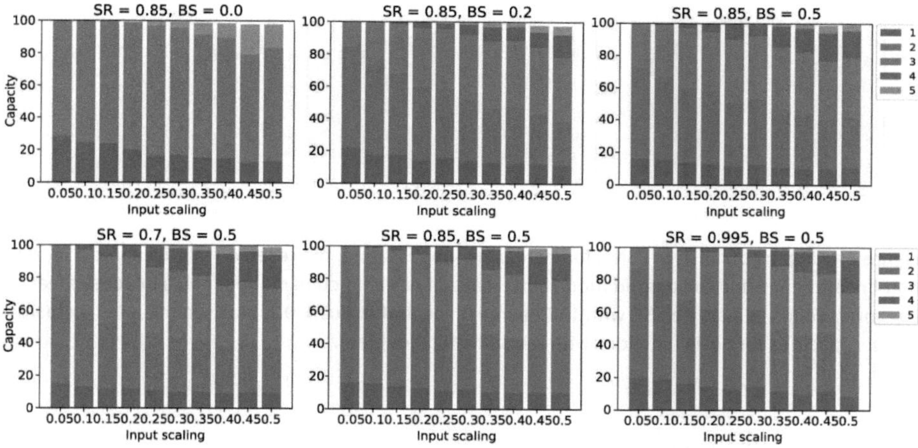

Figure 2.2: Summarized total information processing capacity profiles (for total degrees up to 5) for different settings of the scaling parameters in echo state networks: (top panels) Impact of the bias scaling, showing no contributions of even degree capacities for zero bias and increasing nonlinear contributions for increasing bias; (bottom panels) Impact of the spectral radius, showing increased linear memory as the spectral radius approaches 1.0 (each bar corresponds to a single ESN with 100 neurons, spectral radius and bias scaling are indicated in each panel's title; colors group capacities for basis functions with the same total degree).

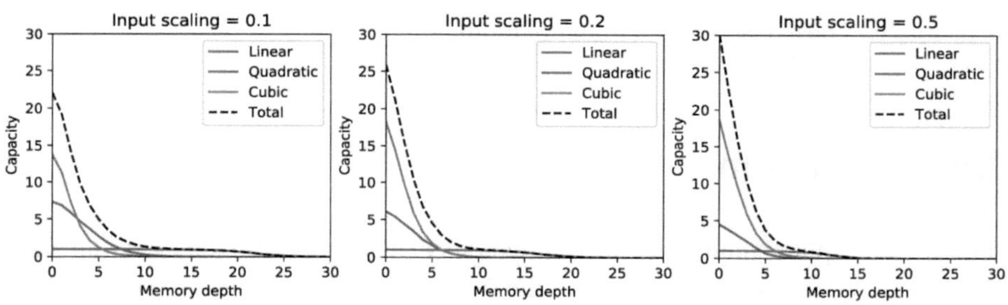

Figure 2.3: Linear and non-linear contributions to the total memory capacity: for increasing input scaling, the reservoir is driven into a more non-linear regime. The total linear memory capacity decreases and the contribution of non-linear capacities to the total capacity increases (each plot corresponds to a single ESN with 100 neurons, spectral radius equal to 0.85, bias scaling equal to 0.05).

polynomial basis functions. As a consequence, the capacities that measure these contributions get very small and, therefore, hard to estimate accurately.

The same nonlinear mixing also dramatically reduces a system's performance under multiple inputs [45], as well as its noise robustness (since each noise source can be considered as an additional input signal). When multiple inputs are present, one needs to extend the Hilbert space model to multiple inputs, i. e., consider all func-

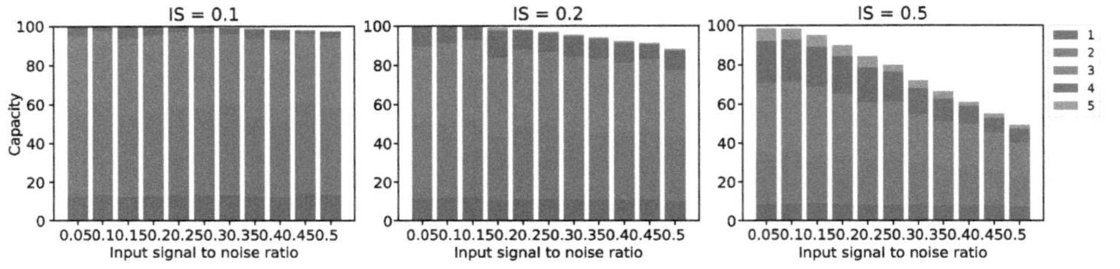

Figure 2.4: Impact of noise in ESNs: for increasing input scaling, the reservoir is driven into a more non-linear regime and the degradation of total capacity as a function of input noise increases (each bar corresponds to a single ESN with 100 neurons, spectral radius = 0.7, bias scaling = 0.5; colours group capacities for basis functions with the same total degree).

tions of the complete (fading memory) input history of all inputs. This includes all higher degree cross products between delayed values originating from different input signals, which also emerge naturally from the nonlinear mixing. Figure 2.4 illustrates this for reservoirs that are driven with a single input, to which an increasing amount of uniformly distributed noise is added. Because each reservoir is driven with a different input power (input scaling), the signal-to-noise ratio is reported in the plots. Whereas the total linear capacity is barely affected by increasing noise, the available nonlinear capacity quickly decreases for more nonlinear reservoir. As more noise gets mixed non-linearly with the information signal, less information about this signal and its history can be extracted by a linear readout.

2.4 Conclusion

The way the input signal(s) of a driven dynamical system perturbs its internal state, affecting not just the present but also the future of this state, can be considered as a form of *computation*. From its origins in neuroscience and engineering, reservoir computing has evolved into an easy and efficient way to harness this computation, in order to perform nonlinear filtering of the input signal(s) or to extract information from them. The larger the number of observed state signals and the more diverse the ways in which they are affected by current and past inputs, the better a system can be used for computation.

In order to make a given system adaptable to a range of tasks, it is beneficial if its dynamical regime can be tuned by a number of global parameters, such as a scaling of the feedback gains or losses, of the input power and of the overall operating point. For systems with a fixed number of observed signals, there exists a trade-off between (linear) memory and nonlinearity. In addition, more nonlinear systems tend to be less robust to noise.

Bibliography

[1] J. von Neumann. "Probabilistic logics and the synthesis of reliable organisms from unreliable components". In: *Automata studies* 34 (1956), pp. 43–99.

[2] S. Mittal. "A survey of techniques for approximate computing". In: *ACM Computing Surveys* 48.4 (Mar. 2016), 62:1–62:33.

[3] Q. Xu, T. Mytkowicz, and N. S. Kim. "Approximate computing: a survey". In: *IEEE Design and Test* 33 (2016), pp. 8–22. ISSN: 2168-2356.

[4] G. Cybenko. "Approximation by superpositions of a sigmoidal function". In: *Mathematics of Control, Signals and Systems* 2.4 (1989), pp. 303–314.

[5] K. Hornik. "Approximation capabilities of multilayer feedforward networks". In: *Neural Networks* 4.2 (1991), pp. 251–257.

[6] C. Horsman et al. "When does a physical system compute?" In: *Proceedings. Mathematical, Physical, and Engineering Sciences/The Royal Society* 470.2169 (2014), p. 20140182. ISSN: 1364-5021. DOI: 10.1098/rspa.2014.0182. URL: http://rspa.royalsocietypublishing.org/content/470/2169/20140182.

[7] K. Caluwaerts et al. "Locomotion without a brain: physical reservoir computing in tensegrity structures". In: *Artificial Life* 19.1 (2013), pp. 35–66.

[8] H. Jaeger and H. Haas. "Harnessing nonlinearity: predicting chaotic systems and saving energy in wireless communication". In: *Science* 304 (2004), pp. 78–80.

[9] H. Jaeger. "The "echo state" approach to analysing and training recurrent neural networks – with an erratum note 1". In: 2010.

[10] W. Maass, T. Natschlager, and H. Markram. "Real-time computing without stable states: a new framework for neural computation based on perturbations". In: *Neural Computation* 14 (2002), pp. 2531–2560.

[11] B. Schrauwen, D. Verstraeten, and J. M. Van Campenhout. "An overview of reservoir computing: theory, applications and implementations". In: *ESANN*. 2007.

[12] M. Lukoševičius and H. Jaeger. "Reservoir computing approaches to recurrent neural network training". In: *Computer Science Review* 3 (2009), pp. 127–149.

[13] M. Lukoševičius, H. Jaeger, and B. Schrauwen. "Reservoir computing trends". In: *KI – Kunstliche Intelligenz* 26 (2012), pp. 365–371.

[14] A. Goudarzi and C. Teuscher. "Reservoir computing: Quo Vadis?" In: *NANOCOM*. 2016.

[15] D. Schiebler. *R and R^2, the relationship between correlation and the coefficient of determination*. http://danshiebler.com/2017-06-25-metrics/. Blog, Posted on June 25, 2017.

[16] L. Buesing, B. Schrauwen, and R. A. Legenstein. "Connectivity, dynamics, and memory in reservoir computing with binary and analog neurons". In: *Neural Computation* 22.5 (2010), pp. 1272–1311.

[17] H. Jaeger et al. "Optimization and applications of echo state networks with leaky-integrator neurons". In: *Neural Networks* 20.3 (2007), pp. 335–352.

[18] H. Jaeger and H. Markram. "On the role of feedback in enhancing the computational power of generic neural microcircuits". In: 2006.

[19] H. Jaeger. "Controlling recurrent neural networks by conceptors". In: *CoRR* abs/1403.3369 (2014). arXiv:1403.3369. URL: http://arxiv.org/abs/1403.3369.

[20] W. Maass and H. Markram. "On the computational power of circuits of spiking neurons". In: *Journal of Computer and System Sciences* 69 (2004), pp. 593–616.

[21] D. Verstraeten et al. "An experimental unification of reservoir computing methods." In: *Neural Networks* 20.3 (Apr. 2007), pp. 391–403. DOI: 10.1016/j.neunet.2007.04.003.

[22] C. Fernando and S. Sojakka. "Pattern recognition in a bucket". In: *Lecture notes in computer science* 2801/2003 (2003), pp. 588–597. DOI: 10.1007/978-3-540-39432-7_63.

[23] K. Nakajima et al. "A soft body as a reservoir: case studies in a dynamic model of octopus-inspired soft robotic arm". In: *Frontiers in Computational Neuroscience*. 2013.

[24] K. Nakajima et al. "Exploiting short-term memory in soft body dynamics as a computational resource". In: *Journal of The Royal Society Interface* 11.100 (2014).

[25] J. Degrave et al. "Developing an embodied gait on a compliant quadrupedal robot". In: *2015 IEEE-RSJ international conference on intelligent robots and systems (IROS)* (2015), pp. 4486–4491.

[26] H. Hauser et al. "Towards a theoretical foundation for morphological computation with compliant bodies". In: *Biological Cybernetics* 105 (2011), pp. 355–370.

[27] H. Hauser et al. "The role of feedback in morphological computation with compliant bodies". In: *Biological Cybernetics* 106 (2012), pp. 595–613.

[28] G. Urbain et al. "Morphological properties of mass-spring networks for optimal locomotion learning". In: *Frontiers in Neurorobotics*. 2017.

[29] A. V. Avizienis et al. "Neuromorphic atomic switch networks". In: *PLoS ONE*. 2012.

[30] M. S. Kulkarni and C. Teuscher. "Memristor-based reservoir computing". In: *2012 IEEE-ACM international symposium on nanoscale architectures (NANOARCH)* (2012), pp. 226–232.

[31] H. O. Sillin et al. "A theoretical and experimental study of neuromorphic atomic switch networks for reservoir computing." In: *Nanotechnology* 24.38 (2013), p. 384004.

[32] J. P. Carbajal et al. "Memristor models for machine learning". In: *Neural Computation* 27.3 (2015), pp. 725–747.

[33] D. Kudithipudi et al. "Design and analysis of a neuromemristive reservoir computing architecture for biosignal processing". In: *Frontiers in Neuroscience*. 2015.

[34] C. Du et al. "Reservoir computing using dynamic memristors for temporal information processing". In: *Nature Communications*. 2017.

[35] M. Dale et al. "Evolving carbon nanotube reservoir computers". In: *UCNC*. 2016.

[36] A. Goudarzi, M. R. Lakin, and D. Stefanovic. "DNA reservoir computing: a novel molecular computing approach". In: *DNA computing and molecular programming* (2013), pp. 76–89.

[37] H. Jaeger. "Tutorial on training recurrent neural networks, covering BPTT, RTRL, EKF and the 'echo state network' approach". In: *Technical Report GMD Report 159, German National Research Center for Information Technology* (2002).

[38] O. L. White, D. D. Lee, and H. Sompolinsky. "Short-term memory in orthogonal neural networks." In: *Physical Review Letters* 92.14 (2004), p. 148102.

[39] K. Vandoorne et al. "Toward optical signal processing using photonic reservoir computing". In: *Optics Express* 16 (2008), pp. 11182–11192.

[40] K. Vandoorne et al. "Experimental demonstration of reservoir computing on a silicon photonics chip." In: *Nature Communications* 5 (2014), p. 3541.

[41] A. Katumba et al. "A multiple-input strategy to efficient integrated photonic reservoir computing". In: *Cognitive Computation* 9 (2017), pp. 307–314.

[42] A. Katumba et al. "Low-loss photonic reservoir computing with multimode photonic integrated circuits". In: *Scientific Reports*. 2018.

[43] D. Verstraeten et al. "Memory versus non-linearity in reservoirs". In: *The 2010 international joint conference on neural networks (IJCNN)* (2010), pp. 1–8.

[44] J. Dambre et al. "Information processing capacity of dynamical systems". In: *Scientific Reports* 2 (2012), p. 514.

[45] M. Hermans and B. Schrauwen. "Memory in linear recurrent neural networks in continuous time". In: *Neural Networks* 23.3 (2010), pp. 341–355.

Andrew Katumba, Matthias Freiberger, Floris Laporte, Alessio Lugnan, Stijn Sackesyn, Chonghuai Ma, Joni Dambre, and Peter Bienstman

3 Integrated on-chip reservoirs

3.1 Introduction

In this chapter, we focus on the use of on-chip photonic reservoir systems using integrated photonics in a Silicon-on-Insulator. We show how passive reservoir computing chips can be used to avoid the need for any nonlinearity inside the network itself. We place special emphasis on developing an integrated analog optical readout, and the repercussions this has on the training algorithm. We focus on how these reservoirs can be used to perform a variety of telecom-related tasks (bit level tasks, nonlinear dispersion compensation, etc.) at high speeds and low power consumption. In addition, we present two different alternative topologies which no longer rely on an explicit node structure inside the reservoir: one is based on a chaotic cavity which mixes the signals, and another is a spatial analog of reservoir computing based on pillar scatterers, that can be used to speed up classification of biological cells.

3.2 Passive reservoir computing

A recent development in the design of physical RC systems is the realization that for certain tasks that are not strongly nonlinear, it is possible to achieve state-of-the-art performance using a completely passive linear network, i. e., one without amplification or nonlinear elements. The required nonlinearity is introduced at the readout point, typically with a photodetector [1]. The work discussed in this chapter is also based on this architecture. Aside from the integrated implementation introduced in [1], the passive architecture has been adapted to the single node with delayed feedback architecture in form of a coherently driven passive cavity [2].

Apart from simplicity from a fabrication point-of-view, a further advantage of such a passive architecture is the reduced power consumption, since the computation itself does not require external energy.

The integrated photonic reservoirs typically studied in the past are limited to planar architectures in a bid to minimize crossings that manifest as a source of signal cross-talk and extra losses. This constrains the design space from which reservoir configurations can be chosen. The swirl reservoir architecture, which is also used in this work, was introduced in [3] as a way to satisfy planarity constraints while allowing for a reasonable mixing of the input signals. A 16-node photonic swirl reservoir is shown

https://doi.org/10.1515/9783110583496-003

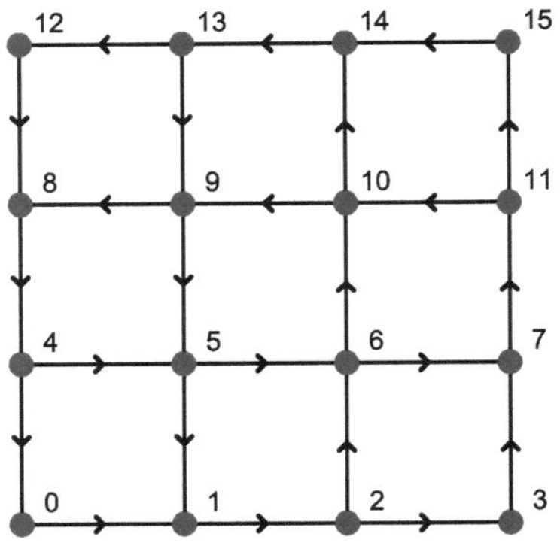

Figure 3.1: Signal flow in a 16-node swirl reservoir architecture. The time-dependent output at each numbered node is linearly combined to result in the answer of the computation. As for inputs, depending on the application the input can be inserted in one or more of the numbered nodes.

in Figure 3.1. One way of looking at this reservoir is that it acts as a giant multipath interferometer, which mixes the input in such a way that it gets converted into a higher dimensional space where a linear classifier will be able to separate the different classes more easily. The input to the integrated photonics reservoir chip could be to a single input node as is in [1] or to multiple inputs, which has some advantages over the former strategy, as is discussed extensively in [4].

In discretized time, the reservoir state update equation is given in a general form by

$$\vec{x}[k+1] = \boldsymbol{W}_{\text{res}}\vec{x}[k] + \vec{w}_{\text{in}}(\vec{u}[k+1] + \vec{u}_{\text{bias}}) \tag{1}$$

where $\vec{u}$ is the input to the reservoir and $\vec{u}_{\text{bias}}$ is a fixed scalar bias applied to the inputs of the reservoir. For an N-node reservoir, $\boldsymbol{W}_{\text{res}}$ is an $N \times N$ matrix representing the interconnections between reservoir components taking into account splitting ratios and losses, with phases drawn from a random uniform distribution on $[-\pi, \pi]$, $U(-\pi, \pi)$. The vector $\vec{w}_{\text{in}}$ is an N-dimensional column vector whose elements are nonzero for each active input node. These input weights are similarly chosen from $U(-\pi, \pi)$.

The output is given by a simple linear combination of the states:

$$\vec{y}[k] = \vec{W}_{\text{out}} \cdot \vec{x}[k]. \tag{2}$$

Our work in [1] experimentally verifies that a passive integrated photonic reservoir can yield error-free performance on the header recognition task for headers up to 3 bit in length with simulations indicating that it should be possible to go up to 8-bit headers (see Figure 3.2). We additionally demonstrated that the passive integrated photonics reservoir can be used for bit level manipulations on digital optical bit streams that could be useful for various telecommunication tasks. The paper [1] also contains more information about the chip design and fabrication procedure.

While the passive reservoir architecture introduced in this section is amenable to various tasks as outlined above, it suffers from major drawbacks due to the inherent limitations of an integrated photonics platform. Particularly, the losses increase with

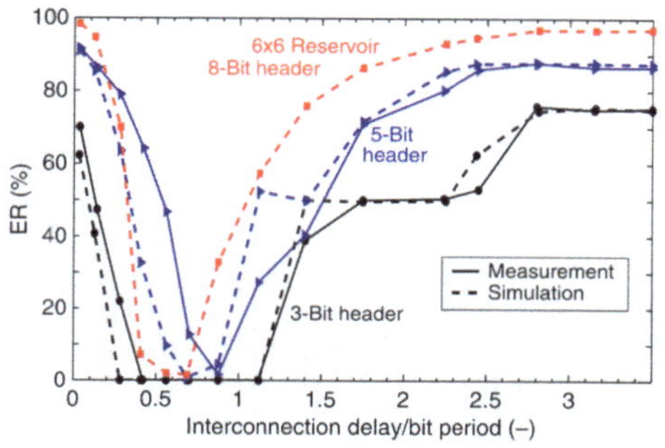

Figure 3.2: Performance of a 6 × 6 swirl passive integrated photonics reservoir on the 3-, 5-, and 8-bit header recognition task [1].

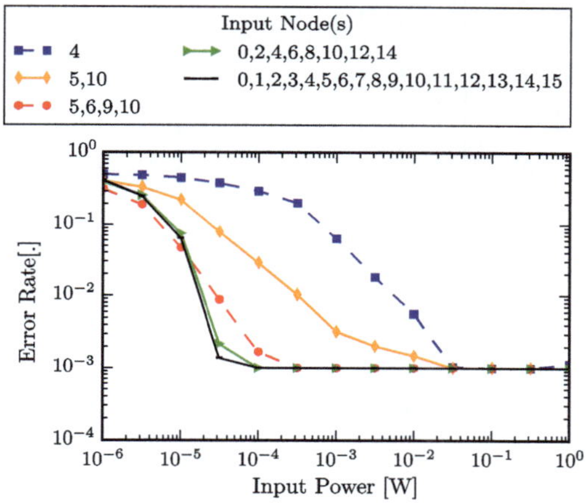

Figure 3.3: Error rate versus total input power for different injection scenarios. The minimum measurable error, given the number of bits used for testing, is 10^{-3}.

the size of the architecture. We have therefore studied techniques and architectures which seek to improve the performance and energy efficiency of this architecture and integrated passive reservoirs in general. We have, for example, found that by carefully selecting the nodes to which the input to the reservoir is provided, we can gain significantly in energy efficiency [4]. As Figure 3.3 shows, we need 2 orders of magnitude less input power if we split the total input power and inject it into the central 4 nodes (5, 6, 9, 10) rather than into any single node. This improvement is due to the increased richness of the reservoir from the varied mixing between multiple copies of the input with different phases, and the more even power landscape that amounts to more efficient mixing as the signal powers are similar across different paths in the reservoir.

In a separate approach, we numerically studied the impact of using multimode rather than single-mode components in the integrated photonics reservoir on its overall energy efficiency [5]. The successful application of this approach strongly hinges on the design of a novel multimode Y-junction with carefully tailored adiabaticity that lowers the losses at combination points in the photonic network constituting the reservoir. As is illustrated in Figure 3.4 for a 36-node reservoir, we can gain up 30 % in per-node power, especially for nodes that are furthest from the input point. This extra

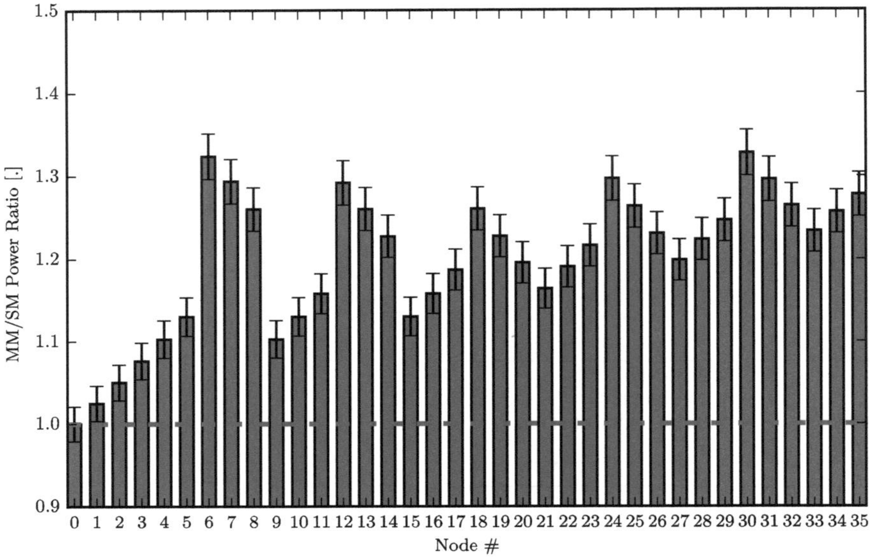

Figure 3.4: Comparison of single-mode and multimode 36 node reservoir average power per node for input to node 0.

power boost could be the difference between being below or above the noise floor at a node.

3.3 Integrated optical readout

3.3.1 Rationale

A drawback of our current experiments is that the linear combination of the reservoir states is still happening in the electrical domain. Indeed, Vandoorne et al. [1] transferred the signal at each node from the optical to the electrical domain using a photodetector, and then sent it through an AD converter. Finally, the required linear combination of signals and readout weights was performed using a microprocessor. See Figure 3.5 for a detailed illustration of the process.

In order to truly reap the benefits of optical computing, signals need to be processed at very high data rates in an energy-efficient way. Considering ecological as well as economical factors, minimizing power consumption is of paramount importance for future computing technologies. From that perspective, the approach pursued in Figure 3.5 and [1] for reading out integrated photonic reservoirs is inefficient, since there is a significant energy and latency cost associated to it. Hence, it is desirable to perform the summing of signals in the optical domain using optical modulators in-

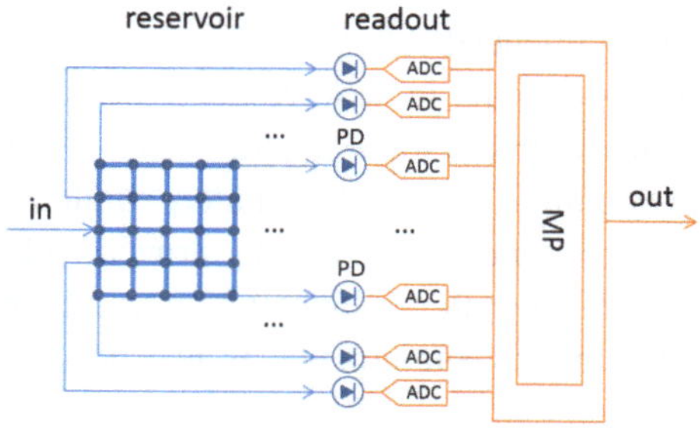

Figure 3.5: Reservoir and readout system of existing RC photonic chip prototypes, with the nodes of the reservoir being collected in the readout section, where optical output signals are converted to electrical signals and then processed to a final output. "PD": photodiode, "ADC": Analogue-to-Digital converter, "MP": Microprocessor. The blue and orange parts represent respectively the optical and electronic signals and components.

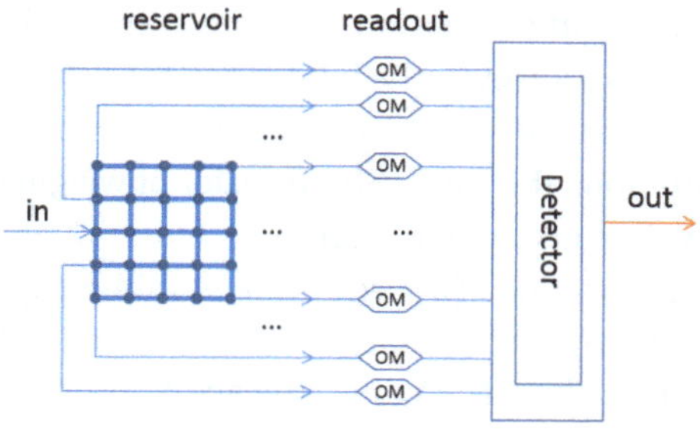

Figure 3.6: Schematic of a fully optical readout. Each optical output signal is modulated by an Optical Modulator ("OM") implementing the weights. The optical outputs are then sent to a photodiode where all signals are summed and then converted to a final electric output signal.

stead of in the electrical domain. Using such an integrated optical readout, only a single photodetector, which receives the coherent weighted sum of all optical signals, is required. A straightforward low-power optical weighting element can take the form of a reverse-biased pn-junction. An even better solution would be to use nonvolatile optical weighting elements, such as the ones that are currently being developed by several groups [6–8]. Figure 3.6 illustrates the concept of a fully optical integrated readout.

3.3.2 Training integrated optical readouts

From a mathematical perspective, an integrated optical readout can be trained in a similar way to readouts in the electrical domain. Of course, one needs to take into account that passive photonic reservoirs make use of coherent light for added richness. This implies that the readout operates on complex-valued signals employing complex-valued weights contrary to the approach in [1], which used real-valued weights on real-valued signals. Nevertheless, in principle, integrated optical readouts can be trained using complex-valued ridge regression [9].

Nevertheless, despite the fact that the ridge regression training algorithm can be transferred to the complex domain, a number of practical challenges occur when training integrated optical readouts. Due to the integration of reservoir and readout on a silicon photonics chip, we lose direct observability of the reservoir states. Observing all states is mandatory though, in order to use classical linear readout training algorithms such as ridge regression and other least-squares approaches. At first glance, a possible solution could be to add a separate high-speed coherent photodetector to each reservoir node, which is only used during training to observe the states. The weights would then be calculated in the electrical domain, while the trained reservoir could still be operated entirely in the optical domain. This is unfortunately challenging due to the fact that high-speed photodetectors tend to be costly in terms of chip footprint. Therefore, such an architecture would not scale well when increasing the number of nodes in the photonic reservoirs to numbers common in classic echo state networks [10].

A second solution could be to train the weights based on simulations of the behavior of a virtual reservoir, using photonic circuit simulation software, which will obviously have full observability of all the nodes. However, the fabrication tolerances of these devices are such that the propagation phase of two nominally identical waveguides could be completely different. This prohibits the successful transfer of weights trained using the idealized simulated reservoir to actual hardware.

A possible resolution of this issue is the application of a pretraining-retraining approach, where a passive photonic reservoir is pretrained in simulation and the trained weights are transferred to an actual reservoir on chip. Thereafter, the actual training error is minimized on the reservoir by fine-tuning the weights of a given integrated optical readout using a black-box optimization approach. Unfortunately, previous simulation studies have shown that such an approach is not feasible, again due to high fabrication tolerances affecting the propagation phase of planar waveguides [11].

A final possibility, which is the one we actually prefer, is to exploit the weighting mechanism of the optical readout of the reservoir to read out all reservoir states through the single photodetector available at the end of the summing structure. Reading out the state variable s_i in response to the training input sequence can be simply realized by setting the weight of that state variable to 1 and all other weights to 0. By presenting the whole training input sequence to the reservoir n times, where n is the number of nodes of the reservoir, the training responses of all nodes can be collected through the single photodetector. By taking the square root of each measured power value, we can approximately invert the nonlinearity of the photodetector and obtain an estimate for the evolution of the light intensity over time at the corresponding reservoir node.

However, since passive photonic reservoirs work with coherent light, it is not sufficient to know only the light intensities at the points predefined as reservoir states: we also need to know the corresponding phase of the light. While the absolute phase of the optical signal inside the reservoir is lost within the photodetection process, the relative phases between the optical state signals influence the power at the detector output. We therefore estimate the phase between two given optical signals within the integrated reservoir by obtaining the evolution of the sum of their states through time as we apply the training signal at the reservoir's input. We now use the phase of one state signal (node) as a reference. Using the evolution of the power of the complex sum

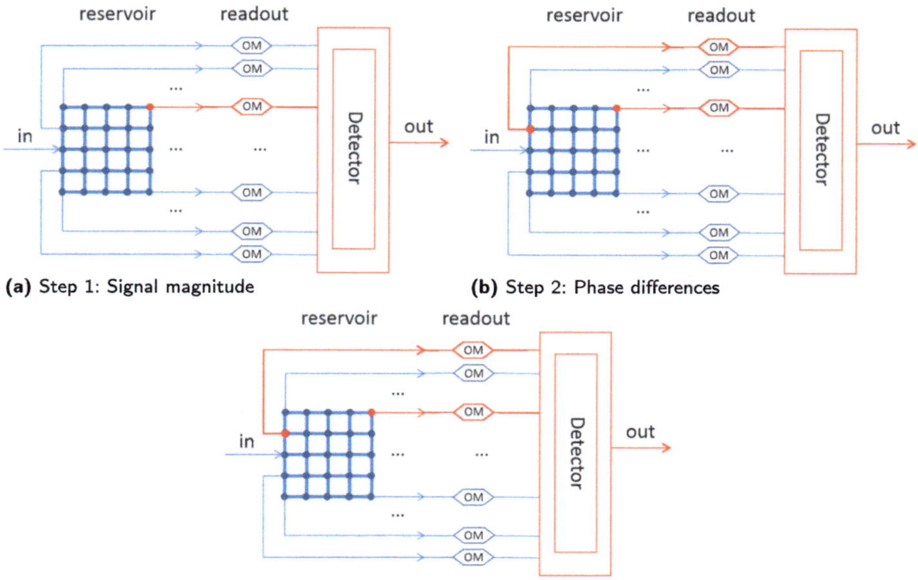

(a) Step 1: Signal magnitude

(b) Step 2: Phase differences

(c) Step 3: Correct sign of the phase differences

Figure 3.7: Illustration of the procedure applied to estimate the time trace of a reservoir state. Weights highlighted in red are set to 1, all remaining weights are set to 0. Step 1: Invert output signal to obtain the magnitude (light intensity) of the activated state. Step 2: Estimate the magnitude of the phase difference for the state in question using the sum of the state and a reference state, as well as the states signal magnitude obtained in step 1. Step 3: Shift the phase of the reference states readout weight by $\frac{\pi}{2}$ (not visible in the diagram), then repeat step 2. Compare the result with the result obtained in step 2 in order to infer the sign of the phase difference between the state in question and the reference state.

between the reference node's signal and each other node signal, as well as the previously determined powers of all individual states, we are able to estimate the relative phase of each node signal with respect to the reference node using basic trigonometric relationships.

The last stage of this calculation consists of an inverse cosine, which is injective, in the sense that there are always two solutions within the range $[-\pi, \pi]$. To discriminate between them, we perform a third measurement between the reference node's signal and each other node's signal, now shifting the phase of the reference node's readout weight by $\frac{\pi}{2}$ and comparing with the phase estimate obtained before. As a result, the whole process requires that we feed the training sequence through the reservoir $3n - 2$ times. See Figure 3.7 for an illustration of the whole process.

Under ideal conditions, this nonlinearity inversion procedure is exact. In order to confirm that, we conduct the following numerical experiment: We train a passive 4×4 photonic swirl reservoir architecture with integrated readout to perform 3-bit

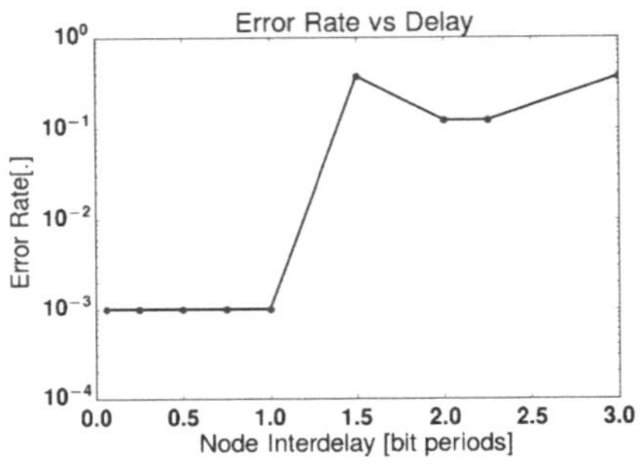

Figure 3.8: Bit error rate versus the time delay between nodes (interdelay) in bit periods for a 4×4 passive swirl architecture with integrated optical readout performing 3-bit header recognition (pattern 101). The minimal detectable error rate is 10^{-3}. Results are averaged over 10 reservoirs with different phase configurations for every data point.

header recognition on a power-modulated digital signal fed into to the reservoir. The integrated optical readout is simulated by inner product multiplication of the complex reservoir states with a complex weight vector and subsequently taking the square of the absolute value of the resulting complex output signal vector. The weight vector is trained on the complex reservoir states using complex-valued ridge regression. Input is fed to the reservoir via node 2 (as indicated in Figure 3.1).

To determine suitable values of the delay time between reservoir nodes (node interdelay) for the architecture, we train our simulated reservoirs with increasing delay time using complex-valued ridge regression. Figure 3.8 shows the achieved bit error rate as a function of increasing interdelay between nodes at a input signal bitrate of 10 Gbps. After having set up this baseline, we exchange the true reservoir states in our setup with the states estimated through the method described above, and find the resulting bit error plot to be identical to the plot in Figure 3.8. A detailed mathematical description of the approach as well as more extensive experiments using a realistic detector model can be found in [12].

Finally, note that although the training of the weights does take some time and still requires the use of an external microprocessor, once the correct weights are identified and set, the entire system can run at high speed without the need for any operations being performed by a microcontroller.

3.3.3 Influence of weight resolution

Weights of the linear combination can be implemented using different approaches, each of which can have a different ability to achieve a fine-grained weight resolution. Reverse-biased PN-junctions, for example, which can be driven by a digital-to-analogue converter, typically provide fine resolution. However, when it comes to alternative nonvolatile weight elements, we can only obtain an intrinsically lower weight resolution because of the physics of their operation, although such elements provide stable tuning with very low power consumption, which is much more energy efficient than reverse-biased PN-junctions. A typical example with such a limitation is

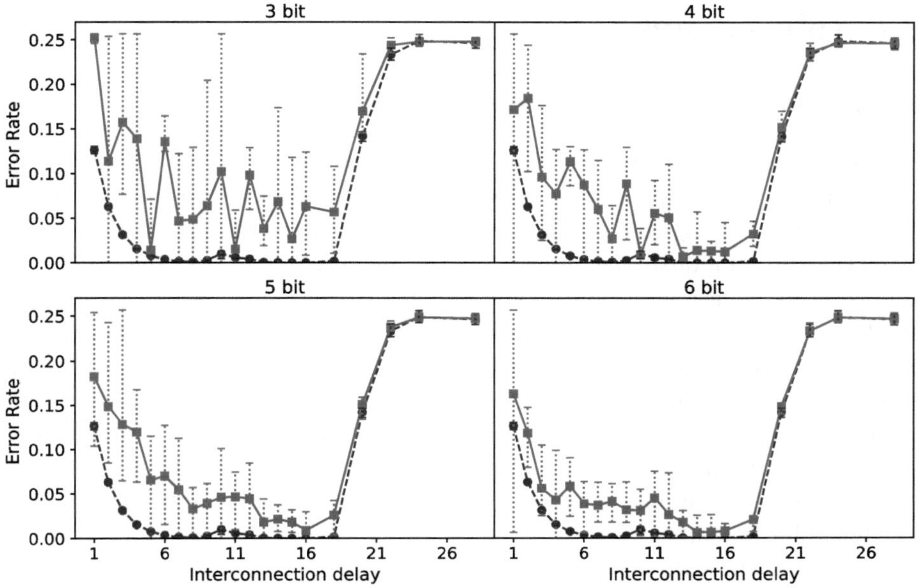

Figure 3.9: Error rate as a function of interconnection delay for different readout weight resolutions (red curves). The blue curves are for the case of infinite weight resolution.

the weighing element based on barium titanate (BaTiO3) [6], an integration of a transition metal oxide material. These elements are typically able to only provide 20 discrete weight levels. Under such circumstances, it is important to investigate how the reservoir computing performance is influenced by such a limited resolution.

To simulate the impact, 4 different weight resolutions (3, 4, 5, and 6 bit) are chosen to study the performance of the reservoir on a 1-bit header recognition task, both for the amplitude and the phase. In the simulation, we use complex-valued ridge regression performed on all the states coming from the reservoir together with the desired output states. In order to achieve simulation results close to realistic situations, different regularization parameters are being used, and the one with the lowest error rate is chosen. To simulate the weight resolution, we simply round each weight that the ridge regression produced to the nearest discretized value. For amplitude, the range of the discrete weights is corresponding to the range of initial infinite resolution weights. As for the phase, the discretized weighting levels are in the interval of $[-\pi, \pi]$.

Simulation results in Figure 3.9 show how the performance of a system changes as the weight resolution decreases compared to a system with infinite weight resolution (blue line). For 6-bits resolution, there is a narrow regime of the interconnection delay where error rates from limited resolution and infinite resolution almost coincide. Moreover, the error bars corresponding to different random initialized reservoirs also decrease in that regime, which indicates a robust behavior.

3.4 Telecom applications

3.4.1 Nonlinear dispersion compensation

Optical technology is at the core of all modern telecommunications for high speed, long- and short-reach applications. Fiber-based technologies are pervasive in different types of networks, from core and metro to data center and local area networks. Moreover, current industrial efforts are towards moving lightwave technology closer and closer to the last-mile end-user and in doing so, take advantage of the massive bandwidth, energy efficiency, and other benefits it provides. These advantages combined with advances in photonics leading to cheaper lasers, modulation technology, optical domain broadband amplification with Erbium Doped Amplifiers, place this technology squarely at the backbone of today's internet information superhighway. However, the various elements constituting fiber-based lightwave networks also contribute to the degradation of optical signals during the generation, transmission, and reception phases [13].

Signal impairments are an inevitability for any kind of communications system. They manifest themselves at the receiver as erroneous detections that need to be addressed. In optical fiber systems, these imperfections can mainly be traced back to dispersion, amplified spontaneous emission at amplification points, attenuation and reflections in fiber links, optical nonlinearities in fibers, timing jitter introduced at O/E and E/O points.

These issues are exacerbated in modern high-speed systems based on coherent modulation formats, where the nonlinearity in the fiber poses serious problems related to error-free propagation. These are typically solved in the electronic domain using advanced DSP post-processing, but such an approach consumes a lot of power and chip real estate. Photonic reservoir computing could provide an alternative here, to undo (part of) these signal impairments already in the optical domain.

To illustrate this, we will first show simulation results related to Nonreturn to Zero (NRZ) signals propagating over a realistic optical link, followed by Binary Phase Shift Keying (BPSK) signals propagating over a link with extreme artificial nonlinearities and intersymbol interference.

NRZ over an optical link

Simulated telecom data is generated with *VPI Transmission Maker v9.2* software, using the setup as in Figure 3.10. The VPI software incorporates realistic models of the signal degradation mechanisms encountered in telecom links, caused by the various optical components and physical phenomena outlined above, and also takes into account how they evolve over the transmission distances considered (which can be long, e. g., in metro and long-haul networks). The generated data is then used for training

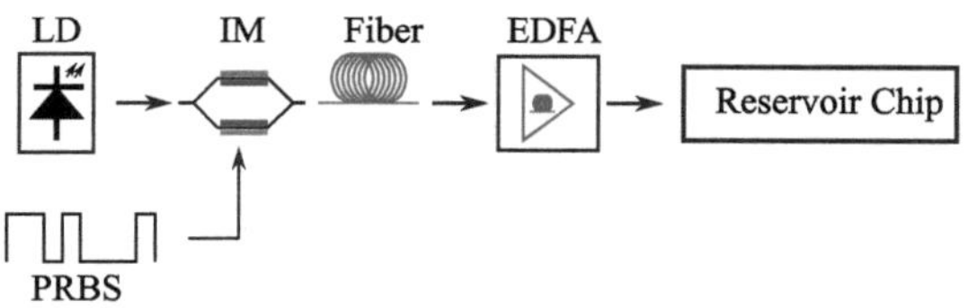

Figure 3.10: Schematic representation of the simulation setup to generate data for the signal equalization task. The input pseudo-random bit sequence (PRBS) signal is modulated onto a laser signal. This is transmitted over a fiber link, after which the data is saved to a file and used as input for the nanophotonic reservoir simulation.

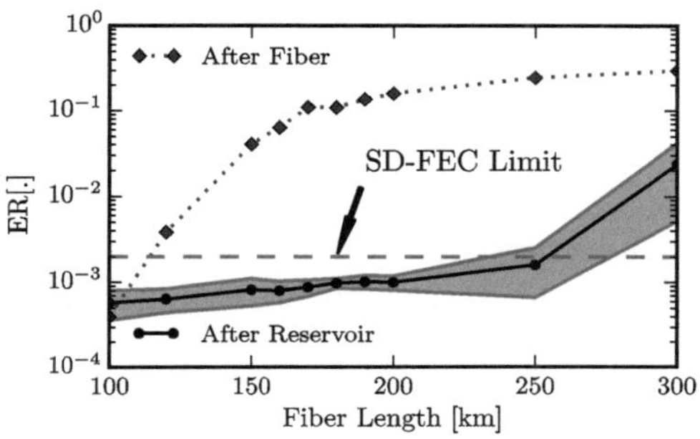

Figure 3.11: Error rates after the fiber link and after the reservoir for lengths up to 300 km. Error margins are also indicated for 5 random initialization of the phases in the reservoir. A Soft Decision Forward Error Correction limit (SD FEC limit) of 0.2×10^{-2} is also shown. Error free operation is possible for all error rate values below this limit.

and testing reservoir designs using in-house circuit simulations and machine learning libraries.

Results are shown in Figure 3.11 for a 10 Gbps link. The results indicate a BER improvement to well below the Soft-Decision Forward Error Correction (FEC) limit of 0.2×10^{-2} for connections containing up to 250 km of fiber. This means that the chip can be used in conjunction with appropriately chosen error correcting codes to achieve error-free communication on the link. Such a design would be suitable for signal equalization in, e. g., metro networks.

BPSK over an artificial channel

In order to show that the techniques described above also work for phase-encoded symbols, we perform a simulation where we send a BPSK signal (i. e., with symbols +1 and −1) over an artificial channel with extreme nonlinearity and intersymbol interference. Modulation speed is 10 GHz. Idealized square pulses are sent through a first-order Butterworth filter with a 3-dB cutoff at 25 GHz. Intersymbol interference is modeled by the following expression:

$$x_{n,\text{total}} = 0.6x_{n-1} + x_n - 0.7x_{n+1}. \quad (3)$$

This means that almost half of the energy of the bit pulse is spread over adjacent time slots.

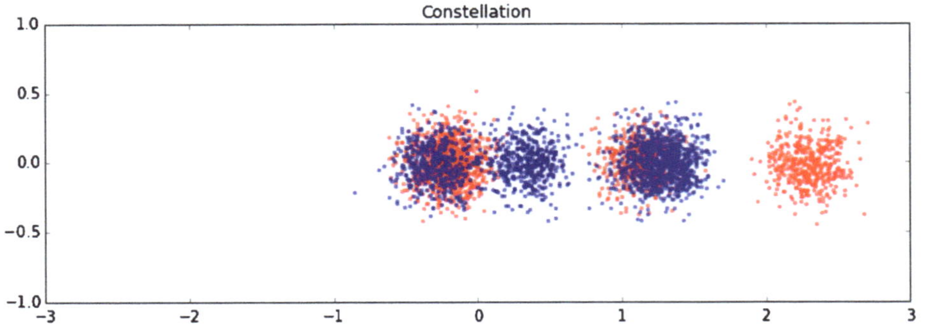

Figure 3.12: Constellation diagram (arbitrary units) of a BPSK signal at the output of an artificial channel with strong intersymbol interference and nonlinearity.

For our artificial channel, we consider a strong 9th order nonlinearity. Given an input x subject to intersymbol interference, we arbitrarily chose to model the output x_{NL} of the nonlinear channel as

$$\tilde{x} = x + 0.036x^2 - 0.011x^3 \tag{4}$$

$$x_{\mathrm{NL}} = \tilde{x} + 0.5\tilde{x}^2 - 0.3\tilde{x}^3. \tag{5}$$

Finally, Additive White Gaussian Noise (AWGN) with signal-to-noise ratio of 20 dB is added.

The constellation diagram at the output of the link is shown in Figure 3.12. Rather than having a blue blob at −1 and a red blob at +1, as would be the case for a much more benign channel, the multiple echoes due to the intersymbol interference and the distortion due to the nonlinearity are clearly visible.

In order to try and undo the transmission impairments, we send that signal through a 7 × 7 silicon photonics swirl network, and train the reservoir to recover the original bitstream, but with a delay of 1 period. The signal at the output of the reservoir is sampled in the middle of the period. After a simple threshold detection, a bit error rate of 4 % is obtained. While not zero, this number has to be compared with a 33 % error rate that is obtained using a tapped filter having access the last 3 samples of the signal (given the size and the losses in the reservoir, this corresponds to a similar memory for both cases). The reservoir clearly has better performance, moreover in a way that is amenable to a direct implementation in the optical domain, without the need for analog-to-digital conversion.

3.4.2 PAM-4 logic

The XOR-task with memory is, due to its strong nonlinearity, a commonly used benchmark task in photonic reservoir computing and in machine learning in general. The purpose of this task will be to train the reservoir to correctly compute the XOR of two symbols appearing at a certain time difference in the datastream. The simplest variation is to XOR the current bit and the previous bit. For increasing time differences between the two bits, the task becomes more challenging in terms of memory needed inside the photonic reservoir. Most often, this task is performed on data encoded with a binary modulation format, but this can be extended to the much more complex case of computing the XOR on a PAM-4 signal, a 2 bit per symbol amplitude modulation format.

The PAM-4 signal used in this example is modulated at 10 GHz, which translates to a bitrate of 20 GHz. Furthermore, a first-order Butterworth filter is used to generate realistic input pulses from a square pulse bitstream. It has the same properties as mentioned in Section 3.4.1. Subsequently, AWGN with signal-to-noise ratio of 30 dB is added. The signal is then low-pass filtered at 25 GHz before downsampling to have a smoother signal for classification.

As the multisymbol XOR task with memory is capable to really challenge the computational capabilities of the photonic reservoir due to its high nonlinearity, it can be expected that using a high number of nodes will be necessary. This will increase the complexity of the reservoir, directly increasing its computational power.

This effect is clearly shown in Figure 3.13, showing a decrease in the symbol error rate (SER) for an increasing number of nodes of the square swirl structure.

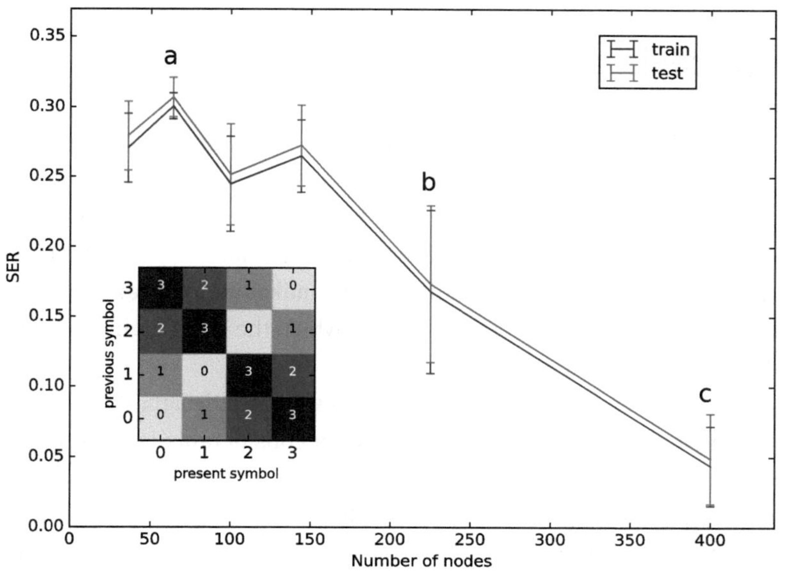

Figure 3.13: Evolution of the SER as a function of the photonic swirl reservoir dimension. The inset shows the truth table for 2-bit XOR with one symbol delay. a, b, and c refer to the histograms in Figure 3.14.

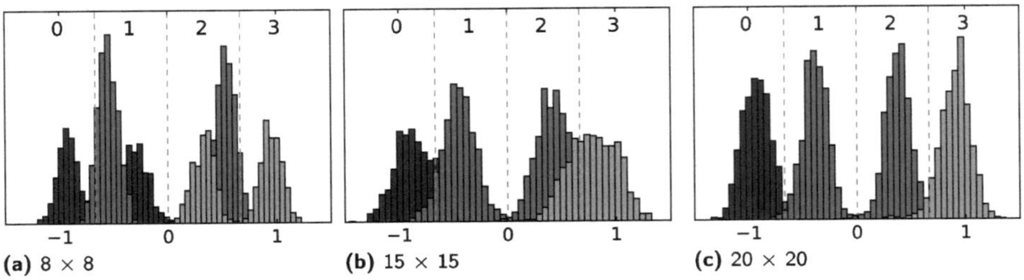

Figure 3.14: Histograms illustrating the classification of the symbols for a photonic swirl reservoir for a given number of nodes, color-labeled with the desired XOR outcome.

As illustrated in Figure 3.14, the 8 × 8 reservoir does not have enough computational power to correctly compute the XOR task, and has a SER (symbol error rate) of 30 %. Similar behavior is observed for smaller reservoirs. From 10 × 10 nodes on, all four levels start to be distinguishable with some overlap remaining. Gradually proceeding to 20 × 20 nodes, the artificial levels in the center that were present for the 8 × 8 reservoir get pushed away to the correct levels for the symbols 0 and 3. This 400-node reservoir is capable of computing the XOR with delay with a SER of less than 5 %.

3.5 Chaotic cavities

So far, the design used for the reservoir has tried to follow the conventional node structure of neural networks quite closely. However, the inherent parallel nature of photonics allows for totally new architectures that depart from this architecture. A possible design consists of a photonic crystal cavity with a quarter stadium shape [14], depicted in Figure 3.15. In this design, the quarter stadium shape makes sure that an input signal gets mixed in a complicated manner [15–17], after which the mixed light leaks out of the cavity along the connected waveguides.

This new design solves a few issues with the more standard design. It allows for a much richer interconnection topology, while needing considerably less chip realestate. The dimensions are 30 μm × 60 μm for a cavity with optimal bitrate around 50 Gbps, while we can in theory go to cavities as small as 7 μm × 7 μm to reach bitrates up to 1 Tbps. On top of that, this photonic crystal design promises very low loss, combined with excellent performance on several benchmark telecom tasks, such as the highly nonlinear XOR task and header recognition tasks, while still accepting bitrates in a wide region of operation.

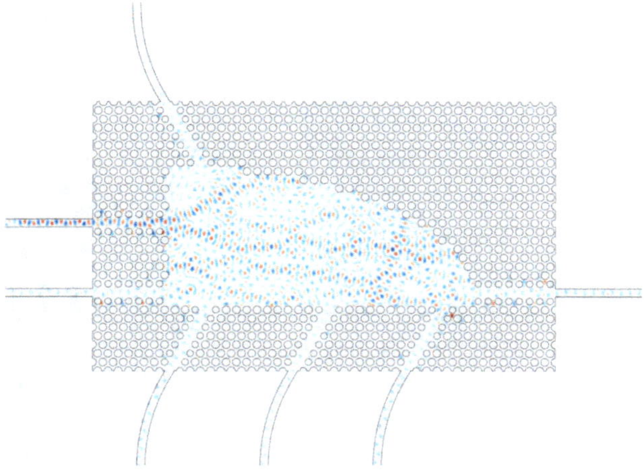

Figure 3.15: A photonic crystal cavity used for reservoir computing. In this case, a single input signal gets mixed inside a photonic crystal cavity. The mixing of the input field can clearly be witnessed by inspecting the field profiles inside the cavity. The mixed light leaks out of the cavity along all the waveguides. By routing this leaked out light to a readout, a reservoir computer can be formed.

3.5.1 Design

While designing a photonic crystal cavity for reservoir computing, several important properties have to be taken into account. First of all, the cavity needs to mix the input fields in a complex way. As already mentioned before, this can be accommodated by choosing a quarter stadium shape, which is known to foster interesting mixing dynamics. Second, the cavity needs to possess a *fading memory*, i. e. the signal should remain inside the cavity long enough to mix with subsequent input bits, but not *too* long such that it obfuscates the patterns emerging. This fading memory can obviously be controlled by controlling the Q-factor of the cavity, i. e. tuning the quality, pitch and diameter of the holes of the photonic crystal lattice. However, changing the size and number of connected waveguides will also yield a nontrivial effect.

Most of the results presented in this section follow the discussion of [14], which uses a 30 μm×60 μm cavity, with 7 connected waveguides: 1 input and 6 outputs routed to a readout. The dimensions of this cavity were specifically chosen to work for the aforementioned photodetector model with cutoff at 25 Gbps. However, since this cutoff is not steep, the reservoir computer remains working up to 50 Gbps and over.

3.5.2 Method

Doing a full reservoir simulation of a photonic crystal cavity is not trivial. First of all, making statements about error rates down to 10^{-3} requires to find the response of the cavity to bit streams with about 10^5 bits, which would be completely impossible if we would limit ourselves to pure FDTD (Finite-Difference Time-Domain) simulations. Instead, the approach visualized in Figure 3.16 is preferred. In this method, the response in the form of an electric field and a magnetic field of a single bit is recorded

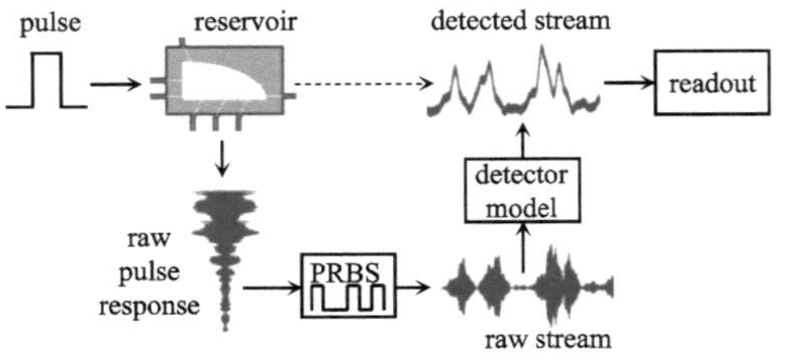

Figure 3.16: Block diagram of the simulation for a photonic crystal cavity reservoir computer. The figure was taken from [14].

from a FDTD simulation. The resulting fields are coherently added together according to a pseudo random bit stream, after which the resulting raw stream is sent through the detector model. Note that this is just the "bit-level" equivalent of the impulse response method, where the response of an arbitrary system is found by convolving the function with the response to an ultrashort impulse. Here, we chose to work with the "bit-level" response instead of the true impulse response because of numerical rounding errors.

3.5.3 XOR Task

As a first benchmark task, a PRBS of 10^5 bits with an input power of 1 mW is sent through one of the photonic crystal waveguides (the top waveguide on the left in Figure 3.15). The light gets mixed inside the cavity and finally, the responses of the other six waveguides are computed. On this recorded output, the readout weights are trained to follow the XOR function between two subsequent bits with an as low as possible mean squared error. After the ideal weights are found, the BER is calculated as the difference between the predicted bits and the target bits. Since we use 10^5 bits in our simulations, the general guideline is to crop the BER at 10^{-3}, i. e., 2 orders of magnitude higher than the lowest BER one can find in the simulation [18]. This procedure, shown in Figure 3.17, can be repeated at different bitrates, after which one can determine the operation range of the cavity.

Interestingly, the reservoir can be made to follow the target function at a wide range of different bitrates, by just changing the readout weights. As can be seen in Figure 3.18(a), we get errorless performance between 25 Gbps and 67 Gbps. What's more, even though we are working with photonic crystals, the reservoir operates in a quite wide wavelength range: 1510 nm–1600 nm, with an exception around 1560 nm, where we probably hit a stop band of the waveguide.

3.5.4 Header recognition

For applications in telecom, performing header recognition is often more useful. By just changing the readout, this photonic crystal cavity can perform this task as well.

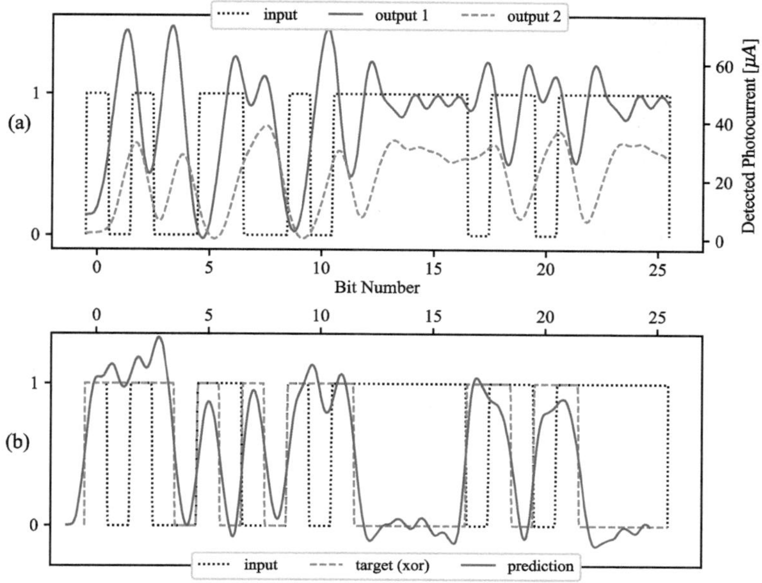

Figure 3.17: (a) Waveforms detected at two of the exit waveguides as the result of a 50 Gbps input. These waveforms get sampled once per bit. (b) After performing a linear combination of the output waveforms, the prediction follows the desired target function, in this case XOR.

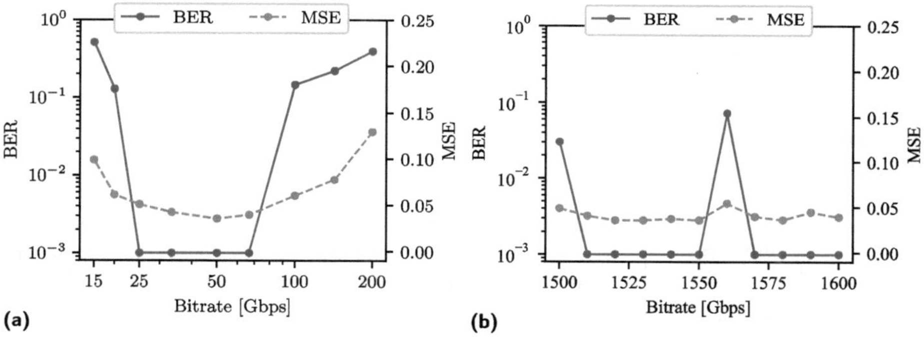

Figure 3.18: (a) There exists a larger span of bitrates the reservoir can work with between 25 Gbps and 67 Gbps at 1550 nm. This region of operation is much wider than for the conventional swirl reservoir. (b) Except for a single outlier at 1560 nm, the cavity operates under a broad band of wavelengths at 50 Gbps.

Concretely, when searching for headers of length L in a random bit stream of 10^5 bits, each bit location was labeled according to the header formed by the bit at that location and the $L-1$ previous bits. This procedure is shown in Table 3.1. Linear Discriminant Analysis (LDA) [19] was then used to find a different weight matrix for each of the different classes.

Table 3.1: Labeling a random bit stream for different header lengths L. Each position in the bitstream gets a class label according to the binary representation of the current bits and the $L-1$ previous bits.

L	...	1	0	1	1	0	1	1	1	...
2		...	2	1	3	2	1	3	3	...
3			...	5	3	6	5	3	7	...
4				...	11	6	13	11	7	...

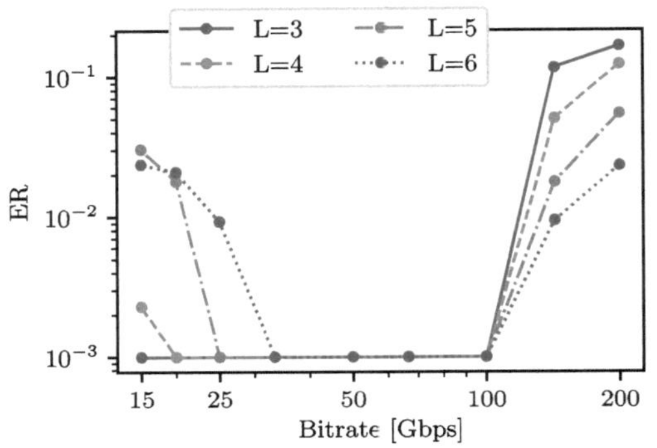

Figure 3.19: By sweeping over the bitrate to find the operation range, we find that the reservoir can distinguish headers up to a header length of $L = 6$ bits, up to 100 Gbps.

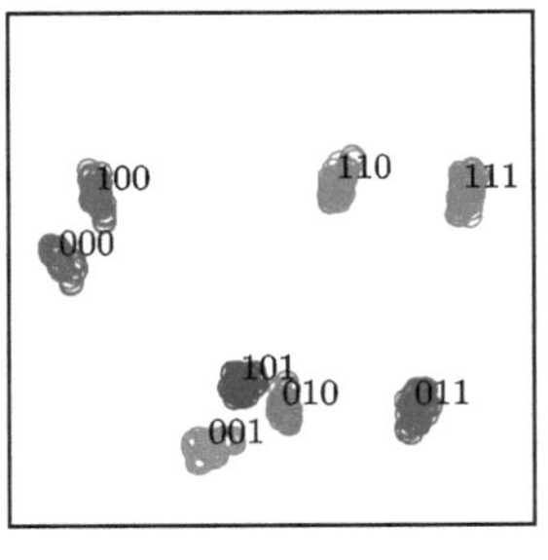

Figure 3.20: We can visualize the separation of three bit headers by projecting on the two primary LDA axes. We see nice separation for all different headers, while similar headers are located closer together.

As can be seen in Figure 3.19, also for header recognition tasks, the cavity works within wide region of operation. We also clearly see that longer headers work better at higher bitrates. This is unsurprising, as for longer headers, one needs to keep more bits in memory, therefore, the bitrate needs to be higher to accommodate this.

One advantage of choosing LDA for obtaining the weight matrices, is that it allows to make a projection from the 2^L-dimensional header-space to a lower dimensional space, in which we can see the separation of the headers visually, as can be seen in Figure 3.20.

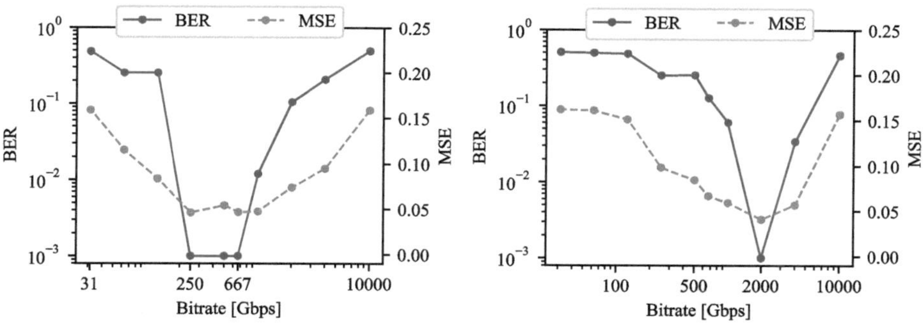

(a) Cavity size $8\,\mu m \times 10\,\mu m$

(b) Cavity size $4\,\mu m \times 6\,\mu m$

Figure 3.21: By reducing the size of the cavity, bitrates of higher than 1 Tbps can in theory be achieved. This is of course only possible if one drops the original detector model.

3.5.5 Q-factor and time scale of the reservoir

In the above simulations, we have always chosen a $30\,\mu m \times 60\,\mu m$ cavity with 7 connected waveguides. In fact, this choice for the dimensions and the number of waveguides was rather arbitrary and one could argue that changing the number of connected waveguides and the size of the cavity will have an important effect on the performance of the reservoir. For example, changing the size of the cavity alone will have an important effect on the Q-factor, and thus on the operating range of the reservoir. In Figure 3.21, the XOR performance at different bitrates is shown for two smaller cavities, under the assumption that we have photodetectors that can reach these bitrates. However, this means that in theory, one can achieve reservoir computing at bitrates higher that 1 Tbps on a chip footprint smaller than $10^{-6}\,cm^2$!

Another form of optimization is changing the number of exit waveguides, as this will inevitable also have a profound effect on the Q-factor of the cavity. This form of optimization is far from trivial, as removing exits will likely also decrease the complexity of the tasks the reservoir can solve. To quantify this effect, we can look at the performance of the reservoir on the XOR task for a range of number of exits.

When the light source is turned off, the power in the cavity with 7 connected waveguides decays exponentially with a slope $m = -0.070\,ns^{-1}$. This yields for the Q-factor at $\lambda = 1550\,nm$:

$$Q = -\frac{2\pi c}{\lambda m} = 16400. \tag{6}$$

Looking at the Q-factor for several variations of the cavity with fewer connected waveguides, we see, as we would expect, that the Q-factor decays harmonically with the number of connected waveguides (see Figure 3.22). We also clearly see that the threshold for the number of connected waveguides for the XOR task lies at 6 waveguides (1 input and 5 outputs).

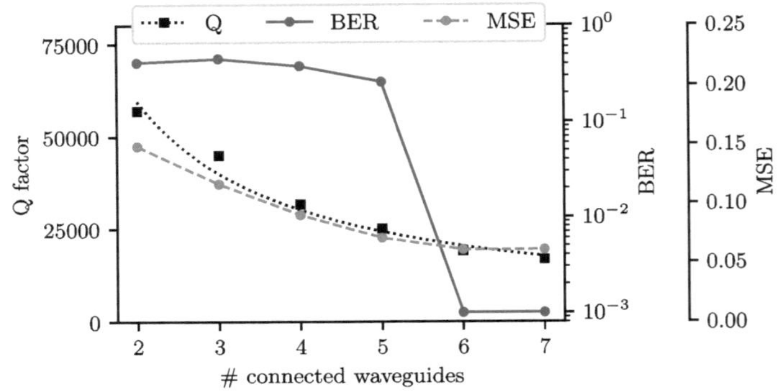

Figure 3.22: The Q-factor decays harmonically with the number of connected waveguides. BER and MSE decrease as well.

3.6 Pillar scatterers for cell identification

3.6.1 Cell sorting with digital holographic microscopy

The sorting of biological cells is of key importance in several biomedical applications, like diagnostics, therapeutics, and cell biology. However, an accurate classification and separation of different cell types is usually expensive, time consuming and often requires alterations of the samples due to the use of labels, e. g., fluorescent tags, that may hinder subsequent analyses [20]. For these reasons, the development of label-free, high-speed, automated, and integrated cell sorting solutions is of particular interest. Among several options, the employment of digital holographic microscopy in microfluidic flow cytometry is a promising candidate. In this technique, the classification is carried out through the analysis of the interference pattern (hologram) projected by the cell when illuminated by monochromatic light. The hologram is acquired by an image sensor and contains information on the 3D refractive index structure of the cells [21]. The large amount of information contained in a cell hologram enables nontrivial analysis and classifications. On the other hand, the computational cost of elaborating such a complex source of information by reconstructing the image from the hologram is a major hindrance to an increase in the cell sorter throughput, e. g., by parallelization of the process.

An important reduction in the required computing power can be achieved by bypassing the reconstruction of the cell image and directly processing the acquired hologram with a machine learning algorithm that carries out the classification task [21, 22]. However, a further improvement of the classification simplicity and performance is desired in order to fully exploit the potential of this implementation.

3.6.2 Dielectric scatterers for an extreme learning machine (ELM) implementation

The basic conceptual structure of reservoir computing, i. e., an untrained recurrent nonlinear network that improves the performance of a trainable linear readout, can be also applied to time-independent signals such as images. In this case, memory in the nonlinear part is not required and the whole system is usually called extreme learning machine (ELM) [23].

Proof-of-concept via FDTD simulations

We provided a proof of concept, based on FDTD simulations, of an integrated photonics application of ELM for fast and label-free classification of biological cells [24]. In this application, a passive optical stage comprising a collection of silica pillar scatterers embedded in a silicon nitride slab waveguide is used to process the light forward-scattered by a cell when illuminated via a green monochromatic source (Figure 3.23). The diffraction pattern projected by the cell and by the dielectric scatterers is acquired by a 1D image sensor whose pixel outputs are used as input for a linear classifier (logis-

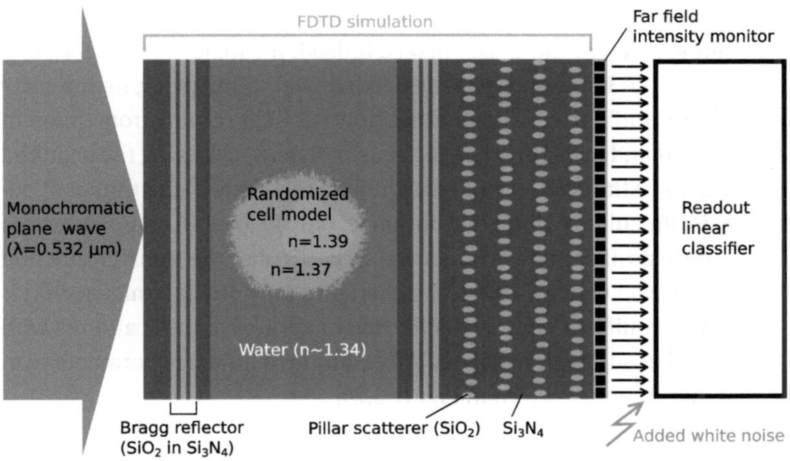

Figure 3.23: Schematic of the classification process. A monochromatic plane wave impinges on a Fabry–Pérot optical cavity composed by Bragg reflectors and containing a microfluidic channel with a cell in water ($n_{H_2O} \sim 1.34$), which has a low refractive index contrast ($n_{cytoplasm} = 1.37$, $n_{nucleus} = 1.39$); the forward scattered light passes through a collection of silica scatterers ($n_{SiO_2} \sim 1.461$) embedded in silicon nitride ($n_{Si_3N_4} \sim 2.027$) and organized in layers; the radiation intensity is then collected by a far-field monitor, which is divided into bins (pixels); each pixel value is fed into a trained linear classifier (namely logistic regression) that consists of weighted sums (one per class) of the pixel values. The weights are trained so that the sum exceeds a certain threshold value only if the corresponding input class is recognized.

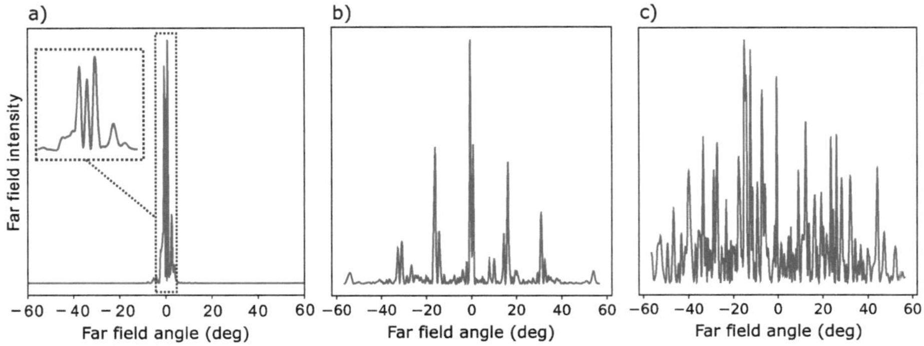

Figure 3.24: Far-field intensity profiles of the light scattered by a cell: (a) without the presence of scatterers the interference pattern is relatively simple and smooth, most of the intensity is confined between −6° and 6°; (b) considering 4 layer of scatterers with a slight random displacement (amplitude of 25 nm) the far-field intensity is distributed around periodically placed peaks, most of the field stays between −40° and 40°; (c) considering 4 layers of scatterers with a larger random displacement (amplitude of 150 nm) the far-field intensity is distributed in a complex pattern mostly between −60° and 60°.

tic regression) implemented in the electric domain. Such a configuration represents an ELM system where the light scattered by the cell is the input signal while the dielectric scatterers and the image sensor play the role of a random nonlinear network whose output nodes are represented by the sensor pixels. Indeed, the image sensor performs a nonlinear function of the impinging phase-encoded[1] signal by giving its intensity as output. The scatterer configuration determines how the light coming from different regions of the cell is split and overlapped on the sensor display, changing the acquired interference pattern. Since the scatterer stage includes a relatively high number of pillars and, therefore, is difficult to completely optimize, only its overall complexity was tuned to maximize the classification performance. In particular, the transfer function complexity of the scatterer stage was tuned by changing only one parameter, that is the amplitude of a slight uniform random displacement applied to the scatterers with respect to their ordered position along layers. An example of how this parameter can modify the interference pattern is given in Figure 3.24.

In order to properly train the readout classifier and to test its performance once trained, a sufficient number (thousands) of diffraction pattern samples had to be computed and provided to the training algorithm. Randomized cell models were employed to create reasonable variability in the diffraction pattern acquisition. In [24], two different classification tasks were considered. The first is based on average nucleus size and aims to distinguish between "normal" cells (small nucleus) and "cancer" cells

1 The cell structure information is mainly encoded in the phase of the scattered light, since cell absorption is usually negligible.

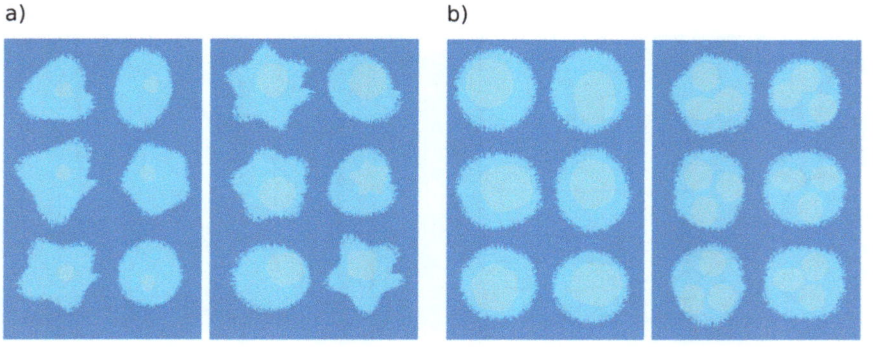

Figure 3.25: Examples of cells automatically generated by the employed randomized models. (a) Comparison between generated examples of "normal" cell and "cancer" cell, with different nucleus size. (b) Comparison between generated examples of "lymphocyte" and "neutrophil," with different nucleus shape.

(bigger nucleus, Figure 3.25(a)). The names in quotation marks were chosen because of the common tendency of cancer cells to show evident irregularities in nucleus size [25]. The second task is based on nucleus shape (the average nucleus area is kept constant) and aims to distinguish between "lymphocytes" (big quasi-spherical nucleus) and "neutrophils" (nucleus divided in 3 lobes, Figure 3.25(b)). The names in quotation marks refer to two among the most common white blood cells that are present in human blood. These two cell models are, physically and biologically speaking, only rough representations of their real counterparts when flowing in a liquid medium. Indeed, the employed models are the result of a trade-off between computational cost and closeness to reality. Such a trade-off is legitimated by the fact that the goal of the work is not to provide absolute references for real applications but, instead, to investigate a relative difference between the classification performance with and without using scatterers. Further details on the employed cell models and machine learning approach are presented in [24].

Results

Let us consider a green laser source ($\lambda = 532\,\text{nm}$) and let us compare the classification error on the test samples when no scatterers are present and when, instead, 4 scatterer layers are employed. The resulting error rate values for different numbers of pixels and for different noise levels (Figure 3.26) show that the use of scatterer layers allows for a significant error rate reduction (up to $\sim 50\,\%$ for nucleus size classification and higher for the nucleus shape classification), provided that a sufficient number of pixels and a low enough noise level are considered. The increased sensitivity of classification performance toward added noise level when scatterers are used is ascribed to the fact

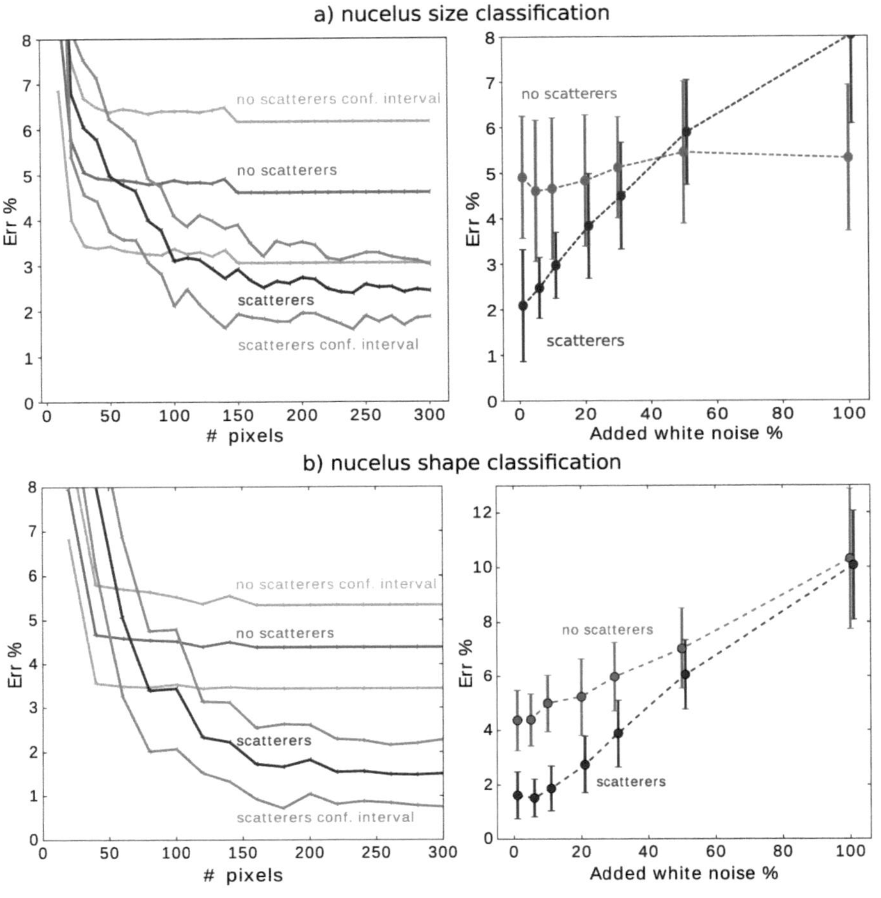

Figure 3.26: Comparison between test error rates corresponding to the absence (in red) and the presence (in blue) of 4 layers of scatterers: (a) "normal" and "cancer" cell classification based on nucleus size; (b) "lymphocyte" and "neutrophil" cell classification, based on nucleus shape. A green laser source ($\lambda = 532$ nm) is employed. On the left: test error rate as a function of the number of employed pixels, with 5 % added white noise. The darker and the lighter versions of the two line colors respectively represent the mean value and the confidence interval (of 2 standard deviations) over the 20 sample sets generated for validation. On the right: test error rate (averaged on the values obtained considering the numbers of pixels 250, 260, . . . , 300) as a function of the added noise percentage. The plots show that the scatterers presence allows for an error rate reduction up to ~ 50 % (or higher for the classification of nucleus shape), provided that a sufficient number of pixels and a low enough noise level are considered.

that the scatterers' presence unfolds the cell diffraction pattern into a higher number of components (Figure 3.24(c)) that may be important for classification. Thus, it is probable that some of these components have low intensity with respect to the average pattern intensity and are therefore easily overcome by high relative levels of noise.

3.6.3 Nonlinearity of phase sensitivity

If there seems to be a limit to the improvements obtainable via the use of scatterers when a green coherent source is used, further simulations considering an UV laser ($\lambda = 337.1$ nm) show that this limit is easily overcome by decreasing the source wavelength [24]. An explanation of this effect can be provided by the following simplified argumentation.

Let us neglect for a moment the light deflection due to the cell refractive index structure and let us consider only the phase shift of the light along all the possible fixed paths (here labeled with n) through the cell to one pixel on the screen. At the top of Figure 3.27 a schematic illustration representing 3 of these paths is shown. Let us state that the light along a path n has unitary initial amplitude and zero initial phase (the reasoning is independent from the initial conditions) and that it accumulates a phase shift θ_n through the cell. Moreover, let us say that its amplitude is reduced by a factor A_n and its phase is increased by ϕ_n along the path to the pixel excluding the

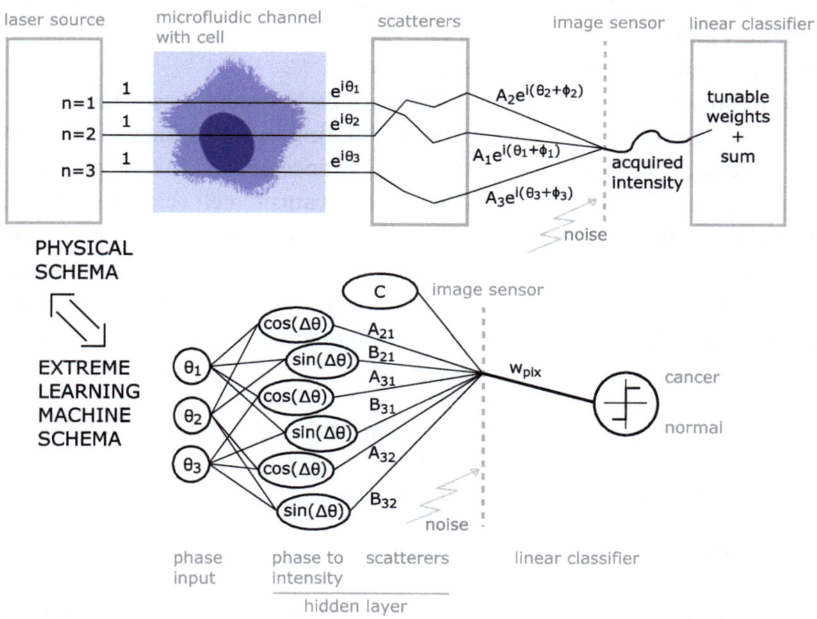

Figure 3.27: Two equivalent schematics of the proposed classifying system. At the top, a physical schema shows an example of amplitude and phase evolution along 3 optical paths that end up impinging on the same pixel of the image sensor. The acquired light intensity is then weighted and summed by a linear classifier. At the bottom, a diagram (under the form of a neural network architecture) represents the corresponding mathematical operations on the light phase accumulated through the cell refractive index structure (see equation (7)). For simplicity's sake, the light deflection due to the cell presence is neglected, and thus the amplitudes A_n and the factors A_{nm}, B_{nm}, and C are considered as constants with respect to the inputs θ_n.

path inside the cell. Thus, the complex amplitude of the radiation impinging on the pixel is $\sum_n A_n e^{i(\theta_n+\phi_n)}$ and the acquired intensity I is proportional to

$$I \propto \left|\sum_n A_n e^{i(\theta_n+\phi_n)}\right|^2 = C + \sum_{m<n} [A_{nm}\cos(\theta_n - \theta_m) + B_{nm}\sin(\theta_n - \theta_m)] \tag{7}$$

where C, A_{nm}, and B_{nm} (that can also account for the presence of scatterers) are constants with respect to θ_n but depend on A_n and ϕ_n. These dependencies are omitted as the phase shifts θ_n are the only actual inputs of our classifying system, neglecting the light absorption in the cell. Equation (7) shows that the phase shift to intensity transfer function on the pixel can be written as a linear combination of all the possible sine and cosine functions whose argument is the phase shift difference between two of the considered optical paths (bottom of Figure 3.27). Note that if the deflection of the light path due the cell's presence was also considered, we would have a richer dependence on θ_n in the right-hand side of equation (7) (C, A_{nm}, and B_{nm} will also depend on θ_n). Nevertheless, the sines and the cosines in equation (7) would still be present and the following argument would still be relevant. It is important to note that, in this representation, the only role of the scatterer layers is to improve classification performances by providing more suitable weights A_{nm}, B_{nm}, and C.

Let us now consider, for instance, the difference $\Delta\theta$ between phase shifts corresponding to a path through the nucleus and a neighboring path that instead does not intersect the nucleus. Let us call this phase shift difference $\Delta\theta_n$ in the case of a "normal" cell (smaller nucleus) and $\Delta\theta_c$ in the case of a "cancer" cell (bigger nucleus). We can intuitively say that if the readout linear classifier is able, for example, to detect the difference ΔI between the intensity contributions produced by $\Delta\theta_n$ and $\Delta\theta_c$, respectively, among the other intensity contributions, the system can be successfully trained in carrying out the classification task. From equation (7) follows that an estimate of this critical intensity difference is given by

$$\begin{aligned} &\Delta I \propto A[\sin(\Delta\theta_c) - \sin(\Delta\theta_n)] + B[\cos(\Delta\theta_c) - \cos(\Delta\theta_n)] \\ &\text{with} \quad \Delta\theta_c = \frac{2\pi D_c}{\lambda}(n_{\text{nucleus}} - n_{\text{cytoplasm}}) \\ &\text{and} \quad \Delta\theta_n = \frac{2\pi D_n}{\lambda}(n_{\text{nucleus}} - n_{\text{cytoplasm}}) \end{aligned} \tag{8}$$

where A and B are constants, $D_c \sim 2.5\,\mu\text{m}$ and $D_n \sim 1.2\,\mu\text{m}$ are the average diameter of the "cancer" and the "normal" cell model, respectively, λ is the wavelength of the considered radiation, $n_{\text{nucleus}} = 1.39$ and $n_{\text{cytoplasm}} = 1.37$ are the refractive index of the nucleus and of the cytoplasm in the employed cell model. Let us stress that we expect bad classification performances if the system has a too linear or a too random response. It can be noted that in equation (7) these two undesired conditions may be ascribed to $\theta_n - \theta_m \ll \pi$ (linear regime) and $\theta_n - \theta_m \gg \pi$ (periodic regime), respectively. When we focus on distinguishing between different nucleus sizes, equation (8) have to be considered. In particular, if $\lambda = 0.532\,\mu\text{m}$ we have $\Delta\theta_c \sim 0.6$ and $\Delta\theta_n \sim 0.3$, which are quite smaller than π. By looking at the expressions for these two differences, it can be noticed that they can be increased by lowering the wavelength. This implies the need to employ an UV laser, which is usually significantly more expensive than its green counterpart and could damage the illuminated cell. A more feasible solution consists of increasing the effective optical path length through the cell by inserting it in an optical cavity. Indeed, a cavity would make the impinging light pass, on average, more than once through the cell.

3.6.4 Combination of dielectric scatterers and optical cavity

Equation (8) suggests that the classification improvement due to the use of dielectric scatterers can be enhanced by inserting the cell in a properly designed optical cavity. Indeed, this allows to increase the nonlinearity of the nucleus signal representation in the acquired intensity pattern. In practice, in the FDTD simulation design 2 Bragg reflectors were placed at the 2 external sides of the microfluidic channel, orthogonally to the light beam direction, creating a Fabry–Pérot cavity (Figure 3.23). The employed Bragg reflectors are each composed of 3 layers of SiO_2 with a width of (455 ± 10) nm in a Si_3N_4 cladding. The error in the layer width was implemented by adding a random value sampled from an uniform probability distribution between −10 nm and 10 nm. It approximately accounts for fabrication errors. The distance $D = 21.02\,\mu\text{m}$ between the reflectors was chosen so that the portion of light passing through and near the nucleus of the cell is resonant. This was done by monitoring the light intensity inside the cavity for different values of D. Note that such a tuning was relatively easy to perform because the cell acts as a weak converging lens, providing an additional light confinement along the microfluidic channel direction.

The reflectivity R of the reflectors is also a crucial parameter, since it controls the cavity Q-factor and, therefore, controls both the sensitivity of the resonance to intracavity optical path lengths and how long the light stays, on average, inside the cavity. This means that by tuning R a trade-off has to be achieved between how much the phase shift due to the selected resonant cell part is increased and how much the corresponding resonance is stable. Indeed, if the cavity Q-factor is too high, the resonance strength might be strongly influenced by uninteresting small details of the cell structure or by fabrication errors. As previously explained (Subsection 3.6.3), the cavity should be designed so that the average light phase shift differences due to the optical feature of interest corresponding to the considered classes are roughly between $\pi/2$ and 2π. In particular, the reflectors employed in the simulations (composed of 3 layers) have a satisfying reflectivity of $\sim 56\,\%$, while it turned out that similar reflectors with 4 and 5 layers have a too high reflectivity, respectively of $\sim 73\,\%$ and $\sim 85\,\%$. Fi-

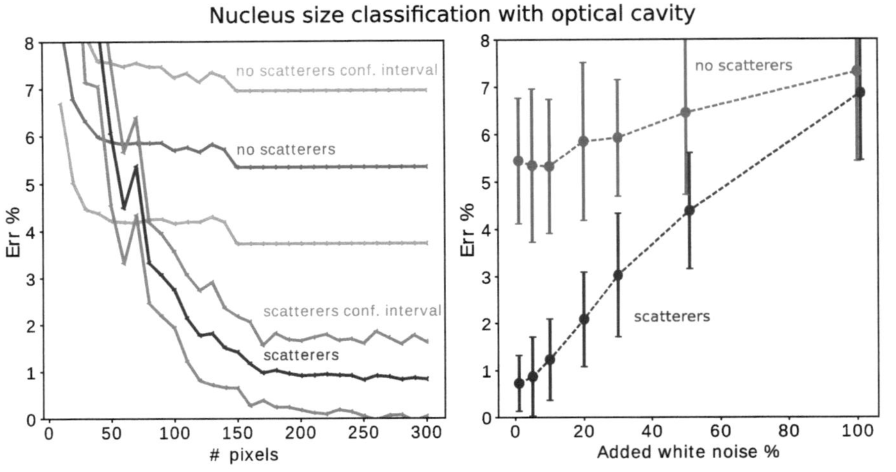

Figure 3.28: Comparison between test error rates corresponding to the absence (in red) and the presence (in blue) of 4 layers of scatterers, employing a green laser source ($\lambda = 532$ nm) and the described integrated optical cavity. On the left: test error rate as a function of the number of employed pixels, with 5 % added white noise. On the right: test error rate (averaged on the values obtained considering the numbers of pixels $250, 260, \ldots, 300$) as a function of the added noise percentage. These graphs are to be compared with Figure 3.26(a): the employment of an optical cavity significantly enhances the classification improvement obtained with the use of scatterers.

nally, the introduction of the optical cavity implied that the simulation time had to be properly increased to 1.2 ps, while the other simulation and machine learning details were as previously described.

Results

Figure 3.28 shows the results of nucleus size classification obtained employing a green laser source ($\lambda = 532$ nm) and the described integrated optical cavity. A substantial improvement is observed with respect to the correspondent case without cavity (Figure 3.26). In particular, the classification improvement due to the use of scatterers is increased by factor 5 for sufficiently low but still plausible noise levels ($< 10\,\%$). At these noise levels, the results are similar to what was obtained with an UV light source (see [24]), without the drawback of possible cell damage. For higher noise levels, an increased sensitivity to noise pushes the classification error rate to significantly higher values. Furthermore, it should be stressed that an additional advantage arising from the use of an optical cavity is that it can be designed to increase the intensity pattern sensitivity toward specific optical path lengths, making the optical features of interest more evident to the readout classifier with respect to other competing ones.

3.7 Conclusion

In this chapter, we reviewed integrated photonic reservoir computing chips, with a special emphasis on passive reservoir structures. We presented simulation results that show that this paradigm can be used successfully to perform a variety of tasks (bit level tasks, nonlinear dispersion compensation, etc.) at high speeds and low power consumption. In addition, we presented a spatial analog of reservoir computing based on pillar scatterers and a cavity, that can be used to speed up classification of biological cells.

Bibliography

[1] K. Vandoorne et al. "Experimental demonstration of reservoir computing on a silicon photonics chip." In: *Nature Communications* 5 (2014), p. 3541.

[2] Q. Vinckier et al. "High-performance photonic reservoir computer based on a coherently driven passive cavity". In: *Optica* 2.5 (2015), pp. 438–446. ISSN: 2334-2536.

[3] K. Vandoorne et al. "Parallel reservoir computing using optical amplifiers". In: *IEEE Transactions on Neural Networks* 22 (2011), pp. 1469–1481.

[4] A. Katumba et al. "A multiple-input strategy to efficient integrated photonic reservoir computing". In: *Cognitive Computation* (Apr. 2017), pp. 1–8. ISSN: 1866-9956.

[5] A. Katumba et al. "Low-loss photonic reservoir computing with multimode photonic integrated circuits". In: *Scientific Reports* 8.1 (2018). ISSN: 20452322.

[6] S. Abel et al. "A strong electro-optically active lead-free ferroelectric integrated on silicon". In: *Nature Communications* 4 (2013), p. 1671.

[7] C. Ríos et al. "Integrated all-photonic non-volatile multi-level memory". In: *Nature Photonics* 9.11 (2015), pp. 725–732.

[8] B. Van Bilzen et al. "Production of VO 2 thin films through post-deposition annealing of V 2 O 3 and VO x films". In: *Thin Solid Films* 591 (2015), pp. 143–148.

[9] A. Hoerl and R. Kennard. "Ridge regression: biased estimation for nonorthogonal problems". In: *Technometrics* 12.1 (1970), pp. 55–67.

[10] H. Jaeger and H. Haas. "Harnessing nonlinearity: predicting chaotic systems and saving energy in wireless communication." In: *Science (New York, N.Y.)* 304 (2004), pp. 78–80. ISSN: 0036-8075.

[11] M. Freiberger et al. "On-chip passive photonic reservoir computing with integrated optical readout". In: *Rebooting computing (ICRC), 2017 IEEE international conference on*. 2017.

[12] M. Freiberger et al. "Training passive photonic reservoirs with integrated optical readout". In: *Unpublished, submitted for review to Neural networks and learning systems, IEEE transactions on*.

[13] I. Djordjevic, W. Ryan, and B. Vasic. *Coding for optical channels*. Boston, MA: Springer US, 2010. isbn: 978-1-4419-5568-5.

[14] F. Laporte et al. "Numerical demonstration of neuromorphic computing with photonic crystal cavities". In: *Optics Express* 26.7 (2018), pp. 7955–7964.

[15] H.-J. Stockmann and J. Stein. "Quantum chaos in billiards studied by microwave absorption". In: *Physical Review Letters* 64 (19 May 1990), pp. 2215–2218. DOI: 10.1103/PhysRevLett.64.2215. URL: https://link.aps.org/doi/10.1103/PhysRevLett.64.2215.

[16] M. Sieber et al. "Non-generic spectral statistics in the quantized stadium billiard". In: *Journal of Physics. A, Mathematical and General* 26.22 (1993), p. 6217.

[17] C. Liu et al. "Triggering extreme events at the nanoscale in photonic seas". In: *Nature Physics* 11.4 (2015), pp. 358–363.

[18] M. Jeruchim. "Techniques for estimating the bit error rate in the simulation of digital communication systems". In: *IEEE Journal on Selected Areas in Communications* 2.1 (Jan. 1984), pp. 153–170. ISSN: 0733-8716.

[19] A. J. Izenman. "Linear discriminant analysis". In: *Modern multivariate statistical techniques*. Springer, 2013, pp. 237–280.

[20] D. R. Gossett et al. "Label-free cell separation and sorting in microfluidic systems". In: *Analytical and Bioanalytical Chemistry* 397.8 (2010), pp. 3249–3267.

[21] L. Lagae et al. "High throughput cell sorter based on lensfree imaging of cells". In: *2015 IEEE international electron devices meeting (IEDM)*. Dec. 2015, pp. 13.3.1–13.3.4. DOI: 10.1109/IEDM.2015.7409689.

[22] B. Schneider et al. "Neural network for blood cell classification in a holographic microscopy system". In: *2015 17th international conference on transparent optical networks (ICTON)*. July 2015, pp. 1–4. DOI: 10.1109/ICTON.2015.7193315.

[23] G.-B. Huang, Q.-Y. Zhu, and C.-K. Siew. "Extreme learning machine: theory and applications". In: *Neurocomputing* 70.1 (2006). Neural networks, pp. 489–501. ISSN: 0925-2312. DOI: 10.1016/j.neucom.2005.12.126. URL: http://www.sciencedirect.com/science/article/pii/S0925231206000385.

[24] A. Lugnan, J. Dambre, and P. Bienstman. "Integrated pillar scatterers for speeding up classification of cell holograms". In: *Optics Express* 25.24 (Nov. 2017), pp. 30526–30538. DOI: 10.1364/OE.25.030526. URL: http://www.opticsexpress.org/abstract.cfm?URI=oe-25-24-30526.

[25] D. Zink, A. H. Fischer, and J. A. Nickerson. "Nuclear structure in cancer cells". In: *Nature Reviews. Cancer* 4.9 (2004), p. 677.

Daniel Brunner, Julian Bueno, Xavier Porte, Sheler Maktoobi, and Louis Andreoli

4 Large scale spatiotemporal reservoirs

4.1 Introduction

A core advantage of optical systems is their possibility to process information contained in a 2D-plane in parallel. Numerous schemes take profit from this concept, among others, the 4*f* optical-correlator [1] or the parallel implementation of optical matrix multiplications [2, 3]. Taking the application of an optical matrix product further enables the optical realization of a Hopfield network. There, diffraction implements a predetermined coupling matrix based on the analytical relationship introduced in Section 1.2.5. Using photorefractive crystals, diffraction combined with phase-conjugation can even implement deep neural networks with a hardware-based gradient back-propagation learning routine [4]. Most of these concepts rely on volume-holograms to implement a fully-optimized coupling matrix. This approach creates exceptional space-bandwidth products, yet also results in limitations such as a detrimental cross-talk between weights [5]. Thanks to the RC concept's relaxed requirements on the internal coupling matrix, the impact of these limitations is of reduced importance. Furthermore, control over all individual entries of the connectivity matrix is not required and simpler concepts can be leveraged to implement coupling fully in parallel. We will create large-scale networks of discrete photonic emitters based on diffractive coupling [6] and will implement learning in photonic hardware based on a digital micro-mirror device (DMD) [7]. We, too, illustrate avenues to optically inject information into such systems [6]. With that, we realize the connections between the input, hidden and readout layers all-optically based on fundamentally parallel concepts. We illustrate diffractive coupling in two different systems: one uses the pixels of a spatial light modulator (SLM) as nonlinear elements, the other is based on a commercial array of vertical-cavity surface-emitting lasers (VCSELs).

4.2 Diffractive coupling

From its principles, a reservoir is a spatiotemporal dynamical system comprising a complex and recurrent network of nonlinear nodes. Spatiotemporal dynamics can be mapped onto multiple *spaces*, including additional temporal [8] or spectral [9] coordinates. An implementation of nonlinear networks in physical space has the fundamental advantages of efficiently exploiting the parallelism of optical components. Other than in electronics, where the physical realization of each network connection requires a dedicated wire, in optics a single element can provide such functional-

https://doi.org/10.1515/9783110583496-004

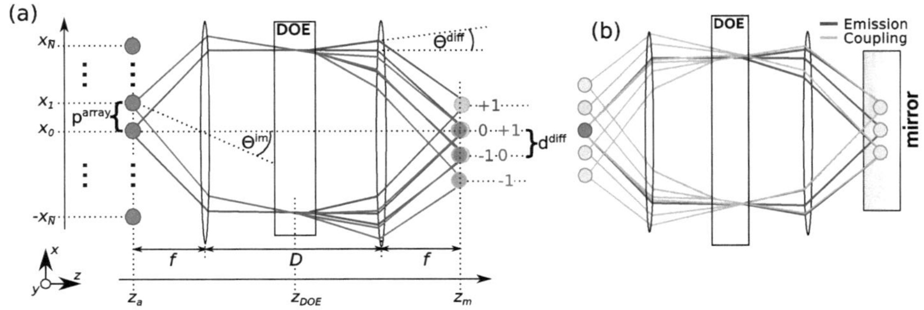

Figure 4.1: (a) Imaging two emitters via a 2f-imaging arrangement, including between both lenses a DOE creating three discrete diffractive orders of identical amplitudes. The focal distance is f, Θ^{im} the angle between the principle rays of neighboring emitters, and Θ^{diff} the angle between diffractive orders. For the right set of parameters, the distance between diffractive orders in the image plane equals the distance between neighboring emitters: $p^{array} \approx d^{diff}$. (b) Placing a mirror in the image plane of the second lens results in (i) double passing through the DOE, and (ii) diffractive orders located at the positions of neighboring emitters.

ity for the entire network. For the case of coupling an array of discrete optical emitters spaced at period p^{array}, the coupling mechanism has to precisely match the array's period [6]. We assume a 2D array of optical nonlinear elements with positions given by $\mathbf{r}_{i,j} = p^{array} \cdot (i,j)$, where $i,j \in \{-\tilde{N},\tilde{N}\}$ assign physical node positions to integer indices. This results in an array consisting of $N = (2\tilde{N})^2$ nodes. In Figure 4.1(a), we schematically illustrate the principle of spatial multiplexing discrete positions by combining imaging and diffraction. The schematic illustration is restricted to the x and z-dimensions as properties along y are identical to x and can therefore be omitted for illustration-simplicity. Emitters are located at z_a, one focal distance f in front of the first lens or microscope objective (MO), which is followed by the diffractive optical element (DOE) located at z_{DOE}. Following classical optical design principles, the DOE should ideally be located at a distance f behind the first and in front of the second lens [10]. In practice, this is often not feasible, for instance due to geometric restrictions by mechanical components, but most importantly due to the use of MOs. For the sake of minimal aberrations, the use of infinity-corrected MO is strongly recommended, even required. Yet, these typically posses a small, in many cases even negative back-focal distance. Placing the DOE in the Fourier-plane is therefore impractical or impossible. Finally, a second lens or MO, located at $z_a + f + D$ creates an image-plane located at z_m.

As illustrated in Figure 4.1(a), the result is the distribution of the original emitter's optical field across various diffraction orders at $z = z_m$. For a first simple analysis of the imaging system, we assume the DOE to be a simple transmission diffraction grating with a periodic phase-modulation along x and y. Based on the analytical solution for diffraction by a grating and an optical ray-treatment of an aberration-free lens, we obtain

$$\Theta_i^{\text{im}} = \tan^{-1}\left(\frac{i \cdot p^{\text{array}}}{f}\right) \tag{1}$$

$$\Theta_{i,m}^{\text{diff}} = \sin^{-1}\left(\sin\Theta_i^{\text{im}} + m\frac{\lambda}{p^{\text{DOE}}}\right), \tag{2}$$

which are the two relevant angles describing imaging (Θ_i^{im}) and diffraction ($\Theta_{i,m}^{\text{diff}}$) in the system's collimated space. Parameters are wavelength λ and the period of the DOE's phase or amplitude modulation p^{DOE}. The same relationships would be found in the y-direction using integer j.

In order to understand the optical mechanism underlying coupling by diffractive-imaging, Figure 4.1(a) shows the process for an emitter located in the center of the array, $i = 0$ and $x_0 = 0$, and its nearest neighbor $i = 1$ located at $x_1 = p^{\text{array}}$. For both emitters, the grating creates a family of diffractive orders in z_m and, for coupling between their optical fields to be established, some of these orders need to overlap in space. The criteria for coupling between diffractive order $m = +1$ of emitter $i = 0$ and order $m = 0$ of emitter $i = 1$ is then given by

$$\sin\Theta_{0,+1}^{\text{diff}} = \frac{\lambda}{p^{\text{DOE}}}, \quad \tan\Theta_1^{\text{im}} = \frac{p^{\text{array}}}{f} \tag{3}$$

$$\Theta_{0,+1}^{\text{diff}} = \Theta_1^{\text{im}}. \tag{4}$$

Equations (3) and (4), respectively, give the imaging angle for emitter $i = 1$'s imaging angle (Θ_1^{im}) and the +1-diffraction order's angle for emitter $i = 0$ ($\Theta_{0,+1}^{\text{diff}}$). The second lens or MO images both respective wavefronts, and hence, for overlap of their optical fields at z_m, the propagation angle of both fronts needs to be identical, creating the condition of equation (4). Solving this set of equations is straightforward, and for the right set of system parameters coupling is created between neighbors of the array.

We typically adjust the experimental parameters, i. e., λ, p^{array} or p^{DOE} such that coupling is optimized for the central emitter: $\lambda/p^{\text{DOE}} = \sin(\tan^{-1}[p^{\text{array}}/f])$. The resulting linear superposition of their optical fields is weighted according to the respective diffractive orders' intensities. However, it is clear that this alignment criteria is only exactly satisfied for coupling the pair of emitters initially used to solve equation (4). For pairs of emitters located away from the optimal position, the overlap between neighboring emitters' optical fields will gradually be reduced due to the different trigonometric relationships for imaging and diffraction; see equation (3). As soon as the resulting misalignment between the neighboring optical fields becomes of the order of the emitter's physical size, coupling will be seriously hampered. In consequence, the array size is effectively limited, as will be discussed in Section 4.2.2. Crucially, for the elements located inside the area for which equation (4)'s condition is fulfilled, coupling is implemented fully in parallel and without any additional energy cost besides optical attenuations introduced by the individual optical elements such as DOE or

MOs. In general, the concept does not require both MOs to have identical focal distances. For that case, f in equations (1), (3), and (4) refers to the first MO's focal distance.

If emitters are directly sensitive to optical feedback, i. e., semiconductor lasers such as VCSELs, locating a mirror at $z_m = z_a + 2f + D$ results in an imaging optical resonator, a configuration illustrated in Figure 4.1(b). The reflected optical fields will traverse the DOE a second time, resulting in additional diffraction and distribution of diffractive orders in array-plane z_a that creates a larger coupling range than the one in z_m. Furthermore, in reflection the image formed at position z_a is created by a $4f$-configuration. Consequently, diffractive orders will be arranged around their original emitter's position $\mathbf{r}_{i,j}^{\text{array}}$, creating local coupling including self feedback: emitter (i, j) will be injected by its own field and emitters located in its neighborhood with a range defined by the DOE's particular diffraction pattern. Finally, the second lens can in principle be removed, creating feedback according to $2f$-imaging, and hence coupling of emitter (i, j) to emitters located in the neighborhood centered at $(-i, -j)$. It is important to highlight that RC does not imply self-coupling, yet from a nonlinear dynamical point of view the resulting behavior of the network will most certainly be different. Finally, $W_{i,j}^{\text{DOE}} \neq W_{j,i}^{\text{DOE}}$ would allow for a network structure according to a directed graph. For arrays of electro-optical emitters, a direct coupling approach as previously discussed for semiconductor lasers will typically not be sufficient; unless conversion from optical to electrical signal and vice versa is carried out at the site of the emitters. While such a situation is created by optical light valves [11] and advanced arrays based on integration of detectors and lasers [12], most arrays of electro-optical nonlinear elements including electro-optical SLMs will not provide this functionality. For these systems, the diffracted image will need to be recorded with an electronic spatial sensor—typically a camera, which in turn must be coupled to the electro-optical emitter array.

4.2.1 Coupling matrix

For the detailed characterization of a network coupling matrix established by diffractive-coupling (W^{DOE}), we first employed the SLM-based setup 1 [7]. Figure 4.2 illustrates such an experimental scheme. The field of an illumination laser (Thorlabs LP660-SF20, $\lambda = 661.2\,\text{nm}$, $I_{\text{bias}} = 89.69\,\text{mA}$, $T = 23°\text{C}$) is adjusted to s-polarization via polarization-controlling paddles and collimated by a lens (L1, Thorlabs AC254-035-B-ML), creating a plane-wave illumination of the SLM. A second lens (L2, Thorlabs AC254-200-B-ML) is positioned such that the illumination is focused at the back-focal plane of the first MO (MO1, Nikon CFI Plan Achro 10X). This microscope objective receives the s-polarized illumination that is reflected by the polarizing beam splitter (PBS) cube. Between the PBS and MO1, we locate a half-wave plate ($\lambda/2$) which is aligned such that the illuminated SLM (Hamamatsu X13267-01) is operated in intensity

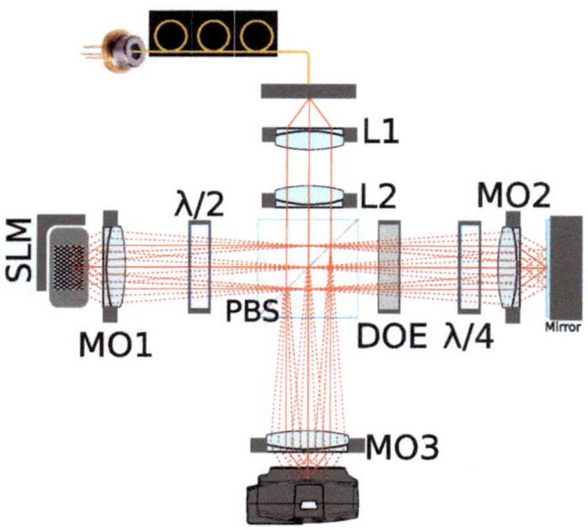

Figure 4.2: The surface of a SLM is illuminated by a laser's plane wave. A $\lambda/2$-plate adjusts the SLM to operate in intensity modulation, and the signal reflected off the SLM is filtered by a polarization beam splitter (PBS). In the PBS's transmission direction, a diffractive optical element (DOE) introduces diffractive orders that are imaged onto a camera after a double pass through a $\lambda/4$-plate.

modulation mode. Instead of the standard configuration of a 3-paddel fiber polarization controller we employ 5 paddels which significantly increases the extinction ration at the PBS. After reflection off and potential polarization modification by the SLM, the p-polarized fraction of light is transmitted through the PBS, where it passes the DOE (HOLOOR MS-443-650-Y-X) and a quarter-wave plate ($\lambda/4$). A second microscope objective (MO2, Nikon CFI Plan Achro 10X) creates an image-plane, in which we located a broadband highly-reflective dielectric mirror. Following reflection, light again passes the DOE, and due to the double-pass through the $\lambda/4$-plate p-polarization is converted into s-polarization, resulting in near 100 % reflection at the PBS. There, the twice-diffracted optical fields are imaged on a Camera (CAM, Thorlabs DCC1545M) by yet another microscope objective (MO3, Nikon CFI Plan Fluor 4X). Crucially, in contrast with the previous section, the location within the emitter array is now only identified by a single index i, which is straightforward to convert into an 2D index of (i, j) via concatenation. We opted for this simplification in order to avoid three-dimensional tensors to describe the coupling between emitters.

In this characterization, after interaction with each of the SLM's pixels, the p-polarized fraction of the reflected optical field is imaged onto the camera in a $4f$-configuration, including double-diffraction by the DOE. Each SLM pixel is controlled via its gray-scale value (GS), hence the 8-bit SLM's state vector is $\mathbf{x}^{\text{SLM}} \in \{0, 1, \ldots, 255\}$, and the resulting optical field aligned in p-polarization for pixel i is given by

$$E_i = E_i^0 \cos\left(\frac{2\pi}{\kappa_{\text{SLM}}}(x_i^{\text{SLM}} + \theta_i^0)\right), \tag{5}$$

where E_i^0 is the optical field illuminating pixel i, $\kappa_{\text{SLM}} = 244.6 \pm 1.6$ the conversion between pixel gray scale and polarization angle in radians; gray scale offset $\theta_i^0 = 11.1 \pm 1.1$ is a device related constant. Uncertainties given for κ_{SLM} and θ_i^0 correspond to the standard deviations measured across all pixels. After double-passing the DOE, the signal registered on the camera is given by

$$x_i^C = \alpha \left| \sum_j^N W_{i,j}^{\text{DOE}} E_j \right|^2, \tag{6}$$

where $\mathbf{x}^C \in \{0, 1, \ldots, 255\}$ is the 8-bit camera-state and $\alpha = \frac{GS}{I^{\text{sat}}} \cdot \text{ND}$. The camera's saturation intensity is given by I^{sat}, and ND the transmission efficiency through the full setup, including neutral density filters selected such that the dynamical range of the camera is best exploited and overexposure is avoided. Most importantly, $\mathbf{W}^{\text{DOE}}$ is the network's coupling matrix created by the DOE. Camera-state $\mathbf{x}^C$ is linearly rescaled in size such that its dimensions match the number of active SLM pixels, resulting in $\tilde{\mathbf{x}}^C$. This additional step is required since (i) the optical image is magnified by 2.5 due to the magnification ratio between MO1 and MO3, and due to (ii), the different pixel sizes of SLM (12.5 μm) and camera (5.2 μm).

In agreement with our alignment condition given by equation (4), the illumination wavelength λ was chosen such that in combination with the DOE and MO1 the spacing between diffractive orders (p^{diff}) matches the SLM's pixel-pitch ($p^{\text{array}} = 12.5\,\mu$m). For obtaining W^{DOE}, $\tilde{\mathbf{x}}^C$ was recorded once for each pixel $i \in \{1, 2, \ldots, 2025\}$ being switched to its maximal p-polarization configuration ($x_i^{\text{SLM}} \sim 110$) while all other pixels set to minimal p-polarization ($x_{j \neq i}^{\text{SLM}} \sim 50$). This procedure was carried out for all pixels, once with and once without the DOE, enabling us to first confirm that, without the DOE, pixels exclusively experience self-coupling. Moreover, this procedure also allowed for the normalization of W^{DOE}. As the DOE with 3×3 diffractive orders is operated in double pass, the final diffraction is a convolution of the diffraction pattern with itself, on average resulting in a 5×5 coupling pattern. Figure 4.3 shows the experimentally obtained coupling matrix W^{DOE} for a network of 2025 (45×45) nodes. Upon inspection of the insets, showing a more detailed zoom of W^{DOE}, one can see strong variations in local connectivity strengths. This is due to each pixel illuminating a DOE area comparable to the DOE's lowest spatial frequency. As this area shifts slightly from pixel to pixel, the intensity distribution between diffractive orders varies. This inherently creates the heterogeneous photonic network topology needed for computation according to the RC concept [13].

In single-pass and for the optical wave traversing the DOE with a diameter significantly larger than p^{DOE}, the resulting pattern is the 3×3 configuration of diffractive orders. Combined, these diffractive orders account for approximately 70 % of the entire optical intensity, which in addition is very uniformly distributed across the 9-diffractive orders. One could therefore approximate the average coupling for a single pass configuration as a 2-dimensional square window function, which is one for each entry within the 3×3 coupling window and zero otherwise. This assumes that local diffraction pattern fluctuations, consequence of the small beam diameter for individual pixels, average out across the network, hence resulting in the DOE-diffraction

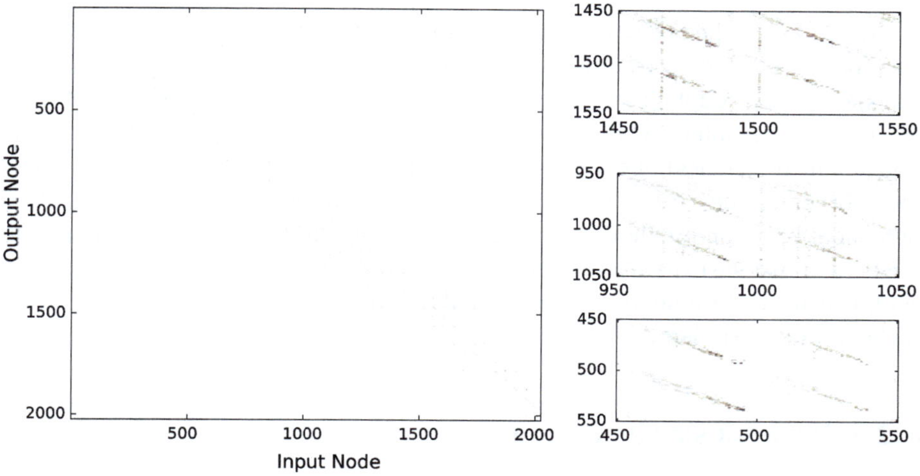

Figure 4.3: Diffractive coupling matrix W^{DOE} of a network consisting of 2025 nonlinear photonic nodes. The contrast in the large panel was artificially enhanced to reveal the coupling structure. The smaller panels (original contrast) on the right show local coupling strengths, revealing complex connection weight distributions, while also revealing the local nature of the coupling.

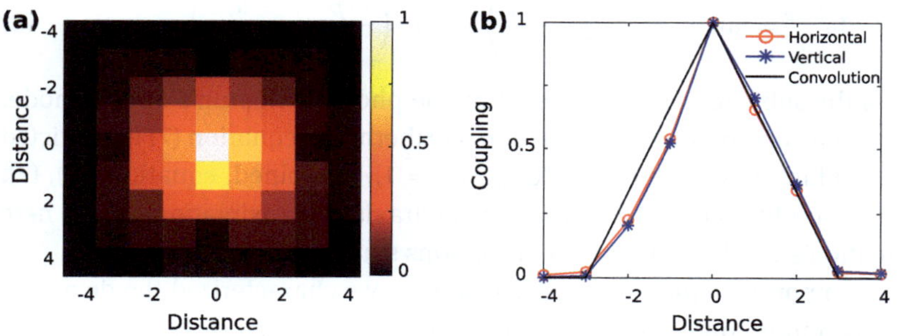

Figure 4.4: (a) Normalized, average coupling strength against coupling distance. The dominating coupling term corresponds to self-coupling, and coupling takes place within a radius of ≈3. (b) Horizontal and vertical profiles through the center position of data shown in panel (a). Average coupling is highly symmetric and experiences a linear decay with distance. This topology can be excellently explained by the convolution of two step functions with a width of 3, creating a triangular distribution. This is the result of double passing the DOE, and hence convoluting the 3×3 coupling matrix with itself.

pattern obtained for illuminating many DOE-periods p^{DOE}. For the double-pass configuration, such an assumption results in the convolution of this 2D window function with itself, creating a pyramidal distribution now consisting of 5×5 nonzero entries. In Figure 4.4, we show the globally averaged coupling properties obtained from our experiment. For that, we took the images $\tilde{\mathbf{x}}^C$ previously recorded for the coupling matrix characterization, and select an area of 9×9 entries centered around the activated pixel, i. e., hence the position of the zero order. Subarrays for all pixels are summed and normalized, and the resulting average network coupling is shown in Figure 4.4(a). As expected, the coupling strength at the center is strongest, corresponding to the average zero-order intensity after the DOE-double pass. In Figure 4.4(b) we show the horizontal (red circles) and vertical (blue stars) profile-cuts through the center of panel (a) data. The experimentally obtained results excellently match the expected pyramidal coupling profile with an overall width of 5 at its base. For coupling radii larger than two the experimentally obtained average coupling is not strictly equal to zero, which we attribute to the non negligible contribution originating to the DOE's higher diffractive orders.

4.2.2 Network size limitations

While already demonstrated to couple several thousand photonic nodes, diffractive coupling's validity depends of positions $r_{i,j}^{\text{array}}$. From equation (3), it follows that overlap between the orders of neighboring emitters is given if

$$\Phi_i^{\text{Node}} > \Theta_{i,\pm 1}^{\text{diff}} - \Theta_{i\pm 1}^{\text{im}} \tag{7}$$

$$\Phi_i^{\text{Node}} = \tan^{-1}\left(\frac{i \cdot p^{\text{array}} + a/2}{f}\right) - \tan^{-1}\left(\frac{i \cdot p^{\text{array}} - a/2}{f}\right). \tag{8}$$

Here, Φ_i^{Node} is the subtended angle covered by the photonic neuron's optical mode, which is emitted by an aperture of diameter a. Furthermore, equation (7) is restricted coupling via the first diffractive orders ($\Theta_{i,m}^{\text{diff}}$, $|m| = 1$). Combined, equations (1), (2), and (7) create a condition corresponding to the paraxial approximation, where here the approximation's validity is limited to deviations smaller than Φ_i^{Node}.

Due to the importance for the concept's validity, we characterized the deviation away from the coupling conditions for a large range of emitter positions r_i. In the experiment, we implemented the setup schematically illustrated in Figure 4.1(a). To emulate the emission of a single-mode optical network node, we used emission from the end of a single-mode optical fiber (Thorlabs TW670R5A2) coupled to the same laser as in Section 4.2.1 with a wavelength of $\lambda = 661.2$ nm. The fiber was shifted along the object plane via a micrometric xy-stage (Thorlabs ST1XY-S/M), and the image was recorded on a CMOS-camera with 2.2 μm pixel size (IDS USB 3 μEye LE). A MAG = 10 microscope objective (Nikon Plan N, NA = 0.25) was used to collimate the fiber's emission, which was imaged onto the camera with a MAG = 4 microscope objective (Nikon N4X-PF, NA = 0.13). Both microscope objectives were separated by $D = 50$ mm, with the DOE located approximately mid-way between both. At each position r_i, we recorded the camera's image, and since in this configuration light passes the DOE only once, we fitted the resulting diffractive orders via 9 Gaussian profiles.

In Figure 4.5(a), we show the experimentally obtained mismatch (stars) between the expected and real positions of diffractive order = −1 for $0 < r_i < 2$ mm on a

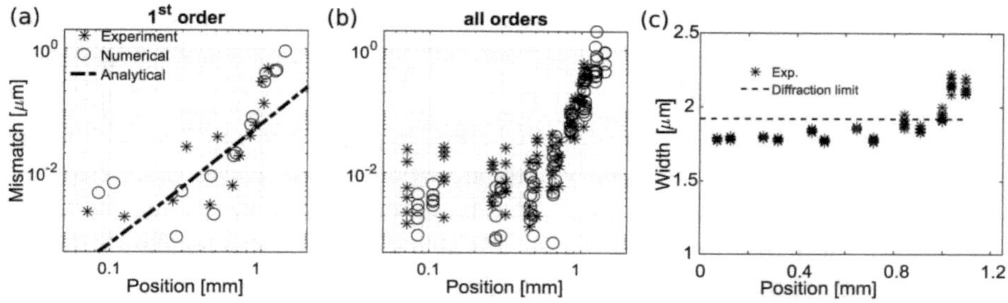

Figure 4.5: (a) Mismatch between the first diffractive order and nominal positions of emitter versus position for the first diffractive order in the horizontal plane. (b) As in (a) but for all orders. The increase in mismatch for distances beyond 1 mm is caused by beam-vignetting. For arrays of 4 mm^2 size, the mismatch remains below 1 μm. (c) For the same area, the imaging system remains diffraction limited, allowing to couple arrays consisting of 40.000 single mode emitters separated by a pitch of 10 μm.

double-logarithmic scale. From the data, it is apparent that the mismatch remains below 0.1 μm for positions located within a circle of $r_i \lesssim 1$ mm. Assuming $p^{\text{array}} = 10\,\mu$m, a typical value for such discrete photonic emitter arrays [7, 14], diffractive coupling is therefore possible for more than 30.000 nodes. However, for $r_i > 1$ mm one can identify a strong increase in the coupling mismatch. Comparison with the analytical results (dashed line) obtained with the help of equations (1) and (2), it is clear that experimental results and analytical solution strongly diverge.

The underlying reason can be understood on the basis of a numerical simulation. For that, we computed the optical propagation of the collimated beam via the method based on the angular spectrum of plane waves, importantly not employing the paraxial approximation [10]. Using phase-retrieval, we determined the phase-profile of the DOE, which was included in the beam-propagation simulation. Finally, we simulated the microscope objectives based on a numerical method employing the Debye-approximation [15]. All relevant parameters like NA and MAG of the microscope objectives as well as the distance between the individual optical elements were taken from the experiment. Results of the numerical simulation (circles in Figure 4.5(a)) excellently agreed with the experimental results in both, the confirmation of diffractive coupling for $r_i \lesssim 1$ mm and the strong divergence away from the analytical solution for $r_i > 1$ mm. Inspection of the z-positions, critical for the beam propagation, identified the cause: for $r_i > 1$ mm, the entrance pupil of the second microscope objective results in substantial beam-vignetting. Diffraction on that pupil strongly deviates the diffractive orders away from their unperturbed positions. We can therefore conclude that even the already excellent current scalability up to networks hosting more than 30.000 nodes is limited by the imaging system, not by the diffractive coupling concept.

Figure 4.5(b) shows the coupling-mismatch of all orders for, both, experimental and numerical results. First, the obtained mismatch confirms the limit obtained for the first order. Second, it reveals that for smaller r_i the experiment exhibits a position resolution limited to below ~40 nm. This excellent accuracy is consequence of the camera's low detection noise allowing for highly accurate fitting. Interestingly, we find numerical simulations are equally limited in position resolution. The limit of ~10 nm obtained there is consequence of the fitting algorithms convergence criteria.

Finally, successful coupling does not only require accurate agreement between the different diffractive order's positions, but also that diffractive imaging does not notably deteriorate the quality of the emitted optical mode. In Figure 4.5(c), we therefore show the width of the individual diffractive orders in the image plane obtained from the experiment (stars). The dashed line corresponds to the diffraction limit of the optical system, demonstrating that our system remains diffraction limited as long as beam-vignetting can be neglected. Imaging performance slightly better than the diffraction limit can be attributed to the uncertainty associated to the single-mode fiber's NA, which in turn causes uncertainty in calculating the collimated beam's width. All together, we can confirm the excellent scalability of diffractive coupling, allowing the creation of single-mode emitter networks consisting of 10s of thousands of elements.

4.3 Networks of vertically emitting lasers

In a second experiment, we investigated diffractive-coupling of a small array of semiconductor VCSEL lasers. In this early stage experiment [6], we implemented a different optical setup to create the $4f$ diffractive coupling. In Figures 4.1 and 4.2, $4f$-imaging is realized via two lenses or MOs, with the DOE and other optical elements included in the collimated-space between. This approach, inspired by designs from infinity-corrected microscopy, has the advantage that aberrations are strongly reduced as all flat optical elements are located in the space where beams are collimated. However, on the down side, one cannot adjust f, which therefore is not a parameter but a constant in the alignment condition of equation (3). For our commercial VCSEL array (Princeton Optronics PRI-AA64-PK-SM-W0975), $p^{\text{array}} = 250\,\mu\text{m}$ and $\lambda \approx (966 \pm 2)$ nm are fixed, and so is p^{DOE} of the commercial DOE (HOLOOR 1803). This only leaves focal distance f as a tunable parameter, and the simplest configuration for such a tunable imaging setup is based on direct imaging with a single lens. In Figure 4.6(a), we schematically illustrate the resulting experimental setup. The VCSEL array was placed at distance $f_1 > f$ in front of the only lens inside the resonator (Thorlabs AL1225-B, $f = 25$ mm). According to the imaging properties of a lens, an image is formed at $f_2 = (f_1^{-1} + f^{-1})^{-1}$ with a magnification $M = f_2/f_1$. We can therefore tune f_1 such that the coupling condition of equation (3) is fulfilled, and then simply place the mirror at f_2 behind the lens. Instead of a simple mirror we used a SLM (Holoeye, LC-R 1080) which facilitated additional control over the coupling matrix beyond the coupling implemented via W^{DOE}.

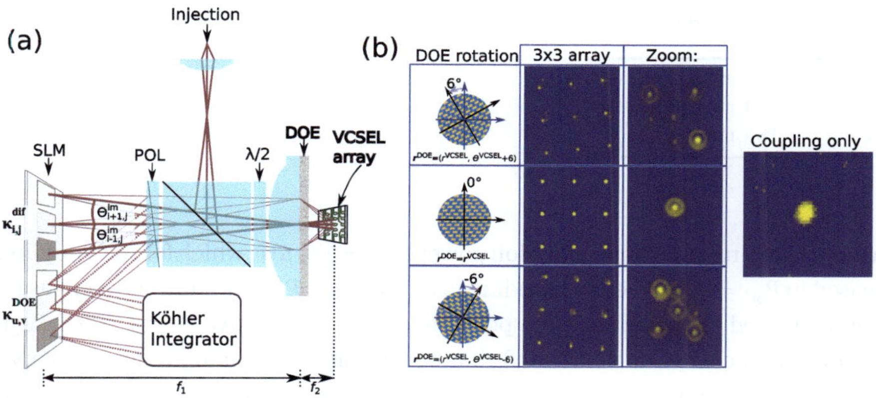

Figure 4.6: (a) Optical $4f$-resonator based on a single lens architecture. The diffraction system's imaging angle can be adjusted to the DOE's diffraction angle by changing distance f_2. The scheme allows for optical injection via an external injection laser and optical feedback is polarization selective. (b) For fine-adjustment, the DOE is mounted on a micrometer controlled rotation stage, and one can see how in the middle column diffractive orders from 9 neighboring lasers excellently overlap. Subtracting images recorded without feedback from the ones with feedback, we can identify the excellent optical quality of the feedback.

A 50/50 beam splitter (BS) created an input port for an external injection laser and allowed imaging of the optical feedback (not shown for simplicity). Behind this BS, a Rochon-prism makes the system's feedback and its output polarization selective. Finally, in the output we included a Köhler-integrator, homogenizing the array's output within a focal spot of ~100 μm. This created spacial overlap between different planar-wave vectors originating from the various VCSELs and, therefore, allowed the detection of the individual VCSEL's properties simultaneously. This step is essential, as otherwise we could not have simultaneously determined global network dynamics but only measure the dynamics of one laser at a time.

Close to the electrically pumped VCSEL's individual bias current threshold ($I_{\text{th}} \sim 0.2$ mA), we obtained a lasing threshold reduction due to self-coupling of ~25 % for the central 3×3 array. Toward elements located further away from the center, this value decreased significantly, and outside the central 5×5 array no such effect could have been reliably determined. This demonstrates that the size of our diffractive network is mostly restricted to the central 3×3 array. We continued and evaluated diffractive coupling by including the DOE and examining modifications to the array's emission power and by coupling-induced network dynamics. For maximizing interaction between the VCSELs in our network, we minimized their spectral detuning by biasing the lasers according to Table 4.1 (T_{array} = 40°C, λ_{array} = 966.92 nm). Two lasers were not pumped: one was not connected by the array-manufacturer, the other could not be tuned sufficiently to come into resonance with the other lasers. Without coupling, we

Table 4.1: Bias current for array emission at 966.92 nm.

2.188 mA	0 mA	0.658 mA
1.77 mA	0 mA	0.908 mA
2.4 mA	1.189 mA	1.411 mA

measured a free-running array emission power of $P_0 = 186\,\mu$W, which for self-coupling increased to $P_{SC} = 195\,\mu$W, corresponding to $P_{SC} = 1.048 \cdot P_0$. Coupling between lasers further increased the array emission power to now $P_{FC} = 205\,\mu$W ($P_{FC} = 1.052 \cdot P_{SC}$). These increases clearly demonstrated that our diffractive-coupling scheme is capable to couple single mode semiconductor lasers if aberrations are kept in check. In order to maximize coupling, careful control of the DOE's rotation angle is essential. In Figure 4.6(b), we show how sensitive the spatial overlap of the individual laser's coupling contributions is in dependence to this alignment parameter. We therefore mounted the DOE in a micrometer controlled rotation stage (Thorlabs CRM1P/M). For correct alignment, we can see that the feedback signal is a single Gaussian-like spot consisting of the individual laser's superpositions. The alignment between the different diffractive orders of neighboring lasers is of such high quality that even the fringes of the Airy-function agree.

Under the bias condition of Table 4.1, each investigated coupling configuration resulted in less relative coupling-induced power increase than the ~25 % obtained close to solitary laser threshold. This reduced relative impact is a direct consequence of the laser network's dynamical state, which significantly differs between the low and the high bias currents. All coupled lasers were biased significantly above solitary threshold and, therefore, very likely exhibit chaotic dynamics [16], which consequently limit the relative power increase.

4.3.1 Network dynamics and optical injection

We continued our analysis by injecting a spectrally aligned external laser into the 7 active VCSELs. By doing so, we demonstrated that diffractive laser networks can be all-optically coupled to external information, essential functionality for them to serve as optical NNs. If the injection laser polarization is aligned to the polarization of the array (s-polarization), we measure a power increase of $P_{CL} = 1.112 \cdot P_{FC}$ due to partially-coherent locking of the array to the external injection laser. When biased according to Table 4.1, the resulting array's emission was strictly linear polarized (p-polarization), which consequently is coupled out of the resonator by the Rochon prism. For successful locking of the array to the injection laser, the array will switch its polarization from p to the s-polarization of the injection laser. We defined the injection induced switch-

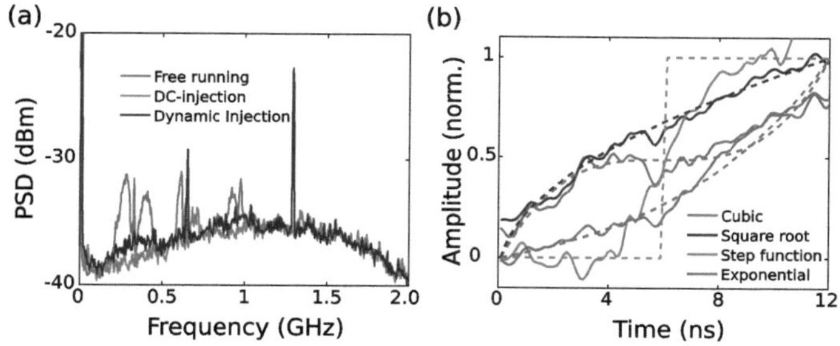

Figure 4.7: (a) Radio-frequency spectra of the coupled free running, DC-injected, and dynamically injected array are shown as red, green, and blue data, respectively. Free running dynamics show clear signatures of the cavity round trip time and of mutual coupling. These dynamics can be efficiently quenched by DC-injection with the external injection laser, whose modulation strongly drives the network. (b) Linear combinations of driven laser dynamics can be used to approximate various nonlinear transformations.

ing contrast Δ_{inj} as

$$\Delta_{\mathrm{inj}} = \frac{I^p_{\mathrm{inj}} - I^{\mathrm{CT}}_{\mathrm{inj}}}{I^p_0}, \tag{9}$$

with I^p_{inj}, I^p_0, and $I^{\mathrm{CT}}_{\mathrm{inj}}$ as the array's p-polarized emission intensity with and without injection as well as the s-polarized injection laser's cross-talk, respectively. Obtaining $I^p_{\mathrm{inj}} = (98.3 \pm 3)\,\mu\mathrm{W}$, $I^p_0 = (294 \pm 3)\,\mu\mathrm{W}$ and $I^{\mathrm{CT}}_{\mathrm{inj}} = (34 \pm 1)\,\mu\mathrm{W}$, the corresponding injection-laser induced polarization switching contrast is $\Delta_{\mathrm{inj}} = 78\,\%$, using a moderate injection laser intensity of ~150 μW per array laser.

Besides modification to the lasers' output power, delayed optical feedback can potentially induce complex network dynamics. These we detected via a fast photoreceiver (FEMTO HSA-X-S-1G4-SI-FS) located in the focal position of the Köhler integrator. In Figure 4.7(a), we show multiple rf-spectra of the network's intensity dynamics for multiple coupling configurations. Red data shows the network's free-running dynamics, revealing multiple broad spectral features around the external cavity's round trip frequency ($\tau^{-1} = 0.65$ GHz). In addition, we can identify strong dynamics at half that frequency, a feature which, for polarization maintaining interaction, is representative for mutually coupled lasers [17]. Hence, dynamics was mainly dominated by the external cavity round-trip frequency ($\tau^{-1} \approx 0.65$ GHz) and its higher harmonics, in addition to the coupling induced component around $(2\tau)^{-1}$. Both, spectrally narrow as well as broadband features are present in the rf-spectrum, indicating a combination of periodic and complex dynamics. According to Table 4.2, bias currents were widely spread, ranging from $3.3I_{\mathrm{th}} \leq I_{\mathrm{bias}} \leq 12I_{\mathrm{th}}$, which was needed for maximizing the array's spectral homogeneity by exploiting the current dependency of the lasers' emission wavelength. Yet, such a large bias current spread induced equally heteroge-

Table 4.2: Diffractive laser network parameters, bias currents as in Table 4.1.

Power w/o coupling	186 μW	P_0
Power self-coupling	195 μW	$P_{SC} = 1.048 \cdot P_0$
Power full-coupling	205 μW	$P_{FC} = 1.102 \cdot P_0$
Coherent locking		$P_{CL} = 1.112 \cdot P_{FC}$
DC-locking fraction		$\Delta_{inj} = 78\,\%$

neous operating conditions for the network's lasers, suggesting complex and diverse dynamics.

We then again injected the laser array with the external injection laser, again with orthogonal polarization alignment. In the resulting rf-spectrum (green data), dynamics are almost fully absent, with only weak dynamical features remaining around τ^{-1}. We therefore successfully quenched the laser network's dynamics by stable locking to the external injection laser. Finally, we investigated the network response to external perturbation. The injection laser was therefore modulated by a Mach–Zehnder modulator (not shown in Figure 4.6(a)) with a frequency of $\nu_{inj} = \tau^{-1}$. Dynamics of the locked network, shown as blue data in Figure 4.7(a), were strongly dominated by the modulation through the injection laser. Crucially, we found strong nonlinear mixing in our laser network: a equally sharp spectral component at twice the injection modulation frequency. As such, we have successfully created a spatiotemporal network of semiconductor lasers. Higher orders too possibly exist, however, these are outside the bandwidth of our fast photo-receiver (1.4 GHz). Additionally, we can drive all network-lasers in parallel at the excellent bandwidth of 0.65 GHz, which results in strong nonlinear mixing of the input information. While still only demonstrated for a small network, this represents the first step toward a fully parallel and ultrahigh speed network of semiconductor lasers serving as a RC.

4.3.2 Function approximation

The final step toward computing with the laser network is to weight and combine the laser's output. For that, we activated all 8 connected lasers of the array's center, biasing them according to Table 4.3. Optical injection was modulated with a sinusoidal intensity and at a frequency of $\nu_{inj} = 33\,\text{MHz}$, whose period was approximately 20 times the network's coupling delay $\tau = 1.3\,\text{ns}$. By choosing such a low modulation frequency, the temporal window during which injection power increased linearly, i. e., between the sin-wave's extrema, spans multiple τ. During this part, the injected signal is approximately a linear ramp, which we used as system input $u(t)$. We recorded the network's response individually for each laser, using the fast photo-receiver and a realtime oscilloscope with a sampling rate of 40 GSamples/s and 16 GHz analog band-

Table 4.3: Bias current for dynamical injection locking and off-line function approximation via VCSEL array.

3.1 mA	0 mA	1.7 mA
2.6 mA	1.24 mA	1.65 mA
3.2 mA	2.0 mA	0.218 mA

width. The network state is then vector $\mathbf{x}(t)$, which contains eight entries—one for each active laser.

We defined multiple target nonlinear transformations $f^T(u(t))$ to be approximated by the laser network's output $y^{\text{out}}(t)$. Based on the $\mathbf{x}(t)$, the system's output is given by

$$y^{\text{out}}(t) = W^{\text{out}}\mathbf{x}(t), \tag{10}$$

where W^{out} are the system's readout weights, which we calculated via the standard matrix inversion techniques introduced in Chapter 2. As targets $f^T(u(t))$, we chose cubic, square root, and exponential nonlinearities plus a step function, shown as red, blue green, and magenta dashed in lines in Figure 4.7(b). In the same figure, we show the resulting outputs $y^{\text{out}}(t)$ in identical color solid lines. The demonstrated system's computation power is certainly limited and relies on off-line procedures. Yet, from the data in Figure 4.7(b) it is apparent that such systems can process data at very high bandwidths. All transformations are realized by the network within 12 ns, using an injection power of ~150 μW per network laser. For a fully implemented optical system, this would correspond to a global clock-rate of 83 MHz, which is close to three orders of magnitude beyond what current implementations can achieve [18]. Also, requiring 150 μW injection power per laser means that over 6000 lasers could be injected with a 1 W single mode injection laser. For such a system, the energy per transformation would be in the order of microjoules for complex transformations potentially realizable with such large networks.

4.4 Reservoir of Ikeda oscillators

Ikeda delay systems have on many occasions demonstrated excellent performance for implementing a photonic RC. The defining feature of nonlinear Ikeda systems is their $\sin^2$ nonlinearity. As in the original publication by Ikeda [19], the nonlinearity is typically created by interference between multiple waves combined with a process sensitive to $|E|^2$. Here, we follow another approach and implemented the Ikeda nonlinearity based on rotation, filtering, and detection of an optical field's polarization. An off-the-shelf system with polarization control over many spatially distributed discrete elements is a SLM. We therefore extended the system originally used in Section 4.2.1.

4.4.1 Experimental setup

The first step toward the physical implementation of a reservoir is realizing a complex spatiotemporal system. Our main objective was to create a system which (i) can readily be scaled up to large numbers of photonic neurons, (ii) can be practically extended by optical input and readout, (iii) is based on fully parallel optical concepts, and finally (iv) is well suited as a proof of concept experiment. Containing over 6000 pixels serving as nonlinear network nodes within an area of 1 mm^2, the SLM system, already utilized for measuring the DOE's coupling matrix W^{DOE}, is excellently scalable. It is an off-the-shelf device, and combined with infinity-corrected microscope objectives the optical setup can readily be extended to realize input and output ports. All optical processes involved in such a system are therefore carried out in parallel. Finally, the device can be fully characterized, allowing for the creation of accurate models, and hence for detailed analysis of fundamental aspects important to a proof of concept, and equally essential for the field's future development.

In Figure 4.8(a), we show the relevant connections and their nomenclature in our reservoir computer, in (b) the complete experimental setup. In our RC system, panel (a), a single input injects information into the reservoir according to injection weights W^{inj}. As before, the reservoir is internally connected in a recurrent topology with connection weights according to the connectivity matrix W^{DOE}. Finally, a single computational result is provided by combining the network's state according to the weight matrix $\mathbf{W}^{\mathrm{DMD}}$. Following the RC concept, input and recurrent internal weights can be

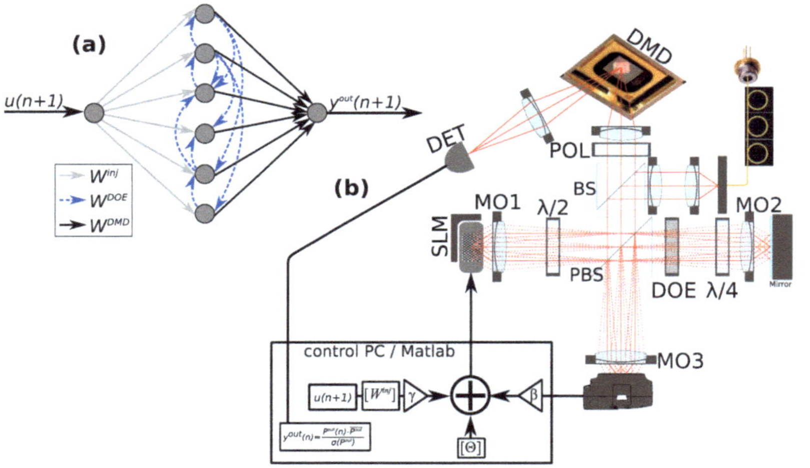

Figure 4.8: (a) Connections implemented optically and electronically in our photonic reservoirs. (b) Experimental setup based on the one introduced in Section 4.2, now extended by the readout weights $\mathbf{W}^{\mathrm{DMD}}$ implemented in the digital micro-mirror device (DMD). The experiment is controlled via a PC, which also implements learning of DMD weights and injects the system with the external information.

chosen randomly [13]. While W^{DOE} is not random (see Section 4.2.1), it certainly has local complexity and is therefore promising to provide the high-dimensional network required by RC.

4.4.2 A driven network of coupled Ikeda oscillators

The experimental system of Figure 4.8(b) is an amended version of the system in Section 4.2.1 and Figure 4.2, making it fit for serving as a reservoir [7]. We included an additional 50/50 beam splitter (BS) in front of the PBS. In reflection, the BS creates the input port for the SLM-illuminating laser and readout of the SLM, and hence of the network state; the readout functionality will be discussed in the following section. All other optical components part of Figure 4.8(b) are identical to the ones introduced in Section 4.2.1.

In order to make the transition from the static system in Section 4.2.1, we introduced a temporal context to our experiment and enabled the injection of external information. Illustrated by the gray box, camera state $\mathbf{x}^{C}(n)$ is recorded and SLM state $x^{\mathrm{SLM}}(n)$ is controlled by MATLAB running on an external computer. As can be seen in Figure 4.8(b), the camera and SLM are connected into a common system, where the first serves as input for the latter. Camera and SLM state vectors are therefore assigned an additional index n, corresponding to integer time of the system. For closing the temporal loop required for a reservoir's recurrent connectivity, we first multiply rescaled camera state $\tilde{\mathbf{x}}^{C}(n)$ with feedback gain β and add phase offset matrix $\boldsymbol{\theta}$ as well as the external information $u(n+1)$, and then send the result to the SLM to create reservoir state $\mathbf{x}(n+1)$:

$$\mathbf{x}^{\mathrm{SLM}}(n+1) = \beta\tilde{\mathbf{x}}^{C}(n) + \gamma\mathbf{W}^{\mathrm{inj}}u(n+1) + \boldsymbol{\theta} \tag{11}$$

$$\mathbf{x}(n+1) = \mathbf{f}(\mathbf{x}^{\mathrm{SLM}}(n+1)). \tag{12}$$

Here, $\mathbf{W}^{\mathrm{inj}}$ is the random injection matrix consisting of elements between 0 and 1, and γ is the injection strength. As $\mathbf{x}^{\mathrm{SLM}}(n+1)$ serves as the argument of the nonlinearity $\mathbf{f}(\cdot)$ provided by SLM pixels and PBS, equations (11) and (12) have the same structure as for the classical RC concept [13]. The only difference is that here the primary network states are optical fields, while the network is updated according to optical intensities. One therefore has to consider that the camera state depends on intensity detection of summed optical fields; see equation (6). Crucially, the only matrix multiplication off-loaded to the control PC is the injection matrix.

We define the reservoir state $\mathbf{x}(n+1)$ as the SLM's intensity in p-polarization, hence detected in transmission after the PBS. The reservoir's dynamical evolution is therefore governed by coupled Ikeda maps according to

$$x_i(n+1) = \alpha|E_i^0|^2\cos^2\left[\beta\cdot\alpha\left|\sum_{j}^{N} W_{i,j}^{\mathrm{DOE}}E_j(n)\right|^2 + \gamma W_i^{\mathrm{inj}}u(n+1) + \theta_i\right], \tag{13}$$

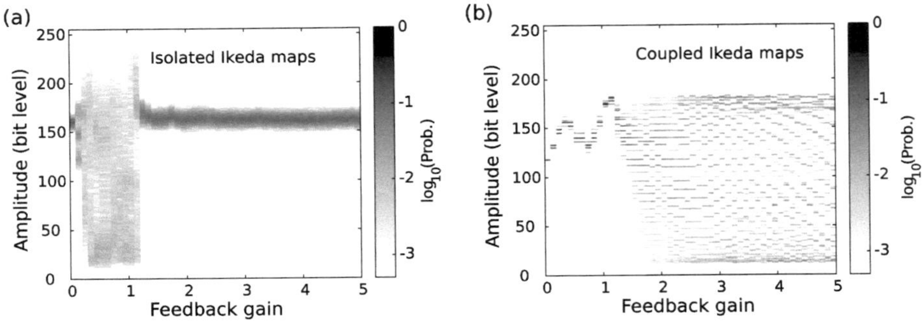

Figure 4.9: (a) Bifurcation diagram of node (22, 25) without coupling to the other network nodes (no DOE). (b) Bifurcation diagram of the same node, now with coupling to other network nodes realized (with DOE).

where $i = 1, \ldots, N$ are the individual photonic neurons. The overall update rate of the entire system is ~5 Hz, which is currently limited by the MATLAB script controlling the SLM. The SLM itself has a maximum frame rate of 50 Hz, which would correspond to the system's hardware limit. Reservoirs of maximally ~2500 nodes can be implemented based on our current experiment. Crucially, this size is not limited by the diffractive coupling concept, as was explained in Section 4.2.2 of this chapter. It is rather the imaging setup's field of view and the size of the illuminating planar wave, both of which can significantly be extended using more compact mechanical structures and replacing L2 of Figure 4.2 by a lens with a larger numerical aperture, respectively.

In Figure 4.9, we show two exemplary bifurcation diagrams for reservoir node (22, 25), recorded within a network of $N = 45 \times 45 = 2025$ nodes. Data of panel (a) was obtained for our system without the DOE, hence for a lattice of uncoupled Ikeda maps. In panel (b), we included the DOE in the beam-path, and W^{DOE} apparently induces strong modifications to node dynamics. Obtaining $\mathbf{x}(n+1)$, however, is not straightforward for the coupled network, since the camera only records the system state after its transformation by W^{DOE}. We therefore carefully characterized the nonlinear function of every SLM pixel thereby obtaining function $\mathbf{f}(\cdot)$, which according to equation (12) we used to translate state $\mathbf{x}^{\mathrm{SLM}}(n)$ into $\mathbf{x}(n+1)$.

An apparent difference between the dynamical behavior is the extended stable state which can be found for the system without nearest diffractive coupling. For $\beta > 1.2$, the system stabilizes after having exhibited chaotic dynamics for smaller feedback gains. The reason behind this unexpected fixed point is the saturation of the node's nonlinear function. For this node in particular, the argument sent to its update, i. e., $x_{1015}^{\mathrm{SLM}}(n+1)$, exceeds 255 for $\beta > 1.2$, and consequently the SLMs introduces an artificial limit, causing a stabilization of the node's dynamics at $x_{1015} = f_{1015}(255)$. Introducing the DOE creates the mentioned inhomogeneities within the network's coupling topology, causing a few individual nodes to experience significantly larger coupling than the majority of the network. As a consequence, the attenuation by the system's ND filters had to be increased by 60 %, and nodes experience average coupling that only saturates for higher β. Finally, the amplitude probability distribution across the coupled node's chaotic state shows an enhanced probability for a few states in the available 8-bit gray scale range. This we attributed to the experimental noise unavoidable during our nonlinear function approximation. The result is therefore not a smooth nonlinear transformation when we approximate the SLM's action. Averaging multiple measurements of the nonlinearity reduced the effect without allowing its full suppression.

4.4.3 Readout weights and photonic learning

The final step to information processing is to adjust the system such that it performs the desired computation, typically achieved by modifying connection weights according to some learning routine. Inspired by the RC concept [13], we constrain learning-induced weight adjustment to the readout layer. The BS introduced in Figure 4.8(b) creates the output-port for our photonic reservoir, which we choose to be of 50/50 splitting ratio to maximize the output power's signal to noise ratio. We focused on a reservoir with $N = 900$ nodes, and since these are spatially distributed we could use a simple lens (Thorlabs AC254-400-B) to image a version of the reservoir's state onto a digital micro-mirror array (DMD, DLi4120 XGA, pitch 13.68 μm). Individual mirrors of the DMD can be flipped between ±12°, and only for −12° the optical signal contributes to the output at the detector (DET, Thorlabs PM100A, S150C); see Figure 4.8. Our physically implemented readout weights are therefore strictly Boolean. Taking profit of orthogonal polarization between the field imaged on the camera and the DMD, our system's output is

$$y_k^{\mathrm{out}}(n+1) \propto \left| \sum_i^N W_{i,k}^{\mathrm{DMD}} \left(E_i^0 - E_i(n+1)\right) \right|^2 . \tag{14}$$

Here, k is the current learning iteration. In the experiment, weight vector $W_{i=1,\ldots,N,k}^{\mathrm{DMD}}$ corresponds to a square matrix of the DMD's reflectivity toward −12°. An image of the DMD's direction-selective reflectivity can be seen in Figure 4.8(b). An imaging magnification of 20 between SLM and DMD, combined with the different pitch for SLM pixels and DMD mirrors, causes a square area of approximately 18×18 micro-mirrors to correspond to the area of a single SLM pixel. The DMD was mounted on a rotation stage (Thorlabs CRM1L/M), allowing the accurate alignment between SLM and DMD axis. Finally, we carefully determined the location of the SLM network state on the DMD's surface, achieving a position-sensitivity of around one DMD micro-mirror using regular test-patterns loaded onto the SLM. In the following, all arrays of 18×18 micro-mirrors are assigned to their corresponding SLM pixel and are switched uniformly.

Noteworthy, weights implemented by the DMD are not temporal modulations as for example required for reservoir implementations in delay systems [20]. Once a certain configuration of W^{DMD} has been obtained, its function could be implemented by passive attenuations in reflection or transmission. Such passive weights are ultimately energy efficient and typically do not result in a bandwidth limitation. In this specific implementation, once trained, mirrors could simply remain in their position, and, if mechanically clamped, would not further consume energy. Finally, readout equation (14) is optically performed for all elements in parallel.

Training our RC corresponds to the optimization of $W^{\mathrm{DMD}}_{i=1,\ldots,N,k}$ such that after $k = 1, 2, \ldots, K$ learning iterations output $y^{\mathrm{out}}_K(n+1)$ approximates the learning target $y^T(n+1)$ [7]. We trained the system to perform a one step ahead prediction of the chaotic Mackey–Glass sequence [21, 13], and hence $y^T(n+1) = u(n+2)$. Parameters of the MG sequence were identical to [22], using an integration step size of 0.1. For the training, we injected 200 points as training signal $u(n+1)$, and from the resulting reservoir output y^{out}_k we removed the first 30 data points due to their transient nature. We furthermore subtracted its mean and normalized by its standard deviation, resulting in signal $\tilde{y}^{\mathrm{out}}_k(n+1)$, with which we determined the normalized mean square error (NMSE) ε_k between $\tilde{y}^{\mathrm{out}}_k(n+1)$ and $y^T(n+1)$.

Modifications to the DMD configuration are simply the inversion of a single set of micro-mirrors for a particular neuron i, and a modification from $\mathbf{W}^{\mathrm{DMD}}_{k-1}$ to $\mathbf{W}^{\mathrm{DMD}}_k$ is rewarded if it resulted in $\varepsilon_k < \varepsilon_{k-1}$. We therefore do not compute a gradient which later is used to precisely adjust $\mathbf{W}^{\mathrm{DMD}}_{k+1}$; our simple scheme modifies the system's behavior (output) and evaluates if modifications have been beneficial. We therefore compute reward $r(k)$ for each learning iteration

$$r(k) = \begin{cases} 1 & \text{if } \varepsilon_k < \varepsilon_{k-1} \\ 0 & \text{if } \varepsilon_k \geq \varepsilon_{k-1}. \end{cases} \tag{15}$$

Of further importance is the particular rule according to which our system selects the entry of W^{DMD}_k to be modified. In the first generation ($k = 1$), the N readout weights $\mathbf{W}^{\mathrm{DMD}}_{k=1} \in \mathbb{Z}\{0, 1\}$ are randomly initialized, the 170 points of $\tilde{y}^{\mathrm{out}}_1$ are measured and ε_1 is determined. For the next ($k = 2, \ldots, K$) learning iterations l_k, points toward the position of the readout weight to be modified according to

$$\mathbf{W}^{\mathrm{select}}_k = \mathrm{rand}(N) \cdot \mathbf{W}^{\mathrm{bias}}, \tag{16}$$

$$[l_k, W^{\mathrm{select,max}}_k] = \max(\mathbf{W}^{\mathrm{select}}_k), \tag{17}$$

$$W^{\mathrm{DMD}}_{l_k,k} = \neg(W^{\mathrm{DMD}}_{l_k,k-1}). \tag{18}$$

Here, rand(N) creates a random vector with N entries equally spaced between 0 and 1, max(·) returns position (l_k) and value ($W^{\mathrm{select,max}}_k$) of its argument's largest entry, and $\mathbf{W}^{\mathrm{bias}} \in [0, 1]$ is randomly initialized at $k = 2$ with entries equally spaced between 0 and 1. Symbol ¬(·) in equation (18) is the *not* operator. Entry $W^{\mathrm{DMD}}_{i=l_k,k=k}$ is therefore inverted, with the particular entry chosen according to the location of the largest entry in $\mathbf{W}_k^{\text{select}}$. Matrix $\mathbf{W}^{\text{bias}}$ has therefore the function of a bias controlling the probability of an entry to be selected. It is updated according to

$$\mathbf{W}^{\text{bias}} = \frac{1}{N} + \mathbf{W}^{\text{bias}}, \quad W_{l_k}^{\text{bias}} = 0, \tag{19}$$

and its entries therefore linearly grow by $\frac{1}{N}$ each learning iteration, while the last inverted readout weight's location l_k is set to zero. Consequently, $\mathbf{W}^{\text{bias}}$ biases learning away from modifying weights whose configuration has recently been optimized. In simulations, such a biased learning rule showed significantly faster learning convergence, which we attribute to the fact that it explores the relevant dimensions of W^{DMD} more efficiently. Finally, before the transition to the next learning step, the configuration of $\mathbf{W}^{\text{DMD}}$ is given by

$$W_{l_k,k}^{\text{DMD}} = r(k)W_{l_k,k}^{\text{DMD}} - (r(k) - 1)W_{l_k,k-1}^{\text{DMD}}. \tag{20}$$

Technically, our exploration strategy resembles a stochastic gradient descent, and equations (15) and (20) reinforce modifications which were found beneficial.

4.4.4 Curb performance-limitation of unipolar systems

On our way toward performance optimization, we identified a challenge already faced in the first realization of optical NNs [23]. In general, NN concepts exploit the full range of real numbers, hence including positive and negative values. In many optical architectures, internal and readout connection weights are positive; the same is true for state variable $\mathbf{x}(n + 1)$. This significantly restricts the system's functionality: multiplication with a negative weight does not only allow to subtract responses from different nonlinear nodes, it also allows to change the symmetry of a node's nonlinear transformation. First evaluations of the learning procedure and prediction of the MG series with our system suffered from limited performance since these limitations were not considered.

We introduced a strategy for compensating the absence of inverting a node's nonlinear transformation by negative connection weights. Considering for a moment a system exclusively consisting of nodes with linear rectifier nonlinearities with exclusively positive connection weights. Without negative weights, such a system would globally not be capable to synthesize a transformation from an input onto output with a negative slope, hence the space of functions the system is able to approximate is severely reduced. The same restriction would apply to our reservoir for the case that all nodes operate along the same part of their nonlinearity.

We therefore harness the nonmonotonous and periodic nature of the SLM's $\cos^2(\cdot)$ nonlinearity. A network node's phase offset $\theta_i|_{i=1,\ldots,N}$ is randomly selected from two

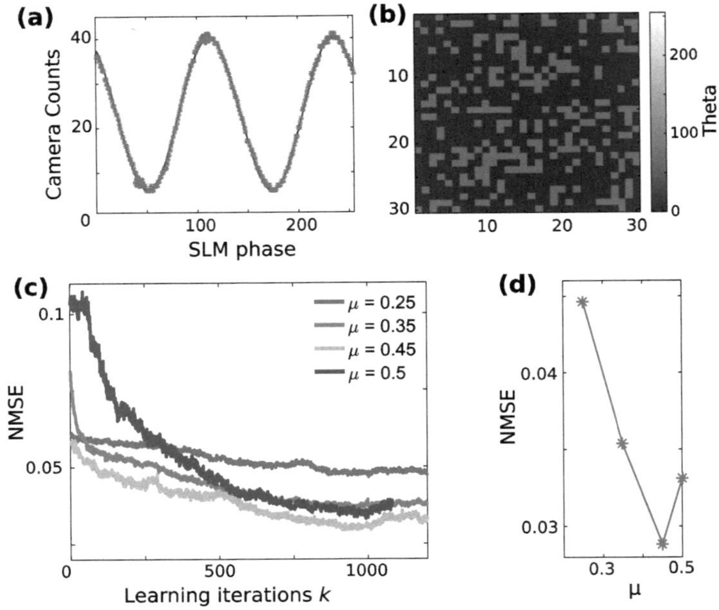

Figure 4.10: (a) Example of the nonlinear function of the photonic reservoir's nodes (red stars). According to a probability of μ, nodes of the network are distributed to operate along their nonlinearity starting from a phase offset close to a local minimum or maximum, blue stars. (b) Random distribution of the phase-offset across the photonic reservoir. (c) Learning curves obtained for different μ, using $\gamma = 0.25$ and $\beta = 0.2$. (d) Best performance obtained at each probability μ, illustrating the strong impact symmetry breaking of the network nodes' responses has on the system's performance.

different values. Locally scanning both offsets for optimal performance, we obtained $\theta_0 = 42 \mathrel{\hat{=}} 0.17\pi$ and $\theta_0 + \Delta\theta = 106 \mathrel{\hat{=}} 0.43\pi$ as the two ideal phase offsets. Nodes therefore are biased close to local minima (θ_0) or maxima ($\theta_0 + \Delta\theta$), which realizes the particular configuration we hypothetically discussed. The consequence are node ensembles which exhibit dynamics operating along their negative, others along their positive slope. We furthermore analyzed the impact of a biased distribution between both values across the network by introducing a probability of μ for a node's offset being $\theta_i = \Theta_0 + \Delta\Theta$.

In Figure 4.10(a), we show learning curves for different values of μ using $\beta = 0.2$ and injection gain $\gamma = 0.25$. Probability-ratios where $\mu = [0.25, 0.35, 0.45, 0.5]$ are shown as blue, red, yellow, and purple data, respectively. In Figure 4.10(b), we give the best performance obtained for each learning curve, revealing a strong impact of our network symmetry breaking. Best performance is found for $\mu = 0.45$, corresponding to a reservoir with an almost balanced distribution between positive and negative slopes. We would like to highlight that changing μ from 0.25 to 0.45 reduces the system's prediction error by approximately 50 %, standing testimony for the effectiveness of our approach. The absence of negative weights in $\mathbf{W}^{\mathrm{DMD}}$, $\mathbf{W}^{\mathrm{DOE}}$ and $\mathbf{x}$ can therefore partially be compensated for by incorporating nonlinear transformations with positive as well as negative slopes. Since such unipolar NNs are a common challenge in many hardware based systems, our result is of high significance for neural network hardware implementations.

4.4.5 System performance

We finally optimized our system's performance by exploring feedback gain β and input scaling γ. In Figure 4.11(a), we show the error convergence under optimized global conditions ($\beta = 0.8$, $\mu = 0.4$ and $\mu = 0.45$) for a training sample size of 500 steps (blue stars). The error efficiently reduces and finally stabilizes at $\varepsilon \approx 0.013$. Considering learning is limited to Boolean readout weights, this is an excellent result. After training, the prediction performance is evaluated further on a sequence of 4500 consecutive data points which were not part of the training dataset. As indicated by the red line in the same panel, the testing error matches the training error. We can therefore conclude that our photonic RNN successfully generalized the underlying target system's properties. The excellent prediction performance can be appreciated in Figure 4.11(b). Data belonging to the left y-axis (blue line) shows the recorded output power, while on the right y-axis (red dots) we show the normalized prediction target signal. A difference between both is hardly visible, and the prediction error ε (yellow dashed line) is small.

There are multiple features in the system's performance we would like to discuss. When repeating learning under identical condition, the system generally converges after comparable numbers of learning iterations k to closely comparable errors. This

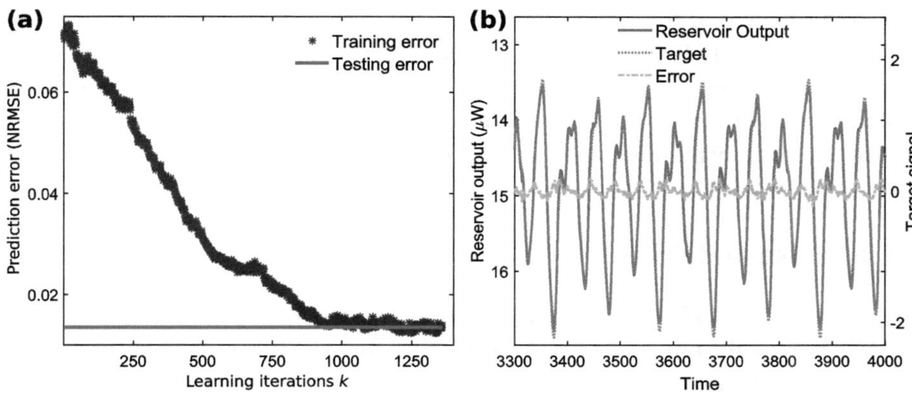

Figure 4.11: (a) Learning performance at optimal parameters ($\beta = 0.8$, $\gamma = 0.4$, $\mu = 0.45$). (b) The photonic RNN's predicted output in μW (blue line) can hardly be differentiated from the prediction target (red dots). Prediction error ε is given by the yellow dashed data.

is of particular interest in light of the significant impact the random initial configuration of $\mathbf{W}^{\mathrm{DMD}}_{k=1}$ has upon NMSE at $k = 1$; see Figure 4.10(c). The initial error before optimization fluctuates by a factor of two, ranging from 6 % to 12 %. Importantly, it is not the optimization starting with the lowest initial error which arrived at the smallest error after learning. This, too, is a feature we have confirmed along many learning experiments. We therefore conclude that (i) our particular learning routine efficiently scans the space of functions available in the photonic reservoir, and (ii) that the final configuration of $\mathbb{W}^{\mathrm{DMD}}_{k=K}$ is not unique. On the contrary, each repetition of the overall learning routing results in a different $\mathbb{W}^{\mathrm{DMD}}_{k=K}$. The system's error landscape likely consists of numerous minima with comparable performance. An alternative explanation is that we are yet far from global optimal performance and deteriorating processes, such as noise, wash-out the error landscape's finer features, resulting in a very broad global optimum. This is supported by our observation that optimal performance is reached rather quickly, and also that often long time-scale parameter drifts deteriorate the system's performance after the optimum has been reached again; see Figure 4.10(c). These will result in systematic modifications to the system's response, also discussed in Section 4.4.6. In a system with multiple local minima, these modifications potentially shift the system across other minima, resulting in a more complex performance impact. Yet we found that such drift appears to be exclusively resulting in monotonous performance reduction. Finally, we observed that learning typically converges after $K \sim N$ learning iterations. If this could be confirmed it would indicate excellent scaling, yet one would first need to investigate the impact the number of reservoir nodes N has on the speed of convergence.

Finally, we down-sampled the injected signals by 3 in order to create conditions identical to [22, 24]. Under such conditions, our error ($\varepsilon = 0.042$) was larger by a factor of 2.2 relative to a delay RC based on a semiconductor laser [22] and by 6.5 relative to a Mach–Zehnder modulator based setup [24]. It however is important to interpret these results in the light of the significantly increased level of hardware implementation in our current setup. In [22, 24], readout weights were applied digitally in an off-line procedure using weights with double precision. In [24], a strong impact of digitization resolution on the computational performance was identified, suggesting that ε can still be significantly reduced by increasing the resolution of $\mathbf{W}^{\mathrm{DMD}}$.

4.4.6 Noise and drifts

The previous section has illustrated that physically implemented analog NNs bring a new feature to the table: various types of noise, for example, in the network's state variable or through the drift of system parameters. Ultimately, network state noise and drift might both be considered perturbations taking place on a different timescale and according to different spectral density distributions. Drift of experimental parameters can either follow white noise if due to interactions with a complex environment, or a

$1/f$ power spectral density distribution if associated to deterioration such as component ageing [25].

To explore the influence of noise acting on timescales of (i) the system's response time as well as (ii) significantly slower such as drifts in the external environment, we measured the reservoir's response every $\Delta T \sim 10$ minutes. As before, input data were $T = 500$ data points of the chaotic MG sequence, resulting in a temporally concatenated reservoir state matrix $\mathbf{X}_t$ with 900×500 entries for a measurement at time t. From this matrix, the first 10 columns were removed due to their transient nature. Other parameters were $\beta = 1.0$ and a uniform phase offset $\Theta = 25$, and measurements were made for $\gamma = 1.0$ as well as for $\gamma = 0.5$. To quantify deviations consequence of noise and drift, we determine the network's response consistency [26]. As consistency C between matrices $\mathbf{X}_{t_1}$ and $\mathbf{X}_{t_2}$ we define the cross correlation at zero lag (CC) between nodes' timeseries and average across the entire network ($C = \overline{CC}(\mathbf{X}_{t_1}, \mathbf{X}_{t_2})$). In order to capture short and long-term effects, we calculate the consistency between consecutive ($C_s = \overline{CC}(\mathbf{X}_t, \mathbf{X}_{t+\Delta T})$) as well as between each recording and the first ($C_l = \overline{CC}(\mathbf{X}_{t=0}, \mathbf{X}_t)$); see Figure 4.12(a). The most apparent feature revealed by the obtained consistencies is the strong influence of injection strength γ. Both, short and long-term consistency for $\gamma = 1$ (blue data) are very high, in both cases above 0.95. Furthermore, there is a clear difference between the short-term (blue circles) and long-term consistency (blue crosses). On average, $\overline{C_s} = 0.993$, around which it oscillated with a period close to an hour. Within windows of reduced parameter drift C_s typically remains at 0.9935, which we take as the system's consistency limit. Long term correlation C_l starts off at a comparable level, yet in the course of two hours drops below 0.98 from where it continues to decrease during the experiment's duration of approximately 16 hours. Such slow timescale can be typically associated to slow modifications and drifts in the system's

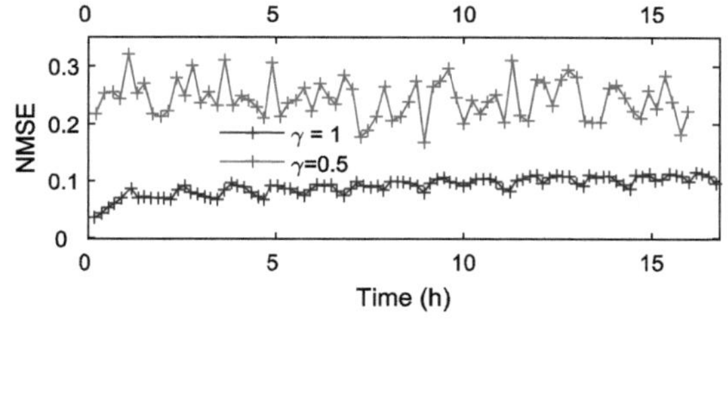

Figure 4.12: (a) Short and long term consistency across the photonic reservoir for different injection strength of the external information. Blue data: $\gamma = 1$, red data: $\gamma = 0.5$. Injection strength has a strong impact on the system's consistency, and under good operation conditions (high consistency) the impact of long-term drift is clearly visible. (b) Short-term consistency gives a lower limit of the prediction error, while drift continuously deteriorates the system's performance away from this limit. For bad consistency, the system's bad performance appears to be solely short-term consistency limited.

environment. We therefore conclude that slow drift has a measurable impact on the system's state, in this case significantly exceeding short-term fluctuations.

Reducing the injection strength to half ($\gamma = 0.5$, red data) has a surprisingly dramatic impact on C_s as well as C_l, which both drop to approximately 0.85. Crucially, one cannot identify any additional long-term (red crosses) relative to short-term (red circles) effects. The system's consistency is therefore limited by contaminations taking place on the fast timescale; drift can be neglected. These measurements highlight an important aspect. Initially, injection strength γ's main purpose was thought to be soliciting a high-dimensional response of the system by inducing excursions through the reservoir's high-dimensional phase-space at $\beta = 1$. It now is clear that, in addition, γ is of importance for the system's stabilization, something already investigated for delay systems [22]. Importantly, the reduction in consistency from 0.993 to 0.85 (factor 20!) strongly exceeds the one of injection strength (factor 2), identifying nonlinear effects as the origin.

Data shown in Figure 4.12(a) identifies the limited stabilizing influence γ has on the impact of slows drifts. Fast timescale noise results in perturbations of the system's trajectory comparable to different initial conditions, which is a consequence of operating the network close to the edge of chaos. Slow-term parameter drift, however, modifies global properties of the system's phase space by systematically modifying the network nodes' responses. Impact of the first can therefore potentially be reduced by a stronger external drive: a reduced fraction of the reservoir's state is susceptible to the "memory" of previous noise's history echoing in the reservoir's state. In general, aspects of computing with networks beyond fixed points become relevant [27]. Increasing the injection strength will on the other hand be of little benefit for suppressing global modifications to the neural network's phase space topology induced by parameter drifts.

The impact these two effects can exert on the computing-performance of photonic or even analogue neural networks is shown in Figure 4.12(b). We compute the readout weights off-line via the common matrix inversion technique, using the first recorded reservoir response. These weights we keep constant for the consecutive recordings of the reservoir state and determine the NMSE based on 100 testing samples. The result therefore shows the impact perturbations on different timescales have upon the system's performance: for a fully consistent system, the NMSE would remain constant. For the consistent reservoir with $\gamma = 1$, we obtain a NMSE $= 0.005$ directly for the response the readout weights were optimized for. Using the reservoir response recorded 17 minutes later already exhibits an increase to NMSE $= 0.035$, from where prediction performance continues to deteriorate. The rate of performance deterioration gradually slows down until performance more or less saturates at around NMSE ≈ 0.1. During hardware learning introduced in Section 4.4.3, a constant competition between system optimization and continuous deterioration due to drift is therefore taking place, and the system's performance is limited as soon as the rate of optimization drops below the rate of deterioration due to drift. One has to keep in mind that consistency and prediction errors were obtained for different parameters suboptimal for prediction.

For the less consistent reservoir at $y = 0.5$, the overall performance is significantly worse. When using the same state for which readout weights were specifically computed, the resulting NMSE is comparable to the one obtained for $y = 1$. However, the error instantly jumps to around NMSE ≈ 0.25 as soon as we compute the output based on the consecutively recorded reservoir states, using the same readout weights. In addition, there are large performance fluctuations, again confirming our interpretation that under these conditions significant trial-to-trial variations are the cause behind the reduced performance. It is therefore not surprising that the long-term drift does not significantly impact the system's prediction performance.

4.4.7 Autonomous system: output feedback

In the original publication of RC by Jaeger [13], long term prediction of the chaotic Mackey–Glass sequence was realized based on an interesting concept: feedback forcing. A feedback-forced architecture uses its own output as future input; see Figure 4.13(a). As learning optimizes the system such that its output approximates the future input value, i. e. $y^{\text{out}}(n + 1) \approx u(n + 2)$, one can substitute the external input via the system's own output and leave the system to evolve autonomously. Switching between the two different inputs is realized by switch S, which during initial training is in position (s_1) such that the reservoir is driven by external data $u(n + 1)$. During prediction, this switch remains in the same position for an initial transient period, but after reaching time $n = n'$ it is toggled to s_2, now connecting the reservoir's input to its own output. Operated in such a way the RC becomes an autonomous and hence self-consistent signal generator with its output-signal specified during training. It is therefore not only an interesting concept for chaotic signal prediction but for any type of complex signal generation. In physical hardware reservoirs, this has so far only been demonstrated using the FPGA [20] controlled delay-system discussed in Chapter 8.

In Figure 4.13(b), we schematically illustrate the hardware realization of this functionality based on our original photonic RC. Switch S is implemented within the MATLAB control routine, which in the feedback forcing position uses the normalized and offset removed version of our systems optical output. The system is trained as in Section 4.4.3 using 500 steps of the Mackey–Glass sequence. After the prediction error converged to values obtained before, we set switch S into position s_2. Unfortunately, at the moment of toggling S from s_1 to s_2, the system enters a second transient and the autonomously evolving photonic RC's output and the target instantly diverge. Our current assumption is that for now the single-step prediction error is still too large, and when switching the difference between $y^{\text{out}}(n+1)$ and $u(n+2)$ induces a too large perturbation.

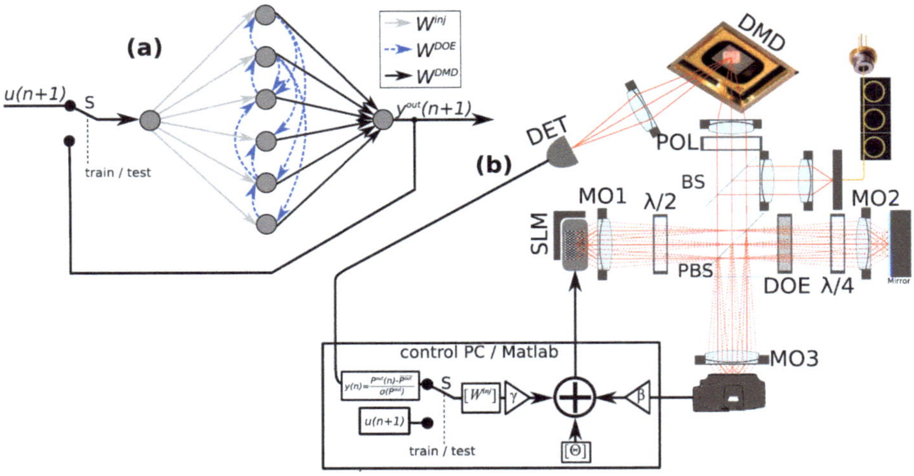

Figure 4.13: (a) Extension of the RC concept, now including the possibility of the system's own output to serve as input. Such feedback forcing is realized by toggling switch *S* after learning. (b) Our experimental realization based on the identical setup as before.

However, we can focus on the system's long-term autonomous dynamics and compare its properties to the ones of the Mackey–Glass sequence. In Figure 4.14(a), we show a section of the original Mackey–Glass time series and our system's output $y^{\text{out}}(n)$ in blue and red, respectively. From a visual inspection, we can see that both outputs share significant similarities. Our autonomous photonic RC is capable to assimilate the nonregular alternations between large and smaller oscillation amplitudes, plus the nontrivial local extrema including the additional small amplitude modulations at those points. In panel (b) of the same figure, we show the autocorrelations of $u(n+1)$ and $y^{\text{out}}(n+1)$, which were obtained based on timetraces with 5000 datapoints. The data demonstrates, first that the period of the nonregular oscillations is well approximated by our autonomous system, in fact within an error of ~10 %. And second, that the system stably approximates the target system as the properties of its output do not change over the time of its free evolution. Finally, in panels (c) and (d) we show the attractors of the original Mackey–Glass signal and of our system's output, respectively, both projected onto their first three dimensions. Attractors were obtained based on the Takens delay-embedding scheme, where we used an embedding delay of $\tau_T = -12$ [28]. While the difference between attractors of panel (c) and (d) are clear, it is also evident that our autonomous system starts to approximate important topological features of the target attractor. Also, one can directly see the impact of the prediction error and potentially of the system's noise: on small distances neighbouring trajectories of our system's output regularly cross. The local structure of the autonomously created attractor is therefore almost entirely lost.

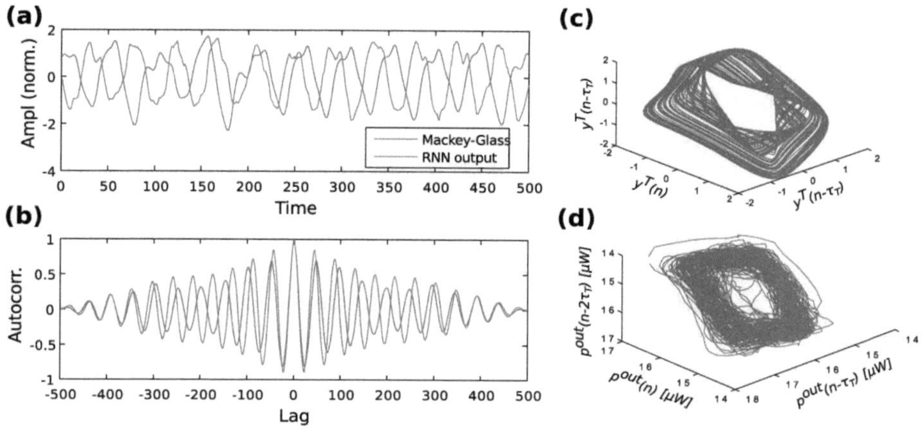

Figure 4.14: (a) Long term dynamics of the photonic reservoir created under feedback forcing (red data), compared to the original Mackey–Glass data. (b) The autocorrelation of both signals shows strongly similar features. The attractor of the original MG sequence, (d), is roughly approximated by the autonomously created photonic reservoir's output, (d).

It is the first time that such a feedback forced photonic RC has been demonstrated based on a spatiotemporal photonic reservoir and weights not implemented via digital, serial hardware. The here reported results prove that the system is capable to autonomously approximate complex temporal signals and evolutions. For future work, multiple improvements to learning should be considered, also carefully investigating the reason behind the current large transient at the moment that switch S is toggled to autonomous operation.

4.5 Conclusion

Large-scale spatiotemporal neural networks are possible and can efficiently exploit the parallelism of optics. After early experiments in classification, we have also shown that these systems are excellent candidates for processing temporal information, such as complex and chaotic signal prediction. Diffractive coupling has demonstrated that it is a powerful tool for the creation of large scale spatio-temporal networks of photonic elements. Based on numerical simulations and analytical considerations, we have shown that the concept is scalable to networks consisting of 10s of thousand photonic elements. Such large scale networks are not only highly attractive for the implementation of photonic neural networks, but also for the fundamental investigations of nonlinear network dynamics and other applications such as potentially coherent beam combining. We experimentally showed that we can couple, both, vertically emitting semiconductor lasers as well as pixels of an electro-optical SLM. With the latter, we have created networks of up to 2025 photonic oscillators, where it is important to highlight that not the diffractive coupling scheme but the imaging setup imposes the size limitation.

In the network, we realized based on the commercially available VCSEL array, we have coupled up to 21 lasers [29]. Nonlinear dynamics of the free-running lasers network clearly showed signatures of mutual coupling. Going beyond just coupling, we simultaneously injected the lasers of our network with an external drive laser. For a constant injection power, we were able to efficiently quench the network's complex dynamics and were able to identify first effects of increased coherence between the lasers consequence of both, coupling and external injection. Using modulation of the injection intensity, we strongly modified the dynamics of the network. Through the presence of the injection modulation frequency's higher orders, we were able to identify nonlinear transformation of the input signal via the network lasers. We recorded such transformations individually, and in an off-line experiment we used these responses to synthesize various target nonlinearities. These results are important first steps to the formation of all-optical neural networks based on large arrays of semiconductor lasers [6, 30]. Such systems could easily be operated at bandwidths exceeding 10s of GHz, which would introduce a shift of paradigm in the implementation of artificial neural networks.

Based on a similar experiment we implemented large scale spatiotemporal networks consisting of up to 2025 Ikeda-maps. Realized using a SLM, the experiment provides access to all system parameters and is therefore excellently suited for proof-of-concepts. Based on this network, we amended the system with an optical output, which can spatially address and weigh individual photonic neurons based on a digital micro-mirror array. This commercially available device realized the reservoir's optical readout weights and we implemented greedy learning [7]. The system is capable to perform one-step prediction of the chaotic Mackey–Glass sequence with excellent performance considering that weights are hardware implemented and restricted to Boolean entries.

Toward achieving this performance, we devised a novel strategy to partially compensate for the network's exclusively unipolar connection weights and states. Taking profit from the SLM's periodic nonlinearity, we divided the network into nodes operating along a positive and a negative section of the nonlinear function. By doing so, we efficiently mitigated one of the challenges typically faced in unipolar systems: the strong reduction of the nonlinear function space's dimensionality available to the system. Such strategies are essential for the future success of the field, where in the hardware implementation of neural networks one undoubtedly will continue to face challenges resulting from restrictions consequence of an implementation substrate's physical properties.

We furthermore took profit from our system's imperfections, which, however, are an undeniable feature of real-world hardware network's noise and drift. These too will most likely be permanent companions of neural networks implemented in hardware substrates. We provided a first detailed study of their impact upon the network's consistency, and in turn on consistency's impact upon the system's computational performance. We found that noise and drift have fundamentally different consequences for computation in such systems, which highlights the importance of careful future investigations into such effects. Of interest would also be if certain learning strategies are advantageous in the light of these degradations unavoidable in physical substrates, and in fact also always present in the human brain.

Finally, we made first strides toward the creation of an autonomous system. Using feedback forcing, we connected the photonic reservoir to its own output after it has been trained for chaotic signal prediction. Following this step, the system autonomously creates a complex time series which bared strong similarity with the original training target. One has to mention that when switching to autonomous operation we found a strong but short transient of bad performance, the origin of which remains to be fully identified.

To conclude, we have demonstrated the feasible creation of large scale photonic networks in various optical substrates. The application to RC demonstrates the enormous potential such systems offer to the field of artificial neural networks. They are fully parallel and global system's bandwidth is not or hardly influenced by the system size. In this aspect, these systems show performance scaling and size superior to others reported for different physical network concepts. Complex coupling and readout weights can be based on passive and constant spatial modulations. These are ultimately energy efficient in their implementation and again do not limit the system's bandwidth. As such, large photonic neural networks can efficiently be implemented and maximally leverage the inherent optical parallelism, and by that its energy efficiency and potential space-bandwidth product.

Bibliography

[1] J. W. Goodman, A. R. Dias, and L. M. Woody. "Fully parallel, high-speed incoherent optical method for performing discrete Fourier transforms". In: *Optics Letters* 2.1 (Jan. 1978), p. 1. ISSN: 0146-9592. DOI: 10.1364/OL.2.000001. URL: https://www.osapublishing.org/abstract.cfm?URI=ol-2-1-1.

[2] R. A. Athale, W. C. Collins, and P. D. Stilwell. "High accuracy matrix multiplication with outer product optical processor". In: *Applied Optics* 22.3 (Dec. 1986), pp. 368–370. ISSN: 0003-6935. URL: http://www.ncbi.nlm.nih.gov/pubmed/22777773.

[3] H. J. Caulfield et al. "Optical implementation of systolic array processing". In: *Optics Communications* 40.2 (1981), pp. 86–90. URL: http://www.sciencedirect.com/science/article/pii/0030401881903333.

[4] K. Wagner and D. Psaltis. "Multilayer optical learning networks". In: *Applied Optics* 26.23 (Dec. 1987), p. 5061. ISSN: 0003-6935. DOI: 10.1364/AO.26.005061. URL: https://www.osapublishing.org/abstract.cfm?URI=ao-26-23-5061.

[5] J. Shamir, H. J. Caulfield, and R. B. Johnson. "Massive holographic interconnection networks and their limitations". In: *Applied Optics* 28.2 (Jan. 1989), p. 311. ISSN: 0003-6935. DOI: 10.1364/AO.28.000311. URL: https://www.osapublishing.org/abstract.cfm?URI=ao-28-2-311.

[6] D. Brunner and I. Fischer. "Reconfigurable semiconductor laser networks based on diffractive coupling". In: *Optics Letters* 40 (2015), p. 3854. DOI: 10.1364/OL.40.003854.

[7] J. Bueno et al. "Reinforcement learning in a large scale photonic recurrent neural network". In: *Optica* 5.6 (Nov. 2018), pp. 756–760. DOI: 10.1364/OPTICA.5.000756. URL: https://www.osapublishing.org/optica/abstract.cfm?uri=optica-5-6-756.

[8] L. Larger, B. Penkovsky, and Y. Maistrenko. "Laser chimeras as a paradigm for multistable patterns in complex systems." In: *Nature Communications* 6 (Jan. 2015), p. 7752. ISSN: 2041-1723. DOI: 10.1038/ncomms8752. URL: http://www.nature.com/ncomms/2015/150714/ncomms8752/abs/ncomms8752.html.

[9] W. Kassa et al. "Towards integrated parallel photonic reservoir computing based on frequency multiplexing". In: *Neuro-inspired photonic computing*. Ed. by M. Sciamanna and P. Bienstman. Vol. 10689. SPIE, May 2018, p. 2. ISBN: 9781510619043. DOI: 10.1117/12.2306176. URL: https://www.spiedigitallibrary.org/conference-proceedings-of-spie/10689/2306176/Towards-integrated-parallel-photonic-reservoir-computing-based-on-frequency-multiplexing/10.1117/12.2306176.full.

[10] J. W. Goodman. *Introduction to Fourier optics*. Roberts & Co, 2005, p. 491. ISBN: 9780974707723.

[11] J. Grinberg et al. "A new real-time non-coherent to coherent light image converter the hybrid field effect liquid crystal light valve". In: *Optical Engineering* 14.3 (June 1975), p. 143217. ISSN: 0091-3286. DOI: 10.1117/12.7971871. URL: http://opticalengineering.spiedigitallibrary.org/article.aspx?doi=10.1117/12.7971871.

[12] S. Lin et al. "GaAs optoelectronic neuron arrays". In: *Applied Optics* 32.8 (Mar. 1993), p. 1275. ISSN: 0003-6935. DOI: 10.1364/AO.32.001275. URL: https://www.osapublishing.org/abstract.cfm?URI=ao-32-8-1275.

[13] H. Jaeger and H. Haas. "Harnessing nonlinearity: predicting chaotic systems and saving energy in wireless communication". In: *Science* 304 (2004), pp. 78–80.

[14] T. Heuser et al. "Fabrication of dense diameter-tuned quantum dot micropillar arrays for applications in photonic information processing". In: *Journal of Applied Physics* 3 (2018), pp. 1–6.

[15] M. Leutenegger et al. "Fast focus field calculations". In: *Optics Express* 14.23 (2006), pp. 4897–4903. ISSN: 1094-4087. DOI: 10.1364/OE.14.011277.

[16] T. Heil, I. Fischer, and W. Elsäßer. "Stabilization of feedback-induced instabilities in semiconductor lasers". In: *Journal of Optics B: Quantum and Semiclassical Optics* 2.3 (June 2000), pp. 413–420. ISSN: 1464-4266. DOI: 10.1088/1464-4266/2/3/331. URL: http://stacks.iop.org/1464-4266/2/i=3/a=331.

[17] T. Heil et al. "Chaos synchronization and spontaneous symmetry-breaking in symmetrically delay-coupled semiconductor lasers". In: *Physical Review Letters* 86 (2001), pp. 795–798.

[18] S. Pichai. *Google CEO Sundar Pichai's I/O 2017 keynote*. 2017. URL: https://www.youtube.com/watch?v=vWLcyFtni6U.

[19] K. Ikeda. "Multiple-valued stationary state and its instability of the transmitted light by a ring cavity system". In: *Optics Communications* 30 (1979), p. 257.

[20] P. Antonik, M. Haelterman, and S. Massar. "Brain-inspired photonic signal processor for generating periodic patterns and emulating chaotic systems". In: *Physical Review Applied* 7.5 (May 2017), p. 054014. ISSN: 2331-7019. DOI: 10.1103/PhysRevApplied.7.054014. URL: http://link.aps.org/doi/10.1103/PhysRevApplied.7.054014.

[21] M. C. Mackey and L. Glass. "Biological populations with nonoverlapping generations: stable points, stable cycles, and chaos". In: *Science* 186.4164 (Nov. 1974), pp. 645–647. ISSN: 0036-8075. DOI: 10.1126/science.186.4164.645. URL: http://www.ncbi.nlm.nih.gov/pubmed/4412202, http://www.sciencemag.org/cgi/doi/10.1126/science.186.4164.645.
[22] J. Bueno et al. "Conditions for reservoir computing performance using semiconductor lasers with delayed optical feedback". In: *Optics Express* 25.3 (2017), pp. 2401–2412. ISSN: 1094-4087. DOI: 10.1364/OE.25.002401. URL: https://www.osapublishing.org/oe/abstract.cfm?uri=oe-25-3-2401.
[23] D. Psaltis and N. Farhat. "Optical information processing based on an associative-memory model of neural nets with thresholding and feedback". In: *Optics Letters* 10.2 (Feb. 1985), p. 98. ISSN: 0146-9592. DOI: 10.1364/OL.10.000098. URL: https://www.osapublishing.org/abstract.cfm?URI=ol-10-2-98.
[24] S. Ortín et al."A unified framework for reservoir computing and extreme learning machines based on a single time-delayed neuron". In: *Scientific Reports* 5 (2015), p. 14945. DOI: 10.1038/srep14945.
[25] V. Kimelblat. "Appplying 1/F noise for drift estimation". In: *Fluctuation and Noise Letters* 10.2 (June 2011), pp. 181–188. ISSN: 0219-4775. DOI: 10.1142/S021947751100051X. URL: http://www.worldscientific.com/doi/abs/10.1142/S021947751100051X.
[26] A. Uchida, R. Mcallister, and R. Roy. "Consistency of nonlinear system response to complex drive signals". In: *Physical Review Letters* 93 (2004), p. 244102. DOI: 10.1103/PhysRevLett.93.244102.
[27] B. A. Marquez et al. "Dynamical complexity and computation in recurrent neural networks beyond their fixed point". In: *Scientific Reports* 8.1 (Dec. 2018), p. 3319. ISSN: 2045-2322. DOI: 10.1038/s41598-018-21624-2. URL: http://www.nature.com/articles/s41598-018-21624-2.
[28] F. Takens. "Detecting strange attractors in turbulence". In: *Dynamical systems and turbulence, Lecture notes in mathematics*. Berlin, Heidelberg: Springer, 1981, pp. 366–381. DOI: 10.1007/BFb0091924.
[29] C. M. Dietrich. "Design and realization of a dynamic semiconductor laser network". MSc thesis. Westfaelischen Wilhelms-Universitaet Muenster, 2015.
[30] D. Brunner, S. Reitzenstein, and I. Fischer. "All-optical neuromorphic computing in optical networks of semiconductor lasers". In: *2016 IEEE international conference on rebooting computing (ICRC)* 1 (2016), pp. 1–2. DOI: 10.1109/ICRC.2016.7738705. URL: http://ieeexplore.ieee.org/document/7738705/.

Silvia Ortín, Luis Pesquera, Guy Van der Sande, and
Miguel C. Soriano

5 Time delay systems for reservoir computing

5.1 Introduction

Delay-based reservoir computers are reservoir computing implementations based on dynamical systems with delay. The dynamical system is often a single nonlinear node. As it will be explained later in more detail, delay-based reservoir computers assign different time-slots of the nonlinear node response to different network nodes, also known as virtual nodes. This time multiplexing of a single nonlinear node's response is equivalent to having a distributed set of virtual nodes along the delayed feedback loop.

The relevance of delay-based reservoir computers relies on the fact that they simplify greatly hardware implementations. As such, the first photonic reservoir computers were based on this approach [1–4].

Here, we provide an overview of the theoretical foundations of delay-based reservoir computers. In addition, we illustrate the main properties of the system with several examples both in numerical simulations and in electronic implementations of the concept. We also examine the main challenges in physical implementations of delay-based reservoir computers.

5.2 Standard reservoir computing

Before getting started with delay-based reservoir computing (RC), we revisit the basic concepts of standard reservoir computing in this section for the sake of completeness. For a more in-depth description of the concept of reservoir computing, we refer the reader to Chapters 2 and 3. Reservoir computing is an implementation of a recurrent neural network with the general idea that the network is split up into several parts. The recurrent part is difficult to train, therefore, another layer (output layer) is added, which is no more than a series of simple linear nodes that interface with the recurrent part [5]. Traditional reservoir computing implementations are generally composed of three distinct parts: an input layer, the reservoir, and an output layer, as illustrated in Figure 5.1.

The input layer feeds the input signals to the reservoir via fixed random weighted connections. These weights will scale the input that is given to the nodes, creating a different input scaling factor for every individual node. The second layer, which is called reservoir or liquid, usually consists of a large number of randomly interconnected nonlinear nodes, constituting a recurrent network. The nodes are driven by

https://doi.org/10.1515/9783110583496-005

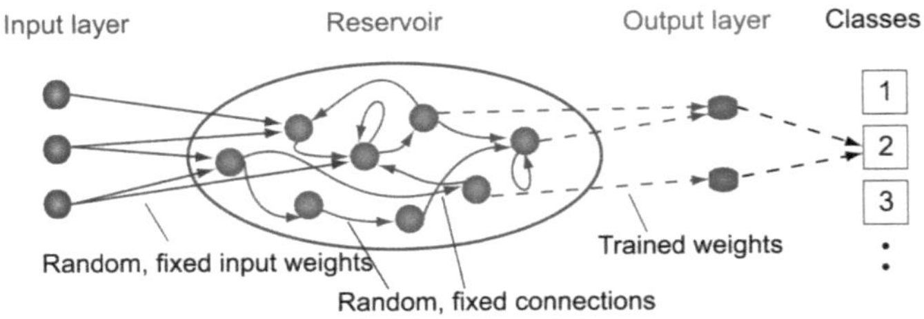

Figure 5.1: Classical reservoir computing scheme. The input is coupled into the reservoir via a randomly connected input layer to the *N* nodes in the reservoir. The connections between reservoir nodes are randomly chosen and kept fixed, that is, the reservoir is left untrained. The reservoir's transient dynamical response is read out by an output layer, which are linear weighted sums of the reservoir node-states. The figure is taken from Appeltant et al. [6].

random linear combinations of input signals. Since every node-state can be seen as an excursion in another state space direction, the original input signal is thus mapped onto a high-dimensional state space. The emerging reservoir state is given by the combined states of all the individual nodes. Contrary to what happens in traditional recurrent neural networks, the coupling weights within the reservoir itself are not trained. They are usually chosen in a random way, globally scaled in order for the network to operate in a certain dynamical regime. Under the influence of input signals, the network exhibits transient responses. These transient responses are read out by the output layer via a linear weighted sum of the individual node-states, with no additional nonlinear transformation happening in the last layer. The training algorithm, which has the goal to find optimum output weights, can thus be drastically simplified to a linear classifier.

The reservoir computing implementation we work with is closely related to echo state networks [7]. In echo state networks, the node-states at time step k are computed according to the following equation:

$$\mathbf{r}(k) = F[\mathbf{W}_{\text{res}}^{\text{res}} \cdot \mathbf{r}(k-1) + \mathbf{W}_{\text{in}}^{\text{res}} \cdot \mathbf{u}(k)]. \tag{1}$$

In this equation, $\mathbf{r}(k)$ is the vector of new node-states at time step k, $\mathbf{u}(k)$ is the input matrix at time step k. The matrices $\mathbf{W}_{\text{res}}^{\text{res}}$ and $\mathbf{W}_{\text{in}}^{\text{res}}$ contain the (generally random) reservoir and input connection weights. The weight matrices are scaled by multiplicative factors in order to get good performance. For the nonlinear function F, often a sigmoidal function, e. g., $F(x) = \tanh(x)$ is chosen. In some cases, feedback from the output to the reservoir nodes is also included.[1] That approach will be used in Chapter 8. In a simplified formulation, the output is a weighted linear combination of the

1 When connections from the output layer back to the reservoir are included, equation (1) becomes: $\mathbf{r}(k) = F[\mathbf{W}_{\text{res}}^{\text{res}} \cdot \mathbf{r}(k-1) + \mathbf{W}_{\text{in}}^{\text{res}} \cdot \mathbf{u}(k) + \mathbf{W}_{\text{out}}^{\text{res}} \cdot \hat{\mathbf{y}}_{\text{out}}(k-1)]$.

node-states and a constant bias value:

$$\hat{\mathbf{y}}_{\text{out}}(k) = \mathbf{W}_{\text{res}}^{\text{out}} \cdot \mathbf{r}(k) + \mathbf{W}_{\text{in}}^{\text{out}} \cdot \mathbf{u}(k) + W_{\text{bias}}^{\text{out}}. \tag{2}$$

In reservoir computing, only the matrices in equation (2) are optimized (trained) to minimize the mean square error between the calculated output values $\hat{\mathbf{y}}_{\text{out}}(k)$ and the required output values $\mathbf{y}_{\text{out}}(k)$.

5.3 Delayed feedback systems

Nonlinear systems with delayed feedback and/or delayed coupling, often simply put as "delay systems," are a class of dynamical systems that have attracted considerable attention, because they arise in a variety of real life situations [8]. They are commonly found in, e. g., traffic dynamics due to the reaction time of a driver [9], chaos control [10, 11], or gene regulation networks where delay originates from transcription, translation, and translocation space [12]. Also in predator-prey models they occur with the time delay representing a gestation period or reaction time of the predators. Sometimes the delay in the system originates from the fact that the previous number of predators has an influence on the current rate of change of the predators [13]. In the brain, delay occurs because of the axonal conduction delay between two neurons [14]. Remote cerebral cortical areas are subject to an entire series of these axonal conduction delays. The total connection delay between these areas could even be tens of milliseconds, but still zero time lag synchronization between remote cerebral cortical areas was observed [15–17]. Delay is found in networks of semiconductor lasers [18] when the signal travels from one laser to the other. Whether it is through free space or via, e. g., an optical fiber, the light needs to cover a certain distance and that requires time. In control systems, the time-delayed feedback originates from the fact that there is a finite time between the sensing of the information and the subsequent reaction of the system under the influence of a control signal. Another example taken from daily life is temperature control of the water coming from a shower. Because of the fact that the water needs to travel a certain distance along the tube between the heating element and the shower head the response to any temperature adjustment of the system is not immediate from the perspective of the user. This could lead to an unstable behavior where the controller increases or decreases the temperature of the water too much due to apparent nonresponsivity of the system.

It has been shown that delay has an ambivalent impact on the dynamical behavior of systems, either stabilizing or destabilizing them [11], with possible emergence of complex dynamics. This has been observed in, e. g., biological systems [19] or laser networks [20]. Often it is sufficient to tune a single parameter (e. g., the feedback strength) to access a variety of behaviors, ranging from stable via periodic and quasi-periodic oscillations to deterministic chaos [21]. In photonics, a normally stable

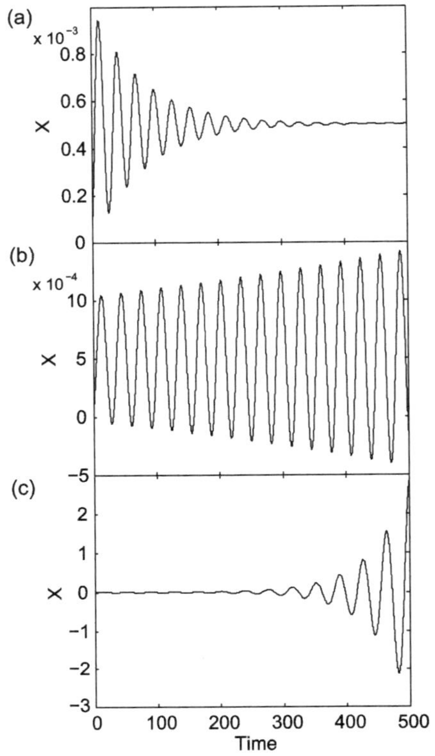

Figure 5.2: Destabilizing effect of delay. Time trace originating from the system given by equation (3) (a) $\tau = 7$, (b) $\tau = 8$, (c) $\tau = 10$. Note the different scaling factors in the vertical axis.

laser source can become chaotic when subjected to feedback even for small feedback strengths. As an example, we take one of the most simple delay systems, given by the equation

$$\dot{x}(t) = -\alpha x(t - \tau), \tag{3}$$

where we choose $\alpha = 0.2$ and τ stands for the time delay. In Figure 5.2, we show the solution of this equation for three different values of τ. When looking at the time trace in Figure 5.2(a) with $\tau = 7$, some damped oscillations can be observed in the transient before the system reaches a constant output value. However, when the delay time is increased to $\tau = 8$, as in Figure 5.2(b), the oscillations are no longer exponentially damped. They increase in amplitude with time. For an even larger τ, equal to 10, this behavior is confirmed with an even stronger growth in amplitude. For this system, the delay clearly has a destabilizing effect.

From the application point of view, the dynamics of delay systems is gaining more and more interest: whereas initially it was considered more as a nuisance, it is now viewed as a resource that can be beneficially exploited. It found applications in chaos-communication [22], and also reservoir computing is an example of benefitting from the delay in the system [6, 1] as presented in this chapter. One of the simplest possible delay systems consists of a single nonlinear node whose dynamics is influenced by its own output a delay time in the past. Such a system is easy to implement since it comprises only two elements: a nonlinear node and a delay loop. When going to more complex situations of several nonlinear nodes being coupled with delay, these systems have successfully been used to describe the properties of complex networks in general. They allow a better understanding of, e. g., synchronization and resonance phenomena [23–25]. Of particular interest for this book is the situation in which only a few dynamical elements are coupled with delay within a certain configuration, e. g., a ring of delay-coupled elements [20].

Mathematically, delay systems are described by delay differential equations (DDE) that differ fundamentally from ordinary differential equations (ODE) as the time-dependent solution of a DDE is not uniquely determined by its initial state at a given moment. For a DDE, the continuous solution on an interval of one delay time needs to be provided in order to define the initial conditions correctly. The general form of a DDE is given by

$$\dot{x}(t) = F\big[x(t), x(t-\tau)\big]$$

with F any given linear or nonlinear function and with τ being the delay time. Mathematically, a key feature of time-continuous delay systems is that their state space becomes infinite dimensional. This is because their state at time t depends on the output of the nonlinear node during the continuous time interval $[t-\tau, t[$. Another interpretation is that a delayed feedback equation leads to a nonrational transfer function, resulting in an infinite number of poles. The dynamics of the delay system remains finite dimensional in practice [26], but exhibits the properties of high dimensionality and short-term memory. Since two key ingredients for computational processing are nonlinear transformation and high-dimensional mapping, delay systems are suitable candidates.

5.4 Delayed feedback systems as reservoirs

The term *reservoir* originally referred to a large, randomly connected fixed network of nonlinear nodes or neurons. However, not all reservoirs are neural networks. Analog physical systems such as the nonlinear behavior of ripples on a water surface have been used for information processing based on the reservoir computing paradigm [27]. Reservoir computing therefore enables the implementation of neuromorphic computing avoiding the need of interconnecting large numbers of discrete neurons.

With reservoir computing, there is no need for reconfigurable connection links within the recurrent network. Such random and fixed connections radically reduce the complexity for a hardware implementation. In previous chapters, it has been shown how recurrent networks for optical reservoir computing have been implemented either on a photonic chip or using diffractive optics. Various photonic techniques can be used to implement optical networks for reservoir computing with a wide range of network topologies. Reservoir computers based on optical-networks, as covered in previous chapters, have many hardware nodes and network degrees-of-freedom, even though they are fixed artificially.

In this section, we will revisit the concept of delay embedded reservoir computing, using only a single nonlinear node with delayed feedback. Thus, from a network perspective, there is only one (hardware) node. Hence, the delay-based approach allows for a far simpler system structure, even for very large reservoir sizes. The advantage of delay-based reservoir computing lies in the minimal hardware requirements as compared to more hardware-intensive systems from previous chapters. In essence, delay-based reservoirs are fixed intrinsically: they take the form of a time-delayed dynamical system with a single nonlinear state variable. Nodes of a delay-based reservoir can be sampled from a spatially continuous medium (i. e., the delay line). These nodes are considered virtual as they are not implemented as components or units in hardware. Nevertheless, delay-based RC has shown similar performance as networked RC, with the advantage that the hardware requirements are minimal as no complex interconnection structure needs to be formed. In photonics, it allows even for the use of hardware that is more traditionally associated with optical communications.

5.4.1 Implementation with a nonlinear node with delayed feedback

The concept of delay-based reservoir computing, using only a single nonlinear node with delayed feedback, was introduced in the early 2010s by Appeltant et al. [6] and Pacquot et al. [28] as a means of minimizing the expected hardware complexity in photonic systems. The first working prototype was developed in electronics in 2011 by Appeltant et al. [6] and efficient optical systems followed quickly after that [29, 1].

In essence, the idea of delay line reservoir computing constitutes an exchange between space and time: what has been done spatially with many nodes is now done in a single node that is multiplexed in time. There is a price to pay for this hardware simplification: compared to an N-node standard spatially-distributed reservoir, the dynamical behaviour in the system has to run at an N-times higher speed in order to have equal input-throughput. Figure 5.3 shows a diagram of a delay-based RC.

Delay-based RC are efficiently implemented with a single nonlinear node (or more general a nonlinear dynamical element) with a feedback loop [6]. As previously stated, in the delay-based reservoir, there is a single node and a collection of virtual nodes (also called virtual neurons) in the delay line. The general equation that governs these delay-systems is

$$T\dot{x}(t) = F\big(x(t), \eta x(t-\tau) + \gamma J(t)\big) \tag{4}$$

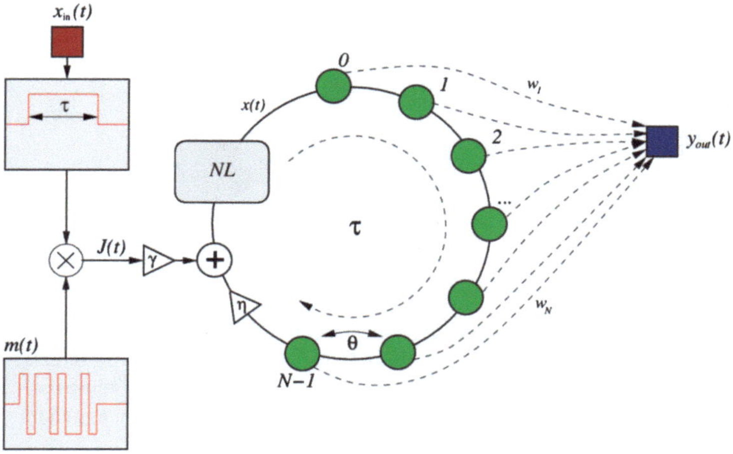

Figure 5.3: Structure of a delay-based reservoir computer. A one-dimensional input signal (in red) is first preprocessed using the masking function $m(t)$. Virtual nodes are defined along the delay line and form the reservoir (in green). The output layer (in blue) is unaltered from the standard RC structure.

where T is the response time of the system, τ is the delay time, $J(t)$ is the masked input, γ is the input scaling or input gain, η is the feedback-strength, and F is a nonlinear function. The masked input $J(t)$ is the continuous version of the discrete random mapping of the original input. To construct this continuous data $J(t)$, the continuous version of the discrete random mapping of the original input ($\mathbf{W}_{\text{in}}^{\text{res}}\mathbf{u}(k)$) is multiplexed in time, as it will be described in the next section.

5.4.2 Time-multiplexing in the delayed feedback approach

The nonlinear (NL) node is subjected to the time-continuous input stream $\mathbf{u}(t)$ or time-discrete input $\mathbf{u}(k)$ (see Figure 5.4), which can be a time-varying scalar variable or vector of any dimension d. The feeding to the individual virtual nodes is achieved by serializing the input using time-multiplexing. In our approach, every time interval of T_{in} (the data injection/processing time) represents another discrete time step. For this, the input stream $u(t)$ or $u(k)$ undergoes a sample and hold operation to define a stream $I(t)$ which is constant during one T_{in}, before it is updated. The resulting continuous function $I(t)$ is related to the discrete input signal $u(k)$ by $I(t) = u(k)$ for $T_{\text{in}}k \leq t < T_{\text{in}}(k+1)$. This procedure is illustrated in the first part of Figure 5.4 for the special case of $T_{\text{in}} = \tau$; function $I(t)$ is also depicted. Thus, in our approach, the input to the reservoir is always discretized in time first, no matter whether it stems from a time-continuous or time-discrete input stream. What is actually injected into the nonlinear node is time-

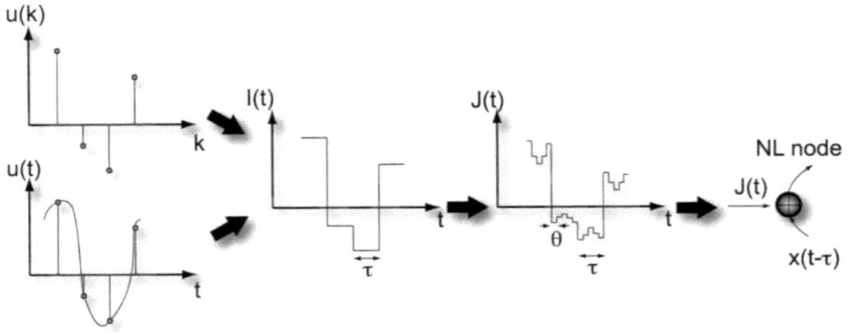

Figure 5.4: Masking procedure. A time-continuous input stream $\mathbf{u}(t)$ or time-discrete input $\mathbf{u}(k)$ undergoes a sample-and-hold operation, resulting in a stream $I(t)$ that is constant during one interval T_{in} before it is updated. In this particular case, $T_{in} = \tau$. The temporal input sequence, feeding the input stream to the virtual nodes, is then given by $J(t) = M \cdot I(t)$. The figure is taken from Appeltant et al. [6].

continuous again, but from this signal no distinction can be made whether the original data points were coming from a discrete or a time-continuous signal.

At the input driving stage, one also introduces a specific input connection structure. In accordance to what happens in traditional neural network reservoirs, every single virtual node can have its proper input scaling factor. In terms of a “classical” reservoir setup, these values correspond to the weights of the connections between the input layer and the reservoir layer. In equation (1), which we repeat here for convenience,

$$\mathbf{r}(k) = F[\mathbf{W}_{\text{res}}^{\text{res}} \cdot \mathbf{r}(k-1) + \mathbf{W}_{\text{in}}^{\text{res}} \cdot \mathbf{u}(k)], \tag{5}$$

these weights were referred to as $\mathbf{W}_{\text{in}}^{\text{res}}$, which is a random $(N \times d)$ matrix in the original concept (we recall that N is the number of virtual nodes and d the dimension of the input). Every input value sent to the time-slot corresponding to a given virtual node is first multiplied by the factor related to that node. This is done to increase variability in the network. However, the delayed feedback system comprises only one physically present nonlinear node that feeds all the virtual nodes in the delay line. Hence, all virtual node-states originate from the same nonlinear transformation and there is no possibility to implement a scaling factor in the virtual node itself. The most convenient option is to imprint coupling weights from the stream $I(t)$ to the virtual nodes by introducing a function $M(t)$, from now on referred to as the input mask, as follows: $M(t) = W_{\text{in},i}^{\text{res}}$ for $(i-1)\theta < t \leq i\theta$ and $M(t + T_{\text{in}}) = M(t)$. This mask function is a piecewise constant function, constant over an interval of θ and periodic, with period T_{in}. Thus, $\theta = T_{\text{in}}/N$ and stands for the temporal separation between the virtual nodes. The values of the mask function during each interval of length θ are chosen independently at random from some probability distribution. When the input signal is one-dimensional, the values to be injected are given by

$$J(t) = I(t) \cdot M(t). \tag{6}$$

The function $J(t)$ is the product of the input and the mask function, as represented in Figure 5.4. When the input consists of d values $I^j(t)$, we generate a separate mask $M^j(t)$ for each input j and subsequently they are all summed together. The value to be injected is then given by

$$J(t) = \sum_{j=1}^{d} I^j(t) \cdot M^j(t). \tag{7}$$

Alternative descriptions of the input mapping procedure in delay-based RC can be found in [30, 31].

5.4.3 Read-out and training in delay-based RC

The output of the nonlinear node is driven by the changes in the input. To relate the states in the delay line to reservoir states corresponding with an input step, the signal needs to be discretized again. The reservoir state comprises the virtual node-states, i. e., the values at the end of each interval θ in the injection time interval T_{in}. For the ith virtual node the kth discrete reservoir state is given by

$$r_i(k) = x(kT_{\text{in}} - (N - i)\theta). \tag{8}$$

Note that this definition implies that the virtual node-state r_i is always read out at the end of the interval θ. Although this is the common procedure throughout this chapter, other choices of sampling position can also yield good results.

Each virtual node r_i is a measuring point or tap in the delay line. However, these taps do not have to be physically realized. Since the x-signal revolves unaltered in the delay line anyway, a single measuring point suffices. After each T_{in}-interval, a new reservoir state ($\mathbf{r}(k) \in \mathbb{R}^{1\times N}$) for the input $\mathbf{u}(k)$ is obtained.

The reservoir states themselves are not the desired outcome of the entire system. A training algorithm is used to assign an output weight to each virtual node r_i, such that the weighted sum of the states approximates the desired target value as closely as possible:

$$\begin{aligned}
\hat{\mathbf{y}}_{\text{out}}(k) &= \mathbf{W}_{\text{res}}^{\text{out}} \cdot \mathbf{r}(k) \\
&= \sum_{i=1}^{N} W_{\text{res},i}^{\text{out}} \cdot r_i(k) \\
&= \sum_{i=1}^{N} W_{\text{res},i}^{\text{out}} \cdot x\left[kT_{\text{in}} - \frac{T_{\text{in}}}{N}(N - i)\right],
\end{aligned}$$

with $W^{\text{out}}_{\text{res},i}$ the weight assigned to the virtual node r_i, x the output of the nonlinear node and $\hat{y}_{\text{out}}$ the calculated approximation of the target. The values of the $\mathbf{W}^{\text{out}}_{\text{res}}$ are determined by a linear training algorithm. The training of the read-out follows the standard procedure for reservoir computing [32, 7]. In this manner, we can map every discrete input step $\mathbf{u}(k)$ onto a discrete target value $\hat{\mathbf{y}}_{\text{out}}(k)$ and this for every k. The testing is performed using previously unseen input data of the same kind of those used for training.

Finally, let us note that as shown in Figure 5.3 and equation (4), the masked input $J(t)$ is scaled by an input scaling factor γ and the feedback $x(t-\tau)$ by a feedback strength η. This is to bias the nonlinear node in the optimal dynamical regime. Optimal values for the input scaling γ and η depend on the task at hand, as well as the specific dynamical behavior of the nonlinear node. Finding the optimal point for these parameters is a nonlinear problem which can be approached by, for example, a gradient descent or by simply scanning the parameter space.

5.4.4 An example: chaotic time series prediction

To compare the approaches of traditional reservoir computing and our delayed feedback system, we demonstrate their function by means of a commonly used benchmark task: chaotic time series prediction. Without going into detail about the exact data processing, we illustrate the different steps and compare the performance. The test data originates from a time series prediction competition, organized as a survey to compare different time series forecasting methods. At that time many new and innovative methods, such as artificial neural networks, emerged to compete with standard prediction methods. In May 1993 in Santa Fe, New Mexico, the NATO Advanced Research Workshop on Comparative Time Series Analysis was held to have an overview of existing methods and their performance [33]. Several time series coming from different systems were provided as a challenge. Here, we consider the set coming from a NH_3 chaotic laser exhibiting dynamics related to Lorenz chaos. A small segment of the input data series is depicted in Figure 5.6, with the laser intensity shown on the y-axis versus the index of the sampled data point.

The goal is to make a one-step ahead prediction, based on the present value of the system and this for all values of the time trace. In our training procedure, both for the case of a reservoir network with many nodes and a delayed feedback system, the time series is fed to the system as examples. The systems will process the input data and nonlinearly transform it. In Figure 5.5, a part of the reservoir states are shown both for a network of randomly connected nodes and for a delayed feedback system, where we consider 400 states in both cases. One time series realization consists of 1000 measurement points. Every point that is fed to the reservoir leads to a change in all 400 node-states of the reservoir, hence 400 series of 1000 points are

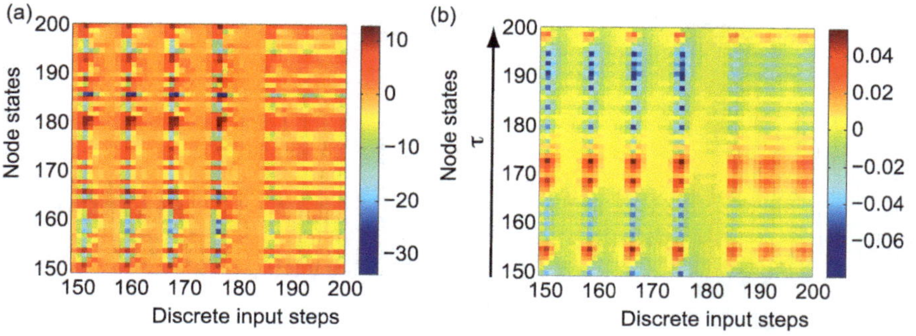

Figure 5.5: Spatiotemporal representation Santa Fe. A zoom is presented of the evolution of the reservoir states of nodes. Feeding in 1000 input steps leads to the construction of 400 reservoir states of each 1000 steps. Here, only 50 input steps are shown for 50 nodes. The state values are shown in color code. (a) Network reservoir approach, (b) delayed feedback reservoir.

recorded as reservoir states. Both systems rely on a different connectivity and configuration, but use the same nonlinear function with identical parameters as network nodes.

Both, in the situation of Figure 5.5(a) and the one of Figure 5.5(b), 400 nodes were used, but only 50 node-states are plotted. In Figure 5.5(a), the reservoir states of a traditional network are depicted. The different node-states are plotted along the y-axis and their evolution in discrete time is given by moving along the x-axis. Figure 5.5(b) shows the states we can obtain with a delayed feedback setup. The representation chosen for this figure is equivalent to the spatiotemporal mapping carried out by the system [34]. Moving along the x-axis gives the evolution in time. Every discrete input step in Figure 5.5(b) corresponds to a jump in time of τ. The general trend of the reservoir states is quite similar for the network and the delayed feedback response. The fact that they both respond in a similar way to identical inputs already gives a first indication that both are able to extract information in an comparable way.

Delay-based reservoirs have achieved performances that are comparable to the state-of-the-art of more traditional reservoir computing approaches for the Santa Fe chaotic time series prediction [35]. In Figure 5.6, the result of the training procedure on these reservoir states is depicted. The crosses correspond to the original target and the black curve is the approximation. Please note that the approximation of the target is also a discrete time series with the same number of samples as the original target. The full lines are present only as a guide to the eyes and do not mean that we only sampled some points of the input or target. For these examples and reservoir parameters, the error, expressed as a normalized mean square error, is 0.0651 for the network approach and 0.0225 for the delayed feedback approach. Lower error values are possible for both the network and the delayed feedback approaches with optimized reservoirs.

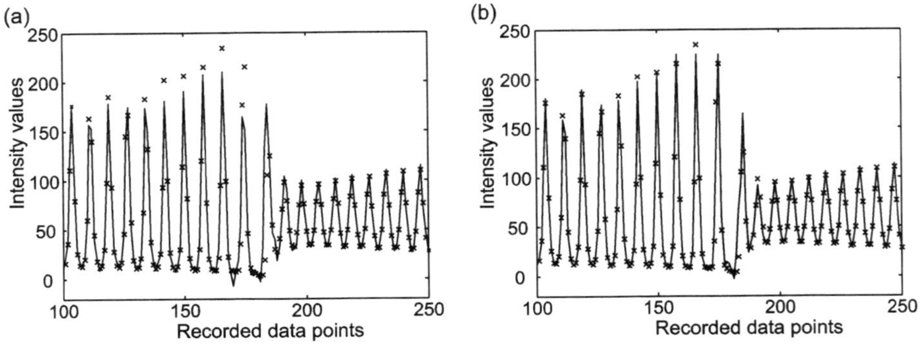

Figure 5.6: Target reconstruction Santa Fe. The crosses represent the sample points of the original target series. The full line connects the approximation of the target. (a) The network reservoir. (b) The delayed feedback system reservoir.

5.5 Interconnection structure of delay-based reservoir computers

In the delayed feedback system with external input as described in the previous section, we can identify four time scales: the separation of the virtual nodes θ, the data injection time T_{in}, the delay time τ, and the response timescale T of the nonlinear node. The data injection time $T_{\mathrm{in}} = N\theta$ is defined by the numbers of virtual nodes N that are necessary to compute a specific task and the node distance θ. The data injection time T_{in}, together with the inherent dynamics of the nonlinear node, controls the connectivity between the virtual nodes. Setting the values of the different time scales, a given interconnection structure is created.

Virtual nodes can be connected in two ways, through the feedback loop, and through the inherent dynamics of the nonlinear node. To create a virtual interconnection between the virtual nodes due to the inherent dynamics of the nonlinear node, the distance between the virtual nodes $\theta = T_{\mathrm{in}}/N$ has to be sufficiently short to keep the nonlinear node in a transient state. If the temporal distance between the virtual nodes θ is smaller than the response time of the system, T, the state of a virtual node becomes dependent on the states of the neighboring virtual nodes [6, 36] (see Figure 5.7(b)). Typically, a number of $\theta = 0.2T$ is quoted [6, 37]. However, there is no reason to assume this could not be task and system bias dependent. If θ is too short, the nonlinear node will not be able to follow the changes in the input signal and the response signal will be too small to measure. If θ is too long, the interconnection structure between neighboring virtual nodes due to the inherent dynamics of the nonlinear node is lost (see Figure 5.8(b)).

The virtual nodes can also set up a network structure via the feedback loop [28, 29]. This can be achieved by introducing a mismatch between the delay time τ and the data injection/information processing time $T_{\mathrm{in}} = N\theta$ ($\tau \neq T_{\mathrm{in}}$) (see Figure 5.11).

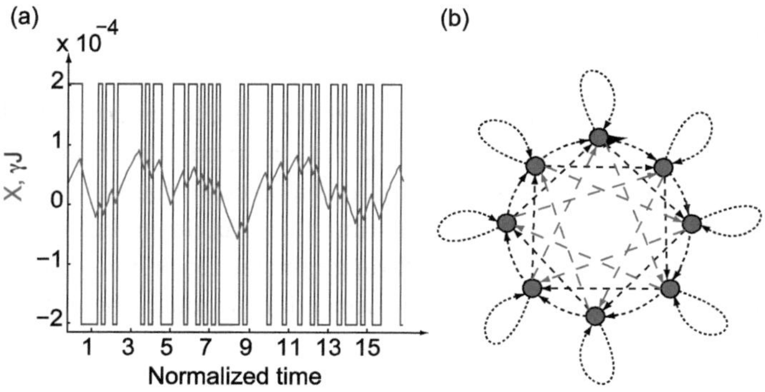

Figure 5.7: Input time trace for small θ and corresponding interaction structure when $T_{in} = \tau$. (a) Input time trace $\gamma J(t)$ (blue) and oscillator output $x(t)$ (red) of our system when the time scale T of the nonlinear system is larger than the separation θ of the virtual nodes $T \gg \theta$ and $T_{in} = \tau$. Here, we choose $T/\theta = 5$. The values on both the x- and y-axis are dimensionless. The mask $M(t)$ takes two possible values. (b) In this case the system does not have the time to reach an asymptotic value. Therefore, the dynamics of the nonlinear node couples neighboring virtual nodes with each other. The figure is taken from the supplementary material of Appeltant et al. [6].

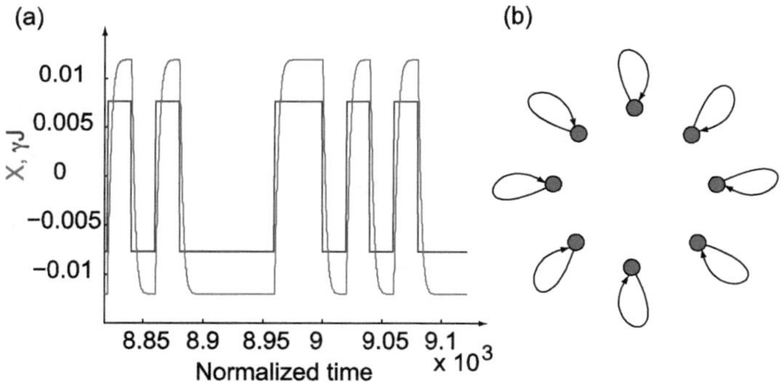

Figure 5.8: Input time trace for large θ and corresponding interaction graph when $T_{in} = \tau$. (a) Input time trace $\gamma J(t)$ (blue) and oscillator output $x(t)$ (red) of our system when the time scale T of the nonlinear system is much smaller than the separation θ of the virtual nodes $T \ll \theta$ and $T_{in} = \tau$. Here, we choose $T/\theta = 0.05$. The values on both the x- and y-axis are dimensionless. The mask $M(t)$ takes two possible values. For this choice of parameters, the system rapidly reaches a steady-state. (b) In this regime, the system behaves like N independent nodes, each of which is coupled only to itself at the previous time step. The figure is taken from the supplementary material of Appeltant et al. [6].

In contrast to traditional reservoirs, where all interactions between nodes take place from one discrete time step to another, the interaction between nodes in delay-based RC occurs through the dynamics of the nonlinear system (usually within the same discrete step) and through the feedback line (usually from one discrete time step to another). For this reason, the connections between the virtual nodes do not quite correspond to the interconnection matrix W_{res}^{res} used for traditional reservoirs in equation (1). Although both types of virtual node connections are not exclusive, in the literature most of the research groups have used either delay-based RC with only the virtual connections created through the inherent system dynamics [6, 37] or connected by the feedback line [28, 29]. For the case in which $T_{\text{in}} = \tau$, we predict good performance when the time scales are related by $\theta \lesssim T \ll \tau$. It is also clear that the operation speed of a delay-based RC, i. e., the data injection time $T_{\text{in}} = N\theta$, depends on θ, so delay-based RC with virtual nodes connected only through the feedback line ($\theta \gg T$) are slower that a counterpart exploiting the virtual connections through the system dynamics ($\theta < T$).

5.5.1 Interconnection structure through system dynamics

When $\theta < T$, the state of a given virtual node at time t depends on the states of the previous virtual nodes due to the noninstantaneous response of the system (T) to the inputs. The strength of this dependency for low-pass filtered systems is an exponentially decaying function of the separation of the virtual nodes.

For the case of $T_{\text{in}} = \tau$ (i. e., the only connection between the virtual nodes are through the system dynamics) and a time series prediction task (NARMA10), it was found that $\theta = 0.2T$ is the best choice for $N = 400$ virtual nodes [6]. This ratio leads to significant coupling between neighboring virtual nodes, as illustrated in Figure 5.7. In Figure 5.7(a), the node output never leaves the transient regime. Since the state of one virtual node depends on the state of the previous ones because of the system dynamics, the equivalent connectivity graph is given by Figure 5.7(b). All nodes are connected to adjacent nodes, with the connection weights exponentially decreasing as we move further back in time. They also experience the self-coupling as $T_{\text{in}} = \tau$ in this case.

The relation between the timescales can be used to establish a more formal link between the traditional formulation of reservoir computing, such as given in Section 5.2 and the virtual interconnections created through the system dynamics. In what follows, we will derive an approximate interconnection matrix $\mathbf{W}_{\text{res}}^{\text{res}}$ describing the coupling between virtual nodes processing information from different input time steps where they are only connected through the system dynamics. For simplicity of notation, in the following we normalize all times with respect to the intrinsic time scale of the nonlinear system T, that is we work in units where $T = 1$. In the following, we consider nonlinear equations of the form:

$$\dot{x}(t) = -x(t) + F\big[x(t-\tau), J(t)\big] \tag{9}$$

with F any nonlinear function and $J(t)$ given by equation (6). We recall that $J(t)$ is constant over each segment with duration θ and equals $W_{\text{in},i}^{\text{res}}u(k)$ over the segment containing virtual node i, with $W_{\text{in},i}^{\text{res}}$ the specific input scaling factor of node i. Assuming a constant value of $F[x(t-\tau), J(t)]$ during the duration θ, solving equation (9) yields

$$x(t) = x_0 e^{-t} + (1 - e^{-t})F[x(t-\tau), J(t)] \tag{10}$$

where x_0 is the initial value at the beginning of each interval θ, i. e., the value for the previous virtual node. In particular, the values of the virtual nodes are given by equation (10) with t replaced by θ. Now returning to the discrete time of input signal $u(k)$, the state of the ith virtual node ($i \in [1, N]$) is reached after a time θ, denoted by $r_i(k)$ (defined in equation (8)). The input to virtual node i at time step k equals $W^{\mathrm{res}}_{\mathrm{in},i} u(k)$. Equation (10) can be rewritten for each virtual node as

$$\begin{aligned} r_1(k) &= r_N(k-1)e^{-\theta} + (1-e^{-\theta})F(r_1(k-1), W^{\mathrm{res}}_{\mathrm{in},1} u(k)) \\ &\ldots \\ r_i(k) &= r_{i-1}(k)e^{-\theta} + (1-e^{-\theta})F(r_i(k-1), W^{\mathrm{res}}_{\mathrm{in},i} u(k)) \\ &\ldots \\ r_N(k) &= r_{N-1}(k)e^{-\theta} + (1-e^{-\theta})F(r_N(k-1), W^{\mathrm{res}}_{\mathrm{in},N} u(k)) \end{aligned} \tag{11}$$

where θ is the separation of the virtual nodes. This equation allows us to recursively compute each virtual node-state at time step k only as a function of the input at the same time step k and virtual node-states at time step $k-1$:

$$r_i(k) = \Omega_i r_1(k-1) + \sum_{j=1}^{i} \Delta_{ij} F(r_j(k-1), W^{\mathrm{res}}_{\mathrm{in},j} u(k)) \tag{12}$$

with

$$\Omega_i = e^{-i\theta}, \quad \Delta_{ij} = (1-e^{-\theta})e^{-(i-j)\theta}, \quad \text{with } i \geq j.$$

This equation is our analogue of equation (1), representing classical reservoirs, and it explicitly describes the state coupling between consecutive time steps through the system dynamics. However, it differs from traditional reservoirs because the nonlinear functions are applied to the states before the summation is taken. Figure 5.9 illustrates this interaction topology by showing interaction strength matrices for two values of θ. The coefficients Ω_i correspond to the values found in the last column, while the diagonal and off-diagonal elements are given by Δ_{ij}. In terms of traditional reservoirs, this can be related to $\mathbf{W}^{\mathrm{res}}_{\mathrm{res}}$ where $T_{\mathrm{in}} = \tau$.

The strongest assumption in this analytical derivation is the fact that the function F is treated as a constant value over the interval θ. To verify whether this approximation is valid we perform a numerical check. While running the reservoir for some random input samples we perturb one of the virtual nodes with a pulse of amplitude 1 and observe how this perturbation is being passed on to other virtual nodes. In this numerical experiment, we choose a Mackey–Glass nonlinearity type to fulfill the role

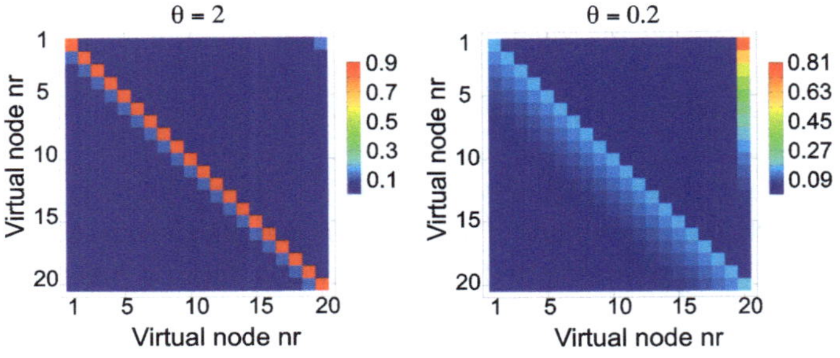

Figure 5.9: Analytical interaction graphs for large and small θ with $T_{in} = \tau$. Interaction graphs for different virtual node separation where we plot the coefficients Ω_i and Δ_{ij} of equation (12) as a matrix using color coding. For large values of θ (left), the diagonal elements are significantly larger than all others, but when θ decreases (right), the exponential tail of the off-diagonal elements and also the connection to the last virtual node of the previous input step become dominant. The figure is taken from the supplementary material of Appeltant et al. [6].

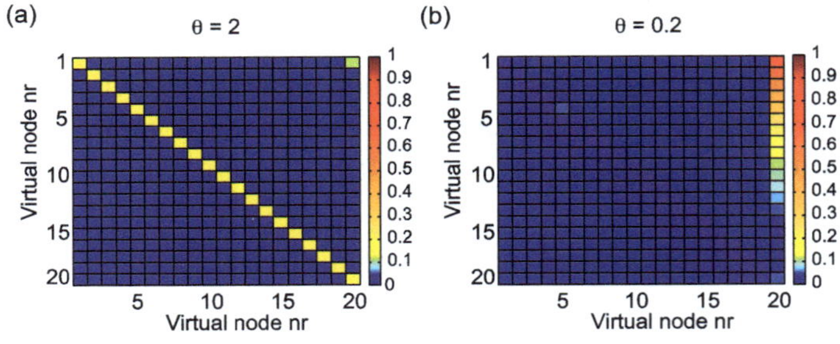

Figure 5.10: Numerical interaction graphs for large and small θ with $T_{in} = \tau$. Interaction graphs for different virtual node separation where we plot the coupling strength between the virtual nodes as a matrix using color coding. For large values of θ (left), the diagonal elements are significantly larger than all others, but when θ decreases (right), the exponential tail of the off-diagonal elements and the also the connection to the last virtual node of the previous input step become dominant.

of the function F. Figure 5.10 shows the interaction strength matrices obtained from numerical simulations. The scaling is expressed in arbitrary units since the obtained values depend on the strength of the pulse and the exact shape of the nonlinear transfer function.

Qualitatively, a confirmation of the analytical result is found. For large values of θ ($\theta = 2$), the self-feedback is the strongest coupling contribution for all virtual nodes. This results in a strong main diagonal in Figure 5.10(a). When setting θ to a small value ($\theta = 0.2$), the effect of the inherent system dynamics becomes more important and the off-diagonal elements are more pronounced. Also the coupling with the last virtual node (last column) is strongly present.

5.5.2 Interconnection structure through the feedback line

Choosing $\theta \gg T$, the state of a given virtual node is practically independent of the states of the neighboring virtual nodes and the connections between the virtual nodes due to the nonlinear node dynamics are negligible (see Figure 5.9). The nonlinear node reaches its steady-state for each virtual node and the reservoir state is only determined by the instantaneous value of the input $J(t)$ and the delayed reservoir state. The system given by equation (9) can then be described with a map:

$$x(t) = F[x(t-\tau), J(t)] \tag{13}$$

with F any nonlinear function and $J(t)$ given by equation (6).

If $\theta \gg T$ and $T_{\text{in}} = \tau$, there is no coupling between virtual nodes and the diversity of the reservoir states is reduced. The behavior in this case is illustrated in Figure 5.8. Figure 5.8(a) shows the injected input (blue) and the corresponding output of the nonlinear node that is sent in the delay line (red). The part of the time trace shown here corresponds to one time-multiplexed input value with a binary mask imprinted on it. Because every mask value is kept constant long enough for the system to reach the steady-state, all node-states with equal mask values are identical. Regardless of the number of virtual nodes that are tapped from the delay line, with this binary mask only two different reservoir state values can be used for computation. Figure 5.8(b) illustrates the equivalent traditional network of nodes in terms of connectivity. All nodes have a self-coupling, induced by the delayed feedback caused by $T_{\text{in}} = \tau$, but they are not influenced by the states of the other nodes in the network.

Virtual nodes can also be connected using the feedback of the nonlinear node if one detunes the input sampling period (T_{in}) to the length of the delay line [28]. This misalignment can be quantified in terms of the number of virtual nodes by using $\alpha = (\tau - N\theta)/\theta$. The topology of the virtual network structure created by this misalignment depends on the value of α. The interaction topology encoded when $\alpha = 1$ (i. e., $\tau = T_{\text{in}} + \theta$) is, for all practical purposes, equivalent to a standard ESN with ring topology [38] (see Figure 5.11). In the case of $1 \leq \alpha < N$, the virtual nodes, $r_i(k)$ can be described when $\theta \gg T$ by

$$r_i(k) = \begin{cases} F(r_{i-\alpha}(k-1) + W^{\text{res}}_{\text{in},i}u(k)) & \text{if } \alpha < i \leq N \\ F(r_{N+i-\alpha}(k-2) + W^{\text{res}}_{\text{in},i}u(k)) & \text{if } i \leq \alpha. \end{cases}$$

5.6 Weights distribution of the input layer

The input layer defines the connectivity between the external input and the reservoir. In traditional RC systems, the connections between the input and the different nodes in the reservoir ($\mathbf{W}^{\text{res}}_{\text{in}}$ in equation (1)) have randomly assigned weights. These weights are typically assigned from a uniform distribution [5].

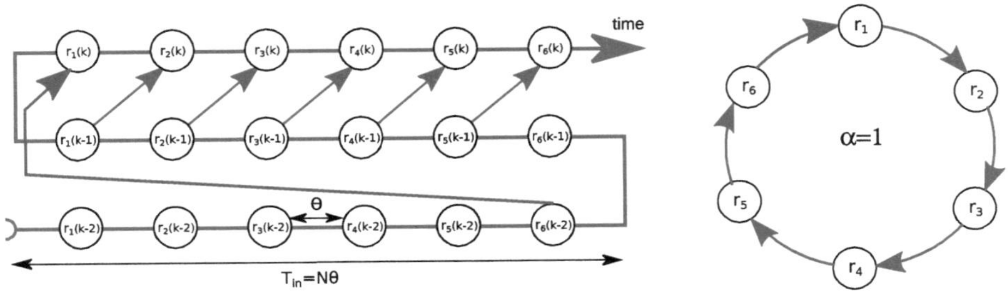

Figure 5.11: Schematic representation of the virtual nodes over the delay line (left) and the corresponding interaction graph (right) when $\tau = T_{\text{in}} + \theta$ ($\alpha = 1$) and $N = 6$. Red arrows indicate the connections at time step $(k-1)$ and blue arrows the connections at the previous time step $(k-2)$.

In delay-based RC, there is only one hardware node and the spatial temporal distribution of the input layer in standard RC has to be performed by time-multiplexing. Thus, as explained in Section 5.4.2, the input mask in delay-based RC is a piecewise constant function (constant over an interval of θ) that is periodically repeated with period T_{in}. The response of the nonlinear system to each piece of the mask function is assigned to the corresponding virtual node. The values of the input mask during each interval of length θ are typically chosen independently at random and define the coupling weights from the input to the reservoir as on the standard RC approach. The input mask performs the random mapping of the input into the reservoir. In addition, the input mask has the important role to maximize the diversity in the responses of the system that can later be used for computation.

In delay-based implementations with $\tau \neq T_{\text{in}}$ and $\theta \gg T$, the input mask values (or weights) are often drawn from a uniform distribution in $[-1, 1]$ since this approach is more closely related with standard RC. Limited research has been done for other distributions of the input weights in this approach. In contrast, the first delay-based RC implementations with $\tau = T_{\text{in}}$ and $\theta < T$ used input weights randomly drawn from a binary uniform distribution [6]. The binary weights are typically randomly distributed in time, although a nonrandom mask construction procedure based on maximum length sequences yields an improvement over random temporal assignments [39]. Later, it was shown that the choice of two-valued input weights is suboptimal in the presence of noise [35, 40] since different virtual nodes end up having similar values that are hard to differentiate in the presence of experimental uncertainties.

For hardware implementations of delay-based RC, it turns out that input weights drawn from either uniform or six-valued distributions that are randomly distributed in time for every piece of the mask induce a larger diversity in the system responses. As a consequence, the use of these distributions of weights yield lower prediction errors in a chaotic time-series prediction task [35, 40].

A final refinement in the choice of input weights has recently been reported in [40]. Focusing on a time-series prediction task and taking $\tau = T_{\text{in}} + \theta$ with $\theta < T$, Nakayama et al. [40] have found optimum prediction errors when the mask is an analog irregular function with the same frequency bandwidth than the nonlinear reservoir system itself. A mask with the required bandwidth can be created following two different procedures, either using a random distribution with a frequency cut-off (colored noise) or using a temporal segment of the intrinsic dynamics of the nonlinear node in the chaotic regime. These results emphasize the importance of the bandwidth of the mask and agree with the findings reported for fully trained hardware systems [41]. In [41], it was shown that the bandwidth of the input mask adjusts to the analog bandwidth of the system when the full system (input, reservoir and output layers) is optimized via back-propagation techniques.

5.7 Computational capacity of delay-based reservoirs

Dynamical systems X driven by time dependent external signals $u(t)$ can process the information contained therein [42]. As it is shown in [42], functions of previous inputs $z(u(t-h), \dots, u(t))$ can be reconstructed from the state of a dynamical system using a linear estimator. This estimator is constructed from N internal variables of the system (see equation (3) in [42]). These N variables provide a high-dimensional space, referred to as a reservoir. The capacity $C[X, z]$ measures how successful the dynamical system X is at computing z (see equations (4)–(5) in [42]).

The total computational capacity of a dynamical system corresponds to the total number of linearly independent functions of the input the system can compute. If the system obeys the fading memory condition [43], the total computational capacity is equal to the number of linearly independent internal variables of the system [42]. In delay-based RC, the internal variables of the system are the virtual nodes, so the total computational capacity of delay-based reservoirs is given by the number of linearly independent virtual nodes. The computational power of delay-based RC is therefore hidden in the diversity of the reservoir states. Neighboring virtual nodes that are connected through the dynamics of the nonlinear node ($\theta < T$) influence each other and have a similar state, yielding to a small diversity in the available reservoir states. As the separation between the nodes increases and $T_{\text{in}} > \tau$, the diversity increases while the virtual connection between the nodes decreases. This behavior can be observed in Figure 5.12, which plots the reservoir states of a delay-based RC for two different values of node separation θ.

The number of linearly independent virtual nodes depends not only on the separation between the virtual nodes but also on the misalignment between T_{in} and τ, i. e., α. When $\alpha < 0$, a number $|\alpha|$ of virtual nodes are not connected through the feedback line with nodes at a previous time and computational capacity is then reduced. Computational capacity is also reduced if $|\alpha|$ and N are not coprimes. In this case, the feed-

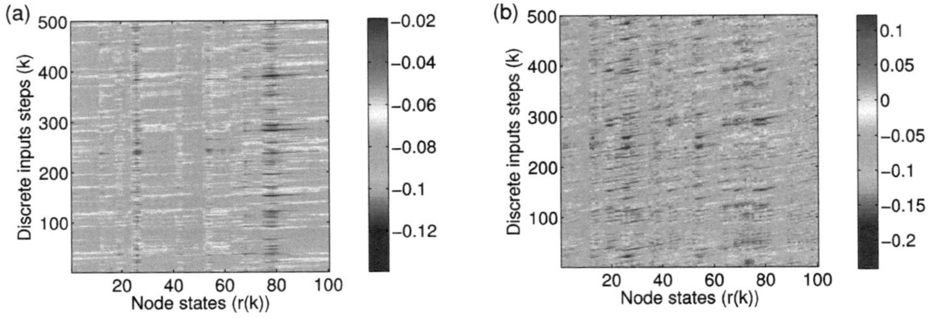

Figure 5.12: Evolution of the reservoir states for a delay-based RC using color coding. The system is governed by equation (9) with $F = \eta F_{\mathrm{sig}}$, where F_{sig} is the sigmoid function, $N = 100$, $\eta = 0.1$, $T = 1$, and $\tau = T_{\mathrm{in}} + \theta$ ($\alpha = 1$). For a small value of θ (a), the state of the virtual nodes are less diverse than for a large value of θ (b). (a) $\theta = 0.2T$. (b) $\theta = 4T$.

back line results in the N virtual nodes forming $\gcd(|\alpha|, N)$ ring subnetworks, where gcd is the greatest common divisor. Each subnetwork has $p = N/\gcd(|\alpha|, N)$ virtual nodes. Virtual node-states belonging to different subnetworks have a similar dependence on inputs and reservoir diversity is reduced. When there are p subnetworks and virtual node connections through dynamics are negligible ($\theta \gg T$), the total capacity of a linear delay-based RC ranges between $(p+1)$ and $2p$ for $0 < \alpha < N$.

In dynamical systems X, when z is a linear function of one of the past inputs, $z(t) = u(t-k)$, the computational capacity corresponds to the linear memory capacity introduced in [44]. The linear memory capacity is a way of estimating the amount of fading memory available in RC systems. Delay-based RC has an intrinsic memory due to its feedback line. This fading memory is essential to perform certain tasks that depend on the context, e. g., time series prediction. As soon as the task requires more memory than the one provided by the system, the performance of the reservoir computer degrades significantly.

The maximum total capacity of a dynamical system is N where N is the number of internal states of the system (the number of virtual nodes in delay-based RC). The total capacity of linear reservoirs (linear function F in equation (9)) is equal to the linear memory capacity. Figure 5.13 shows the linear memory capacity of the linear delay-based RC as the separation between the virtual nodes increases for two different values of the misalignment between the delay and the input, $\alpha = 1$ ($\tau = T_{\mathrm{in}} + \theta$) and $\alpha = 0$ ($\tau = T_{\mathrm{in}}$). In the case of $\alpha = 1$, the linear memory capacity increases with the node separation. Here, larger node separation implies weaker connections through the dynamics and more linearly independence between the virtual nodes. In the limit case of linear reservoir with an instantaneous response to the input ($T = 0$), all the virtual nodes are linearly independent and the total capacity will reach its maximum value, N. As it was shown in Figure 5.11, this topology is similar to the simple cycle reservoir topology introduced in [38]. In contrast, the virtual nodes are only connected

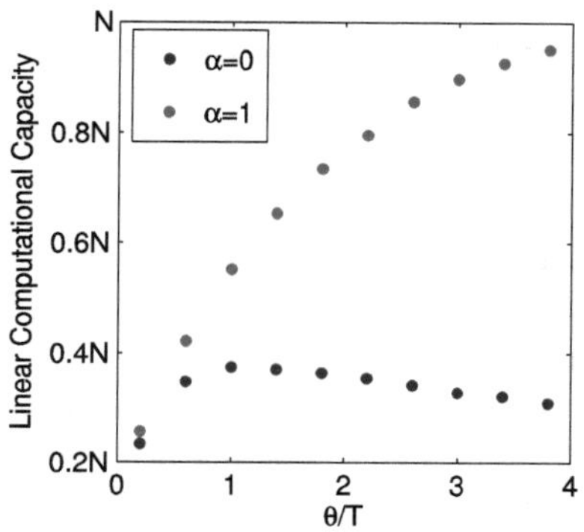

Figure 5.13: Linear computational capacity of a linear delay-based RC with $N = 100$ as a function of the separation between the virtual nodes θ for two different values of α. The delay-based RC is governed by equation (9) with a linear function $F(y) = \eta y$, $\eta = 0.9$, and $T = 1$. The linear computational capacity has been obtained summing until $k = N$.

by the system dynamics in the case of $\alpha = 0$ (see Figures 5.7 and 5.8). As a result, Figure 5.13 shows that the linear memory capacity decreases beyond $\theta > T$ since the delay-based RC is no longer a connected network.

In contrast to linear systems, which only have linear memory capacity, nonlinear dynamical systems have both linear and nonlinear memory. They are therefore capable of nonlinear transformations on the inputs. In this case, however, the total computational capacity is still limited by the dimension of the reservoir. As a consequence, there is a trade-off between the linear memory that dynamical systems possess and their capacity to process the input in a nonlinear way [42]. In this context, a mixture reservoir made of both linear and nonlinear dynamics has been suggested as a workaround to alleviate the memory-nonlinearity trade-off [45].

Hardware implementations of delay-based reservoirs reach a compromise between the computation speed and the available computational capacity when $\theta < T$. This combination of parameters, however, is suboptimal from the point of view of computational capacity.

5.8 Hardware implementations of delay-based reservoirs

Thanks to the versatility of the delay-based RC concept, it can be implemented in very different hardware platforms. The first working prototype was developed in electronics in 2011 by Appeltant et al. [6] and was quickly followed in 2012 by optoelectronic implementations [1, 29]. Moving forward from the optoelectronic implementations, the first all-optical delay-based reservoir computers based on semiconductor optical amplifiers and semiconductor lasers were implemented [3, 4]. A more extensive list of recent hardware implementations of delay-based RC can be found in [46].

The main differences between the delay-based RC experiments are in the nonlinearity of the reservoir and in the relative timing between the input injection time and the delay time. In the case of the all-optical implementations, there are also differences in how the input is injected into the system.

Most hardware implementations of delay-based RC focus on the practical demonstration of the reservoir layer. The input and output layers are emulated off-line on a standard computer. There are, however, first works aiming at the complete implementation of the three layers of RC on analogue hardware. In this way, a proof of concept for stand-alone delay-based reservoir computers has been demonstrated [47].

Optoelectronic and all-optical systems have been widely employed for delay-based RC. A number of classification, prediction, and system modeling tasks have been performed with state-of-the-art results. To name a few, excellent performance has been obtained for speech recognition [1, 48, 4], chaotic time series prediction [1, 35, 4], nonlinear channel equalization [29, 47, 3], and radar signal forecasting [47].

The operating speed of most optoelectronic delay-based RC implementations is in the MHz range, although this kind of setup can even operate at GHz speeds [49]. In the case of all-optical delay-based RC implementations, a photonic reservoir based on a semiconductor laser with feedback has shown unconventional information processing capabilities at Gbyte/s rates [4], one of the fastest reservoir computer up to date. In contrast, most of the electronic implementations are in the range of KHz. These electronic implementations serve as a testbed for the fastest photonic implementations that will be discussed in the following chapters. In particular, the electronic platform allows to explore the particularities of computing with an analog nonlinear system.

5.8.1 An example of an electronic implementation of delay-based reservoir computing

For illustration purposes, in this subsection, we focus on a mixed analog and digital implementation of the delay-based RC concept with a nonlinear analog electronic circuit as a main computational unit [6, 37]. This delay-based RC scheme can be conceptually divided in several distinct blocks, which are schematically shown in Figure 5.14. First, there is an input preprocessing stage where the incoming data are time-multiplexed. A digital-to-analog (DAC) and an analog-to-digital (ADC) converter with 12 bits resolution interface the digital and the analog part, and vice versa. An analog Mackey–Glass electronic circuit [37] is chosen as the nonlinearity in this implementation. A delay element, implemented digitally, provides the required feedback. Finally, at the output post-processing stage, the system output is given by a linear weighted sum of the values of the virtual nodes. The weights are obtained with a simple linear regression during the off-line training procedure.

With the appropriate scaling, the Mackey–Glass system with delay can be modeled, in the presence of a masked input $J(t)$, as follows [37]:

$$\dot{x}(t) = -x(t) + \frac{\eta \cdot [x(t-\tau) + \gamma \cdot J(t)]}{1 + [x(t-\tau) + \gamma \cdot J(t)]^p}, \tag{14}$$

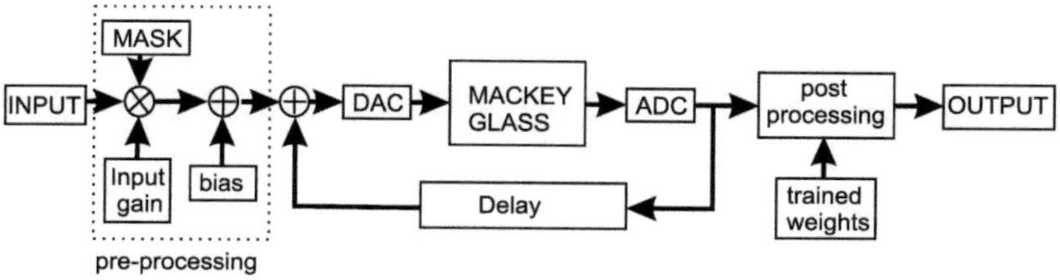

Figure 5.14: Schematic view of the RC implementation based on a single Mackey–Glass nonlinear element with delay. DAC and ADC stand for digital-to-analog conversion and analog-to-digital conversion, respectively. Figure reprinted with permission from Ref. [37], IEEE.

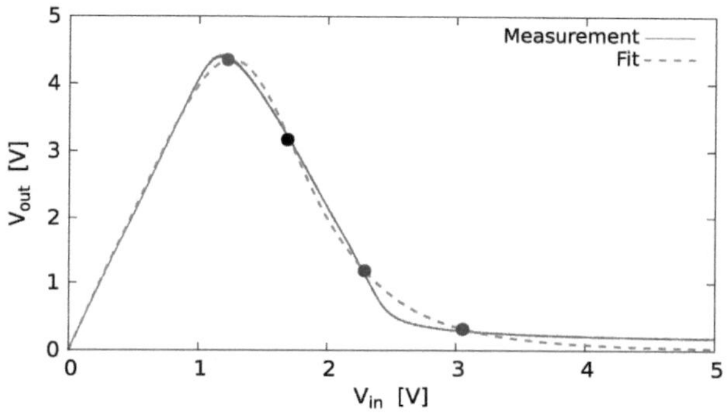

Figure 5.15: Experimental nonlinear function (red solid line) compared to a fit using the Mackey–Glass nonlinearity (green dashed line). The operating points marked with dots of different colors correspond to the solid lines in Figure 5.16. Figure reprinted with permission from Ref. [37], IEEE.

with x denoting the dynamical variable, t a dimensionless time, τ the delay in the feedback loop and η and γ represent feedback strength and input scaling, respectively. Note that for the scaled model $T = 1$. This equation corresponds to equation (9) with a Mackey–Glass nonlinear function. The exponent p can be used to tune the degree of the nonlinearity. Figure 5.15 shows the experimental Mackey-Glass function for this implementation, together with the corresponding numerical fit. In this example, the Mackey–Glass equation fits the experimental nonlinearity with an exponent $p \sim 6$.

We evaluate here the performance of this electronic scheme for the same time-series prediction task that in Section 5.4.4. This task consists on the one-step ahead prediction of a benchmark chaotic time series, the Santa Fe laser time series. For this task, $T_{\text{in}} = \tau$, $N = 400$ and the input mask has six different amplitude levels randomly distributed in time with zero mean. More details about this delay-based RC can be found in [37]. The normalized mean squared error (NMSE) of the one-step ahead prediction over the test set in the experiments (top panels) and the numerical simulations (bottom panels) is shown in Figure 5.16 for different bit resolutions in the ADC. We observe a clear dependence of the NMSE on the number of bits in the output ADC, with a wider region of low NMSE for increasing number of outputs bits. NMSE below 0.05 are obtained for $\gamma \sim 0.3$ and a wide range of feedback strengths when the ADC resolution is larger than 8 bits.

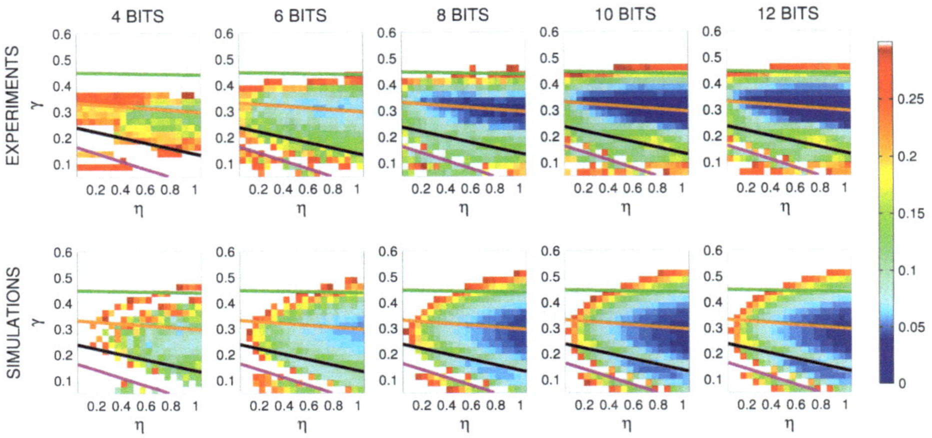

Figure 5.16: Experimental (top) and numerical (bottom) results for the Santa Fe time series prediction test. Color-coded NMSE as a function of the parameters of the system η and γ for different number of bits in the output ADC. The exponent is set to $p \sim 6$, $N = 400$, $\theta = 0.4T$, and $T_{\text{in}} = \tau$. Color lines correspond to the operating points shown in Figure 5.15. The NMSE values are an average over three data partitions. Figure reprinted with permission from Ref. [37], IEEE.

This example nicely illustrates that the nonlinear function of the hardware node plays an important role in the system performance. In delay-based RC, the system output oscillates around the operating point. The operating point and the maximum amplitude of the oscillation determine the effective nonlinear function of the system, i. e., the parts of the nonlinear function that the system really explores. The color lines in Figure 5.16 show the NMSE values when the delay-based RC is operating around the correspondingly colored points in Figure 5.15. For example, when the amplitude of the oscillations is small (low γ) and the operating point is around the inflection point of the numerical Mackey–Glass function (see black dot in Figure 5.15), the response of the Mackey–Glass node is almost linear. In contrast, an operating point at the maximum of the nonlinear function (see pink dot in Figure 5.15) leads to very nonlinear responses of the Mackey–Glass node. Figure 5.16 shows that these operating points lead to large NMSE values, while the lowest NMSE values are obtained when the system is operated around the brown dot in Figure 5.15.

The Santa Fe time series prediction task requires a RC system with fading memory and nonlinear computational capacity. When the operating point is in a very nonlinear region of the Mackey–Glass function (pink line in Figure 5.16), the system does not reach the memory capacity required by the Santa Fe task and the NMSE increases. In turn, when the operating point is in a quasi-linear region of the Mackey–Glass function (black line in Figure 5.16), the system has a large memory but a lower nonlinear computational capacity, and there is a slight increment of the NMSE. As stated in the previous section, there is a trade-off between the memory capacity of a RC system and its computational capacity [42]. In this example, the brown operating point in Figure 5.15 leads to the best compromise between the linear memory and the nonlinear computational capacity.

5.8.2 Challenges in physical implementations of delay-based reservoir computers

The main advantages offered by hardware implementations of RC are high processing speed, parallelism, and lower power consumption compared to digital implementations. However, physical analog systems are affected by noise. The finite signal-to-noise ratio (SNR) reduces the computational capacity [42] and degrades the performance.

Another limitation in hardware implementations of delay-based RC comes from the noninstantaneous response of the system to an input signal. When θ is smaller than the response time of the system, the system's dynamics couples consecutive virtual nodes. These network connections lead to similar virtual node-states, and the computational capacity is degraded. In this case, the covariance matrix of the outputs is ill conditioned with a large ratio between the largest and smallest eigenvalues (condition number). Moreover, this kind of reservoirs are more sensitive to noise [50]. In summary, when the influence of the inherent dynamics is not negligible, the computational capacity is degraded and more sensitive to noise. The influence of the system's dynamics can be mitigated by increasing the node distance θ. Since the information processing rate, given by $T_{\text{in}}^{-1} = (N\theta)^{-1}$, is limited by the system response time, it is desirable from the practical point of view to keep the system as fast as possible. If one desires a high-speed hardware implementation that still exhibits a good computational capacity and a certain degree of noise robustness, it is recommended to use intermediate values of θ but it cannot be arbitrarily small as discussed in Section 5.7.

Role of noise

Several noise sources can be present at the different layers of the RC system. In particular, noise can appear in the reservoir itself, and/or the input and output layers. The noise in the acquisition procedure, i. e., output layer, has contributions from detection noise and digitization noise, which is the strongest noise contribution in the case of mixed analog and digital implementations [35]. Digitization noise originates from the finite resolution of the analog-to-digital (ADC) and digital-to-analogue (DAC) converters, which act as an interface between the analog and digital worlds.

Since performance is degraded by noise, one may try to reduce the digitization noise by oversampling and averaging over the measured response of the system. Signal-to-noise-ratio can be also increased by averaging over several repetitions of the reservoir output. In all of these strategies to reduce the effect of noise, the maximum information processing rate is reduced.

Interestingly, classification tasks are relatively robust against a finite SNR [51]. For a spoken digit recognition task, an all-optical hardware system even outperformed software implementations of RC in terms of speed and accuracy [4]. In contrast, the performance of time-series prediction tasks significantly degrades when the SNR decreases [35]. The difference in sensitivity is due to the different nature of the two tasks. A classification task only requires a winner-takes-all decision that relies mainly on the recognition of the shape of the corresponding digit. This shape is still preserved in the presence of digitization noise. However, time-series prediction actually requires the precise approximation of a nonlinear transformation.

The sensitivity of the reservoir state to noise can have a clear impact on the consistency properties of the system. Consistency relates to the reproducibility of system responses for multiple similar inputs. Computational performance requires reproducible results, making consistency an essential condition for reservoir computing [52]. Different sets of initial conditions due to noise and different noise realizations in the reservoir may result in different temporal evolutions of the dynamics under the injection of the same input. As a consequence, the lack of consistency degrades the performance [40, 52].

In photonic systems based on lasers, spontaneous emission noise is always present in the reservoir. Numerical simulations of all-optical implementations of delay-based RC have shown [53, 54] that time-series prediction task performance degrades for realistic values of the spontaneous emission noise. In all-optical delay-based RC, the computational performance has also been found to be very sensitive to feedback phase [55]. In other words, tiny fluctuations in the precise value of the delay time can have an important impact on the performance. This phase sensitivity can be avoided by modifying the readout layer, such that the read-out weights are optimized from a combination of the reservoir's state and its delayed version [55]. In summary, noise in the output layer is typically the main limiting factor of computing performance for photonic systems [35].

In the following, we show some examples of the performance degradation due the noise, focusing on the cases for which there exist virtual connections through the system dynamics.

Noise in an electronic implementation

Digitization noise in the output layer has been studied in the electronic implementation [51] described in Section 5.8.1. We remind the reader that in this particular implementation $\theta < T$ and there is no mismatch ($T_{\mathrm{in}} = \tau$). The performance for the Santa Fe laser time-series prediction task is evaluated and it is found that performance improves when the bit resolution of the output ADC increases. However, the improvement saturates for resolutions larger than 10 bits (see top panels of Figure 5.16) when other sources of noise become dominant. The maximum SNR attained in this electronic implementation is slightly larger than 60 dB.

Noise in an optoelectronic implementation

The optoelectronic implementation consists of a nonlinear oscillator with delayed feedback [35]. The nonlinear transformation is provided by a Mach–Zehnder modulator (MZM). This system can be modeled, in the presence of a masked input J, as follows:

$$\dot{x}(t) = -x(t) + \eta \cdot (\sin^2[x(t-\tau) + \gamma \cdot J(t) + \Phi] - 0.5), \tag{15}$$

where time and delay τ are scaled to have $T = 1$, x is the dynamical variable, η and γ represent feedback strength and input scaling, respectively, and Φ is the MZM offset phase. It is worth noting that the local properties of the nonlinearity around the operating point can be easily tuned by changing Φ.

We discuss here the implementation in which the number of virtual nodes is $N = 400$, $\theta = 0.2T$ (i. e., virtual connections due to system dynamics are important) and $T_{\text{in}} = \tau$. The performance for the Santa Fe laser time-series prediction task is evaluated for two different masks (a binary mask and a multi-level mask) under different digitization noises. For this system, it has been shown that performance can be improved with a multivalued mask [35].

The influence of the number of output digitization bits on the performance degradation is shown in Figure 5.17 for the two- and six-valued input masks. The prediction errors are consistently lower for the six-valued mask compared to the binary mask over the whole range of digitization bits. Multivalued masks increase the diversity of the reservoir states, and then noise sensitivity is reduced. When a binary mask is used, neighboring virtual node-states tend to be similar due to the short distance between them ($\theta = 0.2T$), and the performance is very sensitive to noise. In the absence of noise, the two types of masks yield the same error for the Santa Fe time-series prediction task.

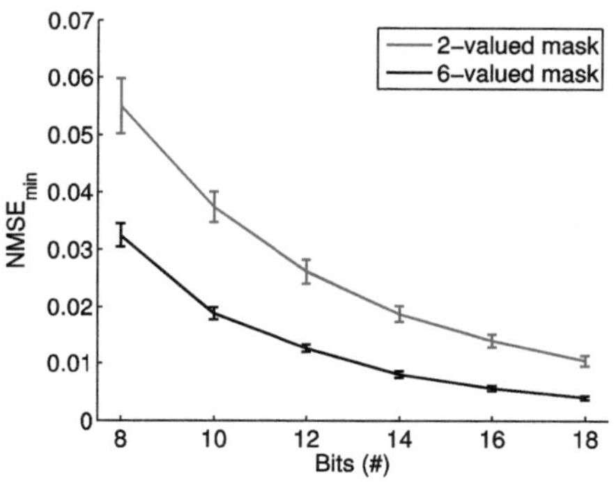

Figure 5.17: Minimum NMSE test prediction error in the Santa Fe laser time-series prediction task for $\eta = 0.8$ and $\gamma = 0.45$ as a function of the number of output digitization bits. The red (black) line corresponds to a binary (six-valued) mask. The error bars correspond to ten different random realizations of the masks. Figure reprinted with permission from Ref. [35], OSA.

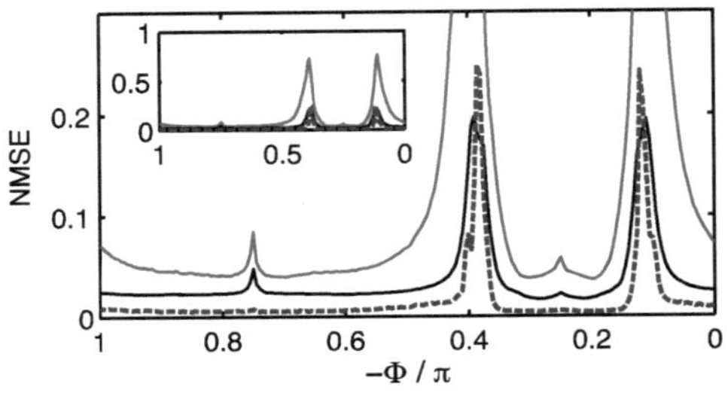

Figure 5.18: NMSE test prediction error in the Santa Fe laser time-series prediction task for $\eta = 0.8$ and $\gamma = 0.45$ as a function of the offset phase ϕ. The red (blue) line corresponds to a binary mask in the presence (absence) of 10 bits digitization noise. The black line corresponds to a six-valued mask in the presence of 10 bits digitization noise. In the absence of noise, a minimum prediction error of about 0.01 is obtained (blue dashed line). Figure reprinted with permission from Ref. [35], OSA.

Figure 5.18 shows the normalized mean square error (NMSE) for the Santa Fe task as a function of the offset phase of the nonlinearity Φ when two-valued and six-valued input masks are used. In the presence of digitization noise (10-bit resolution), the prediction error for a six-valued mask (solid black line) is significantly lower than for the binary mask (solid red line), over the entire parameter range, with a minimum error of about 0.02. The error is not reduced when increasing further the number of discrete values in the input mask.

These numerical simulation results are in agreement with experimental results [35]. The prediction error for the experimental realization of the optoelectronic system is shown in Figure 5.19(b) as a function of the offset phase of the nonlinearity Φ for both the two-valued and the six-valued input masks. The dependence of the error on the offset phase agrees with the numerical results shown in Figure 5.18. A better performance is also obtained with the six-valued mask (solid line) than with the binary mask (dashed line). The minimum prediction error, 0.06 (0.1) for the six-valued (binary) mask, is slightly higher than the numerical results for 8 bits digitization of the ADC. A lowest error of 0.02 can be found when the signal is detected with five times oversampling and subsequent averaging (see Figure 5.19(c)). In this case, the SNR measured at the output layer is equivalent to a 10 bit dynamic range.

The influence of a finite SNR has also been analyzed in an optoelectronic system when the virtual connections through the dynamics are negligible ($\theta = 4T$) and $T_{\text{in}} = \tau + \theta$ ($\alpha = 1$) [30]. We show in Figure 5.20 the memory function obtained from the experiment and the corresponding numerical simulations. The signal to noise ratio of a single measurement is $\approx$ 24 dB. The experimental SNR can be increased to 40 dB by averaging the detection over ten repetitions of the measurement. An excellent agreement is found between numerics and experiments. Experimental memory functions with SNR $\approx$ 40 dB and SNR $\approx$ 24 dB yield a linear memory capacity of 8.5 and 6, respectively. To characterize the linear memory capacity degradation due to noise, we also show the numerical results for the noise-free system, which yield a linear memory capacity around 12. The memory capacity of this optoelectronic system with a SNR of 24 dB is half the one of the noise-free counterpart.

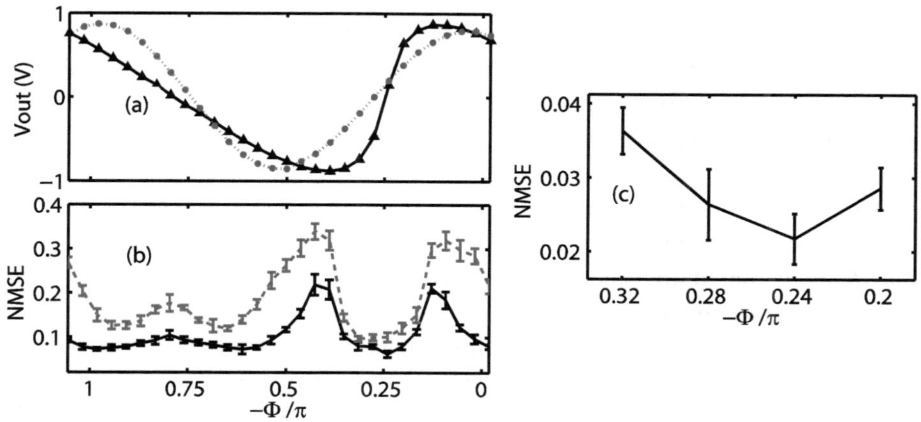

Figure 5.19: (a) Experimentally recorded nonlinearity (dotted line) and operating point (solid line with triangles) as a function of the Mach–Zehnder offset phase for $\eta = 0.8$ and $\gamma = 0$. (b) Test prediction error (NMSE) for the Santa Fe laser time-series prediction task with 400 virtual nodes for two-valued (dashed line) and six-valued (solid line) input masks ($\gamma = 0.45$). (c) NMSE of the predicted time-series for an improved detection with 5:1 oversampling and subsequent averaging obtained for the six-valued mask. Figure reprinted with permission from Ref. [35], OSA.

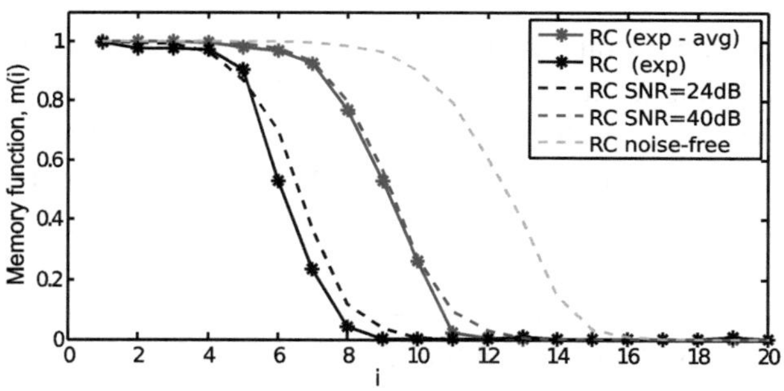

Figure 5.20: The linear memory capacity for the numerical (dashed lines) and experimental (solid lines) realizations of an optoelectronic RC. The system parameters are: $\theta = 4T$, $\alpha = 1$, $N = 246$ virtual nodes, $\eta = 0.9$, $\gamma = 0.3$, and $\Phi = 0.4\pi$. Figure adapted from Ref. [30].

Role of system response time

Delay-based RC consisting of a single nonlinear neuron can be easily implemented in hardware, potentially allowing for high-speed information processing. Since the information processing rate ($1/T_{\text{in}}$) is inversely proportional to the number of virtual nodes (N) and the node separation (θ), the input throughput can be increased by reducing N and θ. However, as shown in previous sections, both the computational capacity (with N being its maximum) and the noise robustness are reduced when the value of θ approaches that of the system response time (T). For this reason, the system response time imposes a limit to the maximum information processing rate.

In this context, parallel-based architectures with k nonlinear nodes reduce the information processing time by a factor of k for the same total number of virtual nodes. It has been shown [56, 57] that for the same $(T/\theta) > 1$ and without mismatch, $T_{\text{in}} = \tau$, performance is improved when different activation functions are used for the nonlinear nodes. In this way, reservoir diversity is increased. However, the hardware implementation becomes more involved than the one of a delay-based RC with a single nonlinear node.

Several strategies have been used to increase reservoir diversity of delay-based RC with a single nonlinear node when $(T/\theta) > 1$ and $T_{\text{in}} = \tau$. First, we have shown in Section 5.6 and earlier in this section that the use of multi-valued input masks increases reservoir diversity and noise robustness.

Another strategy to increase reservoir diversity is to use multiple feedback lines [36, 58]. When the delay times of extra feedback lines are close to but not exact multiples of τ, memory capacity increases and performance improves. In the case of only one extra feedback line with a delay time $\tau_2 = M\theta$ $(M > N)$, the best performance is obtained when M and N are coprimes [59]. In this case, the number of virtual nodes that are mixed together within the history of each virtual node is maximized. Multiple feedback delay lines have been implemented in an optoelectronic system based on nonlinear wavelength dynamics [48]. In this case, 15 delay lines with delay times smaller than $T_{\text{in}} = \tau$ have been used, giving a good performance for a classification task.

Finally, we consider an additional strategy still based on the simple architecture of a single nonlinear node with one feedback delay line. In this strategy, the mismatch α can be used to increase reservoir diversity when $\theta < T$. The value of α has to fulfill the requirement of having no common divisors with N. Otherwise, the computational capacity is reduced due to the formation of subnetworks (see Section 5.7). When $0 < \alpha < N$ has no common divisors with N, all virtual nodes are connected through feedback in a ring. The minimum time steps required for a virtual node to connect through feedback with itself increases with α. Every virtual node connects with itself after $(N + \alpha)$ time steps. For small values of (θ/T), the states of virtual nodes that are separated by less than T are correlated. When the mismatch is increased, virtual nodes are connected through feedback to nodes that are not connected through the inherent system response. Reservoir diversity is then increased and a larger computational capacity is achieved. This is shown in Figure 5.21 for a linear delay-based RC with 97 virtual nodes. The linear computational capacity of the linear delay-based RC increases from 23 ($\alpha = 0$) to 42 ($\alpha = 94$) when $\theta = 0.2T$. If $\theta \gg T$, the total linear computational capacity of the linear delay-based RC is equal to N for $0 < \alpha < N$, i. e., a mismatch $\alpha = 1$ is sufficient and has often been used in delay-based RC systems. Negative values of α have only been used in a system with a microchip laser [60]. In this case, the computational capacity is reduced (see Section 5.7). In addition, a larger $\alpha = 5$ has been used in systems with analogue nonrandom masks [47]. In this case, a large α is necessary to ensure that connected virtual nodes receive a very different version of the input signal.

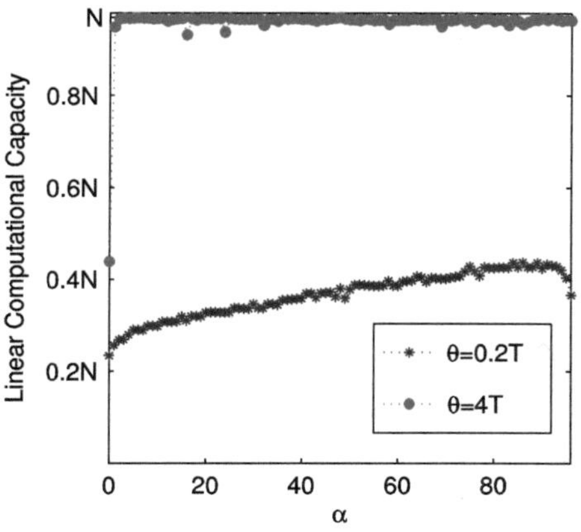

Figure 5.21: Numerical results for the linear memory capacity as a function of the misalignment α for a linear delay-based RC. The system is governed by equation (9) with a linear function $F(y) = \eta y$. The parameters are: $N = 97$ virtual nodes and $\eta = 0.9$.

5.9 Conclusion

Reservoir computing stands out as a simple yet powerful machine learning technique to process sequential data, in which the context is relevant for the information processing. The reservoir is typically nothing else than a randomly connected network with recurrences where the input information is also randomly mapped to the reservoir.

The development of delay-based reservoir computing has contributed to a further simplification of the RC concept in which the connectivity of the reservoir is no longer random but it follows a predetermined (e. g., ring) topology. This agrees with the results presented in [38, 61], in which deterministically connected reservoirs perform as well as random reservoirs. In delay-based RC, the reservoir connectivity can be modified by tuning the relative time-scales of the input sampling period, the feedback line and the system response time.

Most machine learning algorithms are intended to run on software platforms. Although this is also the case for reservoir computing, it turns out that this concept is particularly well suited for hardware implementations due to the random connectivities. The hardware implementation of delay-based reservoir computing has minimal requirements since only a single nonlinear hardware node is needed [6]. The recurrence is provided via a simple delay feedback loop, easy to implement in hardware. The concept of delay-based reservoir computing was initially developed to simplify hardware implementations, with the focus on photonics, and these conceptual simplifications have already allowed for full hardware implementations [47].

Delay-based RC exploits the time-multiplexing technique for the creation of multiple virtual nodes. As a result, the information processing rate is reduced. In this context, frequency multiplexing [62] is a promising way to increase the information processing speed. A combination of time and frequency multiplexing can also be used to reduce the influence of noise by combining several repetitions of the reservoir output in different frequencies.

The development of delay-based reservoir computing approaches, and reservoir computing in general, has greatly benefited from the interactions between experts on machine learning, neuroscience, and dynamical systems theory. We foresee further conceptual developments of this brain-inspired computational paradigm from the joint forces of these different communities. Some of the main challenges remaining ahead are the robustness to noise for the improvement of hardware implementations, the development of deep architectures keeping a minimalistic approach, and the combination of analog and digital systems for high-speed, power-efficient computations.

Bibliography

[1] L. Larger et al. "Photonic information processing beyond Turing: an optoelectronic implementation of reservoir computing". In: *Optics Express* 20.3 (2012), pp. 3241–3249.
[2] Y. Paquot et al. "Optoelectronic reservoir computing". In: *Scientific Reports* 2 (2012), p. 287.
[3] F. Duport et al. "All-optical reservoir computing". In: *Optics Express* 20 (2012), pp. 22783–22795.
[4] D. Brunner et al. "Parallel photonic information processing at gigabyte per second data rates using transient states". In: *Nature Communications* 4 (2013), p. 1364.
[5] M. Lukosevicius and H. Jaeger. "Reservoir computing approaches to recurrent neural network training". In: *Computer Science Review* 3 (2009), pp. 127–149.
[6] L. Appeltant et al. "Information processing using a single dynamical node as complex system." In: *Nature Communications* 2 (2011), p. 468. DOI: 10.1038/ncomms1476.
[7] H. Jaeger and H. Haas. "Harnessing nonlinearity: predicting chaotic systems and saving energy in wireless communication". In: *Science* 304 (2004), pp. 78–80.
[8] T. Erneux. *Applied delayed differential equations*. Springer Science Business Media, 2009.
[9] G. Orosz et al. "Exciting traffic jams: nonlinear phenomena behind traffic jam formation on highways". In: *Physical Review E* 80 (2009), p. 46205.
[10] K. Pyragas. "Continuous control of cahos by self-controlling feedback". In: *Physics Letters A* 170 (1992), pp. 421–428.
[11] E. Scholl and H. G. Schuster. *Handbook of chaos control*. 2nd ed. Weinheim: Wiley-VCH, 2008.
[12] L. Chen and K. Aihara. "Stability of genetic regulatory networks with time delay". In: *IEEE Transactions on Circuits and Systems. I, Fundamental Theory and Applications* 49 (2002), p. 602.
[13] A. Martin and S. Ruan. "Predator–prey models with delay and prey harvesting". In: *Journal of Mathematical Biology*. 43 (2001), pp. 247–267.
[14] H. Haken. *Brain dynamics: synchronization and activity patterns in pulse-coupled neural nets with delays and noise*. Berlin, Germany: Springer Verlag GmbH, 2006.
[15] P. R. Roelfsema et al. "Visuomotor integration is associated with zero time-lag synchronization among cortical areas". In: *Nature* 385 (1997), pp. 157–161.
[16] I. Fischer et al. "Zero-lag long-range synchronization via dynamical relaying". In: *Physical Review Letters* 97 (2006), p. 123902.

[17] R. Vicente et al. “Dynamical relaying can yield zero time lag neuronal synchrony despite long conduction delays”. In: *Proceedings of the National Academy of Sciences of the United States of America* 105 (2008), pp. 17157–17162.

[18] T. Heil et al. “Chaos synchronization and spontaneous symmetry-breaking in symmetrically delay-coupled semiconductor lasers”. In: *Physical Review Letters* 86 (2001), pp. 795–798.

[19] A. Takamatsu, T. Fujii, and I. Endo. “Time delay effect in a living coupled oscillator system with the plasmodium of physarum polycephalum”. In: *Physical Review Letters* 85 (2000), pp. 2026–2029.

[20] G. Van der Sande et al. “Dynamics, correlation scaling, and synchronization behavior in rings of delay-coupled oscillators”. In: *Physical Review E* 77 (2008), p. 055202.

[21] K. Ikeda and K. Matsumoto. “High-dimensional chaotic behavior in systems with time-delayed feedback”. In: *Physica D* 29 (1987), pp. 223–235.

[22] A. Argyris et al. “Chaos-based communications at high bit rates using commercial fibre-optic links.” In: *Nature* 438.7066 (Nov. 2005), pp. 343–346. DOI: 10.1038/nature04275.

[23] S. Boccaletti et al. “The synchronization of chaotic systems”. In: *Physics Reports* 366 (2002), pp. 1–101.

[24] L. M. Pecora and T. L. Carroll. “Synchronization in chaotic systems”. In: *Physical Review Letters* 64 (1990), p. 821.

[25] A. Pikovsky, M. Rosenblum, and J. Kurths. *Synchronization: a universal concept in nonlinear sciences*. Cambridge, UK: Cambridge University Press, 2001.

[26] M. Le Berre et al. “Conjecture on the dimensions of chaotic attractors of delayed-feedback dynamical systems”. In: *Physical Review A* 35 (1987), pp. 4020–4022.

[27] C. Fernando and S. Sojakka. “Pattern recognition in a bucket”. In: *Lecture notes in computer science*. Vol. 2801/2003. 2003, pp. 588–597. DOI: 10.1007/978-3-540-39432-7_63.

[28] Y. Paquot et al. “Reservoir computing: a photonic neural network for information processing”. In: *Proceedings of SPIE photonics Europe, nonlinear optics and applications IV*. Vol. 7728. 2010.

[29] Y. Paquot et al. “Optoelectronic reservoir computing”. In: *Scientific Reports* 2 (2012), p. 287. DOI: 10.1038/srep00287.

[30] S. Ortfn et al. “A unified framework for reservoir computing and extreme learning machines based on a single time-delayed neuron”. In: *Scientific Reports* 5 (2015), p. 14945. DOI: 10.1038/srep14945.

[31] D. Brunner et al. “Tutorial: photonic neural networks in delay systems”. In: *Journal of Applied Physics* 124.15 (2018), p. 152004.

[32] H. Jaeger. “The ‘echo state’ approach to analyzing and training recurrent neural networks”. In: *Technical Report GMD Report, German National Research Center for Information Technology* 148 (2001), pp. 2188–2191.

[33] A. S. Weigend and N. A. Gershenfeld. *Time series prediction: forecasting the future and understanding the past*. Vol. 80. Addison-Wesley, 1993. URL: ftp://ftp.santafe.edu/pub/Time-Series/Competition.

[34] G. Giacomelli et al. “Defects and spacelike properties of delayed dynamical systems”. In: *Physical Review Letters* 73 (1994), p. 1099.

[35] M. C. Soriano et al. “Optoelectronic reservoir computing: tackling noise-induced performance degradation”. In: *Optics Express* 21.1 (2013), pp. 12–20.

[36] L. Appeltant. *Reservoir computing based on delay-dynamical systems*. Vrije Universiteit Brussel/Universitat de les Illes Balears, 2012.

[37] M. C. Soriano et al. “Delay-based reservoir computing: noise effects in a combined analog and digital implementation”. In: *IEEE Transactions on Neural Networks and Learning Systems* 26.2 (2015). DOI: 10.1109/TNNLS.2014.2311855.

[38] A. Rodan and P. Tino. "Minimum complexity echo state network". In: *IEEE Transactions on Neural Networks* 22 (2011), pp. 131–144.

[39] L. Appeltant et al. "Constructing optimized binary masks for reservoir computing with delay systems". In: *Scientific Reports* 4 (2014), p. 3629.

[40] J. Nakayama, K. Kanno, and A. Uchida. "Laser dynamical reservoir computing with consistency: an approach of a chaos mask signal". In: *Optics Express* 24.8 (2016), pp. 8679–8692.

[41] M. Hermans, J. Dambre, and P. Bienstman. "Optoelectronic systems trained with backpropagation through time". In: *IEEE Transactions on Neural Networks and Learning Systems* 26.7 (2015), pp. 1545–1550.

[42] J. Dambre et al. "Information processing capacity of dynamical systems". In: *Scientific Reports* 2 (2012), p. 514.

[43] S. Boyd and L. Chua. "Fading memory and the problem of approximating nonlinear operators with Volterra series". In: *IEEE Transactions on Circuits and Systems* 32.11 (1985), pp. 1150–1161.

[44] H. Jaeger. "Tutorial on training recurrent neural networks, covering BPTT, RTRL, EKF and the 'echo state network' approach". In: *Technical Report GMD Report 159, German National Research Center for Information Technology* (2002).

[45] M. Inubushi and K. Yoshimura. "Reservoir computing beyond memory-nonlinearity trade-off'. In: *Scientific Reports* 7 (2017), p. 10199. DOI: 10.1038/s41598-017-10257-6. URL: https://doi.org/10.1038/s41598-017-10257-6.

[46] G. der Sande, D. Brunner, and M. C. Soriano. "Advances in photonic reservoir computing". In: *Nanophotonics* 6.3 (2017), pp. 561–576.

[47] F. Duport et al. "Fully analogue photonic reservoir computer". In: *Scientific Reports* 6 (2016), p. 22381.

[48] R. Martinenghi et al. "Photonic nonlinear transient computing with multiple-delay wavelength dynamics". In: *Physical Review Letters* 108 (2012), p. 244101.

[49] L. Larger et al. "High-speed photonic reservoir computing using a time-delay-based architecture: million words per second classification". In: *Physical Review X* 7.1 (Feb. 2017), p. 11015.

[50] M. Hermans and B. Schrauwen. "Memory in linear recurrent neural networks in continuous time". In: *Neural Networks* 23.3 (2010), pp. 341–355.

[51] M. C. Soriano et al. "Minimal approach to neuro-inspired information processing". In: *Frontiers in Computational Neuroscience* 9 (2015), p. 68. DOI: 10.3389/fncom.2015.00068.

[52] J. Bueno et al. "Conditions for reservoir computing performance using semiconductor lasers with delayed optical feedback". In: *Optics Express* 25.3 (2017), pp. 2401–2412. DOI: 10.1364/OE.25.002401.

[53] K. Hicke et al. "Information processing using transient dynamics of semiconductor lasers subject to delayed feedback". In: *IEEE Journal of Selected Topics in Quantum Electronics* 19.4 (2013), p. 1501610.

[54] R. M. Nguimdo et al. "Fast photonic information processing using semiconductor lasers with delayed optical feedback: role of phase dynamics". In: *Optics Express* 22.7 (2014), pp. 8672–8686.

[55] R. M. Nguimdo et al. "Reducing the phase sensitivity of laser-based optical reservoir computing systems". In: *Optics Express* 24.2 (2016), pp. 1238–1252.

[56] S. Ortfn, L. Pesquera, and J. M. Gutierrez. "Memory and nonlinear mapping in reservoir computing with two uncoupled nonlinear delay nodes". In: *Proceedings of the European conference on complex systems 2012*. Springer. 2013, pp. 895–899.

[57] S. Ortfn and L. Pesquera. "Reservoir computing with an ensemble of time-delay reservoirs". In: *Cognitive Computation* 9.3 (2017), pp. 327–336.

[58] S. Ortín et al. "Information processing using an electro-optic oscillator subject to multiple delay lines". In: *International quantum electronics conference*. Optical Society of America. 2013, IG_P_7.

[59] P. Nieters, J. Leugering, and G. Pipa. "Neuromorphic computation in multi-delay coupled models". In: *IBM Journal of Research and Development* 61.2/3 (2017), pp. 7–8.

[60] R. M. Nguimdo et al. "Prediction performance of reservoir computing systems based on a diode-pumped erbium-doped microchip laser subject to optical feedback". In: *Optics Letters* 42.3 (2017), pp. 375–378.

[61] A. Rodan and P. Tino. "Simple deterministically constructed cycle reservoirs with regular jumps". In: *Neural Computation* 24.7 (2012), pp. 1822–1852.

[62] A. Akrout et al. "Towards autonomous photonic reservoir computer based on frequency parallelism of neurons". In: *Proc. SPIE*. Vol. 10089. 2017, 100890S.

Laurent Larger

6 Ikeda delay dynamics as *Reservoir* processors

6.1 Introduction

Ikeda dynamics originates from an optical setup proposed in the late 1970s by the Japanese researcher Kensuke Ikeda [1]. It was intended to demonstrate the possibility to generate complex motion such as chaos, in optical systems. Chaotic behavior was indeed harnessed experimentally at that time, being effectively observed in several real world situations in electronics, solid mechanics, fluid mechanics, chemistry, etc., however, not yet in optics. The particular mathematical model describing the Ikeda ring cavity is a delay differential equation involving a sine square nonlinear delayed feedback term. It is very similar to another well-known delay model popularized a few years earlier [2], the Mackey–Glass equation describing the dynamics of blood cells production. The main difference between the two delay models resides in the shape of the nonlinear function, which is a polynomial fraction exhibiting a single maximum for the Mackey–Glass model, whereas a sine function exhibits an infinite number of extrema for the Ikeda model. The Ikeda dynamics, as the slightly earlier model from Mackey and Glass, became a toy model to investigate the richness of delay systems. Beyond the fundamental interest for the understanding of the particular dynamical mechanisms involved in the observed complex behavior, several specifically engineered experiments [3, 4] also triggered some application oriented research. As an example, chaotic motions developed in photonics, together with their synchronization potential, have been useful in the development of cryptographic applications for fiber optics telecommunications [5]. Periodic solutions have also triggered intense research for the design of optoelectronic devices based systems for high spectral purity microwave oscillations intended for radar applications [6]. More recently, similar Ikeda-based optoelectronic setups have been used for the first photonic hardware implementation aimed to demonstrate the capability of complex photonic nonlinear dynamics to efficiently perform machine-learning-based processing of information, according to the Reservoir Computing (RC) concepts [7, 8]. The very first successful hardware implementation of the RC concept was moreover demonstrated one year earlier, involving an electronic circuit also making use of a delayed feedback architecture mimicking the Mackey–Glass model [9]. This chapter is intended to provide a few general concepts related to the particular dynamical complexity of delay dynamics, more specifically the ones based on Ikeda models. Beyond the description of the basic physical properties of such delay systems, we will also illustrate how their intrinsic dynamical features can be interpreted as a way to emulate a virtual neural network, with which the RC concepts can be efficiently implemented.

https://doi.org/10.1515/9783110583496-006

6.2 From the Gedanken experiment to the optoelectronic setup

6.2.1 Ikeda ring cavity, principles of operation

The Ikeda setup is schematically represented in Figure 6.1. It consists of a four mirror optical ring cavity, comprising also a Kerr nonlinear medium inside. Two partially reflecting mirrors are allowing for feeding and extracting light into, and out from, the cavity. Two additional perfectly reflecting mirrors are closing the cavity feedback path. Temporally coherent laser light is assumed, such that the feedback light inside the cavity interferes with the injected light beam. This light interference phenomenon occurs right before entering the Kerr medium. Different interference conditions can be observed, depending on the phase shift cumulated along the cavity (modulo 2π, since thousands of light wavelengths are corresponding to the cavity feedback path). In order to determine this exact interference condition, the optical cavity feedback path can be decomposed into two parts: A fixed part corresponding to the free space propagation, and a variable path originating from the nonlinear phase shift occurring inside the Kerr medium. The latter Kerr phase shift however depends linearly on the intensity level of input light beam to the Kerr medium, i. e., it depends on the intensity level defined by the previously mentioned light interference condition. This intensity is low when the interference is destructive (for a π-phase shift), or high when it is constructive (for a zero-phase shift). At every round trip of the coherent light inside the cavity, a new update of the interference condition might occur, resulting in a new interference intensity level which will be responsible for a new Kerr phase shift, resulting itself in a modified interference condition after the next round trip of the light inside the cavity, and so on.

The observed output light beam appears thus as a continuously changed light intensity, clocked by the round trip time of the light inside the cavity. This round trip time duration is in principle very short, according to the cavity length divided by the extremely fast speed of light in vacuum, typically resulting in a delay of the order of a few nanoseconds. This duration can be typically considered as a time delay separating two successive "updates" of the interference conditions, as it can be observed on the light intensity fluctuations at the output of the cavity. An important dynamical issue to be however considered, is related to the fact this interference condition up-

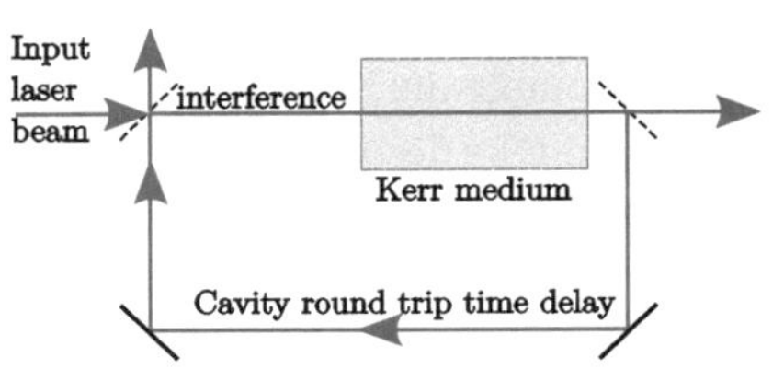

Figure 6.1: Schematic of the Ikeda ring cavity.

date, as well as its actual origin related to the previous round trip Kerr phase change, does not occur instantaneously. Kerr phase change is a light-matter interaction process, definitely occurring at very fast time scales (a few femtoseconds), but not at infinitely fast ones. It is approximately 3–5 orders of magnitude faster than the round trip time. Every Kerr phase change, and thus every resulting interference change and its corresponding light intensity change, occur at finite time scales and in a continuous way, with fastest variation limited by the Kerr medium response time. A self-consistent dynamical state of the Ikeda setup corresponds then to the concatenation of all consecutive infinitesimal temporal changes of the optical phase (or intensity), all over the time interval of a cavity round trip. The resulting dynamics of the phase change, or of its resulting state of interference at the ring cavity input, is actually a continuous time dynamics to be observed as a continuous waveform of either light phase fluctuations, or its consecutive light intensity fluctuations (due to the interference phenomenon). A temporal waveform is permanently flowing inside the optical cavity, one cavity round trip time next to the other. The analysis of the infinite length dynamical waveform will be later on decomposed into finite sub-waveforms, whose duration corresponds to the round trip time delay interval, and whose fine temporal grain fluctuations are ruled – or limited – by the Kerr medium response time.

6.2.2 The all-optical Ikeda setup transposed through an optoelectronic approach

From an experimental viewpoint, the main drawback of the Ikeda setup consists in the actually achievable strength for the nonlinear Kerr phase change, i. e., the phase change span resulting from the minimum and maximum light intensity levels that can enter the Kerr medium. Relatively high power lasers are thus generally required to obtain significant phase change in the Kerr medium, i. e., phase shift of the order of π that is subsequently allowing after the round trips the full scan of constructive and destructive interference conditions. Under this condition only, one can experimentally obtain significantly nonlinear (and complex) dynamics generated in the Ikeda cavity, such as chaotic motions. Such π-phase shifts are unfortunately very difficult to obtain in Kerr media with CW lasers, and most often pulsed lasers are required thus drastically modifying the intrinsic continuous time feature of the original Ikeda idea. Interesting phenomena can nevertheless be obtained, however changing the continuous time dynamics into a discrete time one [10].

Alternative experimental approaches have been rapidly proposed in the literature after Ikeda's original idea. A rather straightforward alternative to the Kerr phase change is the linear Pockels effect, with which a much larger phase change can be obtained (up to several times 2π), with an electrical drive instead of an optical intensity one. This approach is however imposing much slower dynamics, and it is also requiring much longer time delay when one is expecting to keep the same relative ratio between the round trip time (the delay) and the response time for the phase change. Many different experimental solutions can be adopted for the interference phenomenon, e. g., a birefringent interferometer configuration consisting of a birefringent Pockels cell placed between two crossed or parallel polarizers, or a Mach–Zehnder interferometer where one arm involves the electrooptic Pockels effect along a single polarization axis. One then needs a standard photodiode only, in order to detect the intensity of the resulted interference phenomenon. In order to ensure enough feedback gain, some appropriate electronic amplification have to be involved before feeding back the signal onto the electrodes of the Pockels effect medium. Large time delays can be implemented, either through a long enough optical propagation medium in the optical path of the delay dynamics (e. g., optical fiber), or through an electronic delay line placed in the electronic path. Beyond the experimental feasibility, the optoelectronic approach for the Ikeda setup provides nowadays very elegant experimental implementations through the use of modern and broadband Telecom integrated optics devices. This allows to speed up the involved characteristic time scales, thanks to the availability of 10s of picoseconds response time devices. The phase change span, and thus the nonlinear weight of the dynamical process, is moreover still maintained to several πs due to very efficient electro-optic modulation devices. Another practical advantage of the optoelectronic implementation is the availability of various affordable electronic instrumentations (digital oscilloscopes, spectrum analyzers, etc.) allowing for accurate, fast, and easy analysis of the generated temporal waveforms. Time resolutions as fast as a few ps are nowadays available, and the instruments are capable for recording millions of consecutive time delay intervals. Such an efficient instrumentation environment is unfortunately not yet available for femtosecond phenomena, as it would be needed if ultra-fast Kerr media would be involved. Last but not least, in the framework of an information processing system such as a neuromorphic analogue processor based on Reservoir Computing concepts, one has the advantage of an experimental basis that is intrinsically related to an application context derived from information and communication theory. Indeed, through the important development of modern optical telecommunications, photonic systems are benefiting from many high performance and mature technologies dedicated to information processing, filtering, and transmission.

6.3 Modeling and theory

6.3.1 Mathematical model, time scales, motions

According to simple wave physics ruling the ring cavity, and according to the integration of the Maxwell–Bloch equations in the nonlinear light-matter interactions in the Kerr medium, the original Ikeda setup led to the following scalar delay differential

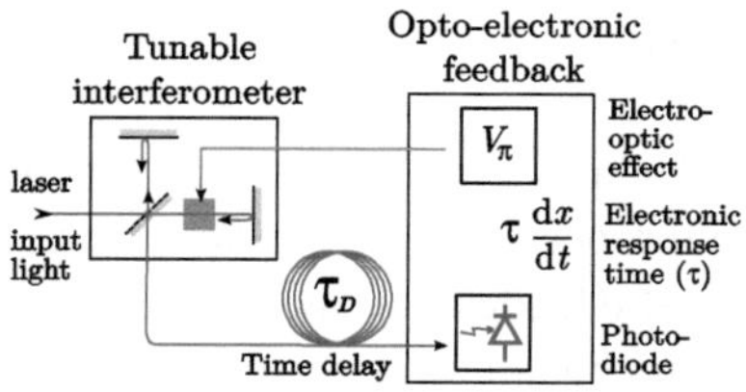

Figure 6.2: Principle of the optoelectronic version for the Ikeda ring cavity.

equation, after a few simplifications:

$$\gamma^{-1}\frac{d\Delta\varphi}{dt}(t) = -\Delta\varphi(t) + A\{1 + B\cos[\Delta\varphi(t - \tau_D) + \varphi_0]\}, \tag{1}$$

where $\Delta\varphi$ is the dynamical variable (phase shift induced by the intensity interference through the Kerr medium), $A = n_2 kLI_0$ is the weight (or gain) for the nonlinear delayed feedback which is directly related to the Kerr effect efficiency (Kerr coefficient n_2, wavenumber k of the laser light, length L of the Kerr medium), B is representative of the contrast of the interference depending on the losses in the cavity, φ_0 is an offset phase related to the static optical path inside the cavity relative to the laser wavelength, γ is the rate of change for the Kerr effect (the inverse being the related response time), and τ_D is the round trip time (or time delay for the dynamics).

The optoelectronic realization of the Ikeda equation, shown in Figure 6.2, follows a very similar mathematical modeling. A simplified normalized equation is usually proposed in the following form:

$$\varepsilon\frac{dx}{ds}(s) = -x(s) + \beta\cos^2[x(s-1) + \Phi_0]. \tag{2}$$

The time variable is here normalized to the delay, introducing the small parameter $\varepsilon = (\gamma\tau_D)^{-1}$ which is representing the finest relative temporal fluctuations within a time delay. The contrast B in equation (1) is usually set to one, since it is rather easy to arrange balanced arms in an interferometer. The weight A of the nonlinear delayed feedback appears now as the factor β of a $\cos^2$-function. Such a writing for the nonlinear function provides a maximum amplitude normalized to unity, and always positive x-values. Parameter β is practically ruled by the different gains involved in the optoelectronic feedback loop: The electrical-to-optical conversion efficiency, corresponding, e. g., to the so-called half-wave voltage V_π involved in an electrooptically tunable interferometer (i. e., the voltage necessary to induce a π-phase shift between the interferometer arms); the optical-to-electrical conversion efficiency, typically the sensitivity of a photodiode; the optical intensity level seeding the interferometer; and the electronic amplification usually needed in the electronic path to properly adjust the voltage level driving the electrode of the electrooptic effect.

The offset phase Φ_0 is a static operating parameter, as φ_0, which can be adjusted independently (through a DC offset voltage applied to a dedicated bias electrode, or any other static tunable optical path).

6.3.2 Linear part of the dynamics

The optoelectronic experimental approach for the Ikeda dynamics conceptually enriched the analysis of Ikeda systems through a signal processing viewpoint, which is complementary to the physics modeling issued from the Ikeda ring cavity design. Indeed, the intrinsic response time of the Kerr effect which is limiting the dynamics, is conceptually equivalent in the optoelectronic approach to the presence of a first-order linear low pass Fourier frequency filter in the electronic path, actually limiting in the same way the bandwidth of the oscillator feedback loop.

The differential equation ruling the nonlinear delayed oscillator, according to this signal processing approach, originates from the linear filter involved in the feedback loop. This leads to a dynamical model obtained through a less conventional approach for physicists, based on theoretical tools brought from signal theory [11]. Equations (1) and (2) can thus be considered as the consequence of a first order low-pass filter. This filter is described in the Fourier domain by the filter function $H(\omega)$, and in the time domain by the corresponding inverse Fourier transform. The latter is known as the impulse response of the filter $h(s)$, i. e., the filter output signal obtained when its input is set to a Dirac distribution:

$$H(\omega) = \frac{1}{1 + i\omega\varepsilon}, \quad h(s) = e^{-s/\varepsilon}\, u(s), \tag{3}$$

where ω is the normalized angular frequency ($2\pi\tau_D f$, f being the Fourier frequency), and $u(s)$ is the Heavyside stepwise function (zero for $s < 0$ and unity for $s \geq 0$). The Fourier analysis of the delay oscillator made through the filter properties, reveals that the cut-off angular frequency of the low-pass filter is $\varepsilon^{-1} \gg 1$. This suggests that many delay modes (integer numbers in the normalized scale in ω) can coexist within the oscillator bandwidth, and they can thus potentially interact through nonlinear mixing present in the feedback loop. Complex motion such as chaos, as well as information mixing, are intrinsically operated by a nonlinear delay dynamics. The temporal description of the delay dynamics can also be reformulated through the use of the impulse response, resulting in an integral writing of the dynamics through a convolution product, instead of a differential equation writing:

$$x(s) = \int_{-\infty}^{s} h(s-\xi) \cdot f_{\mathrm{NL}}[x(\xi - 1)]. \tag{4}$$

The latter equation introduces a more general class of delay equations, simply by extending the linear dynamical process, and keeping the nonlinear delay feedback term. This allows for a more general configuration, in the sense that $h(s)$ is the impulse response related to a linear Fourier filter of potentially any kind. This flexibility is important in the sense it introduces delay dynamics with other kinds of Fourier filters. When dealing with an information processing framework, it is indeed often of interest to select some of the available frequencies, and not others. From a more experimental viewpoint, the optoelectronic setup for the Ikeda dynamics might also involve filtering details that are not of concern in the case of the all-optical Ikeda ring cavity. This was precisely the case when Ikeda-generated broadband chaos was investigated in the framework of its application to secure optical communication. While seeking ultra-fast chaos generation with an optoelectronic chaos, the use of broadband RF amplifiers in the electronic path, necessarily involved a DC nonpreserving device because of the intrinsic low cut-off frequency generally owned by broadband amplifiers (typically 50 kHz–20 GHz instead of DC-20 GHz). This forces the feedback loop to have a bandpass filter feature, which fact led to the discovery of unusual complex and rich dynamics, such as chaotic breathers, low frequency limit cycles, stable period-1 (single delay) limit cycle [12], and also chimera states ([13], see Section 6.4.2). The minimal changes to be adopted in the differential equation in order to take into account the additional high-pass filtering, compared to the conventional low pass case, is to introduce:

- Either an integral term $\delta \int_{s_0}^{s} x(\xi)\,d\xi$ in equation (2) ($\delta \ll 1$ being defined as τ_D/θ where θ is the characteristic response time of the high-pass cut-off filter);
- Or more conventionally, a slow variable $\delta \cdot y$ in the same equation, y being defined through a second differential equation $(dy)/(ds) = x$.

The nonlinear delayed feedback function $f_{NL}[x]$ can be considered with a more generic form for delay differential equations. It can be of any kind in equation (4), whether the ones in equation (1) or (2), or also, e. g., the one involved in the Mackey–Glass model, $\beta \cdot x/(1 + x^n)$.

6.3.3 Feedback and nonlinearity

As presented in the previous section, the delay equations belonging to the Ikeda family have well identified and separated linear and nonlinear parts. The linear part consists in the local time (non-delayed) contribution and can take the form of a linear differential equation. The nonlinear part is concerned by a delayed term. It is worth mentioning that other very popular delay systems show opposite features: in external cavity laser diodes, the local part is nonlinear (laser rate equations), whereas the delayed part is usually a small linear superposition of an external electromagnetic feedback light field [14].

In the presently reported case, nonlinearity occurs in a feedback process. Its role is to bring through the feedback, a possible destabilization of the linear filter (which is usually stable, i. e., in equation (1) one has $\gamma > 0$). The nonlinearity operates in the vicinity of a rest state x_0 defined by a fixed-point equation, $f_{NL}[x_0] = x_0$. It introduces thus an amplitude-gain coupling, since when x_0 is changed by δx, the nonlinearity modifies δx by a factor $f'_{NL}[x_0]$ (the derivative of the nonlinear function evaluated at

x_0). Because of the feedback action, its small amplitude sign strongly matters. Indeed, the effective stability of the nonlinear feedback system (without considering yet any delay) is characterized by a new rate of change $\tilde{\gamma} = \gamma(1-f'_{NL}[x_0])$ (in the case of equation (1) or (2)). This is a simple illustration that negative feedback ($f'_{NL}[x_0] < 0$) generally keeps the system stable when having a small enough gain, but it makes it faster ($\tilde{\gamma} > \gamma > 0$); on the contrary, positive feedback can make it unstable (if $f'_{NL}[x_0] > 1$), as it is also well known from standard proportional control schemes.

The way the rest state x_0 can be set, is thus obviously an important issue for the properties of (delayed) feedback systems. This has a straightforward influence on the amplitude-gain nonlinear coupling, i. e., on the strength of the slope $f'_{NL}[x_0]$. The rest state (and its small amplitude-gain properties, strength, and sign) is typically fixed by setting an offset, which is the static phase parameters φ_0 or Φ_0 in equation (1) or (2). Another important feature for the nonlinear function, is actually the number of extrema it exhibits over the range of amplitudes x that is spanned for a particular solution $x(t)$. This is connected to the notion of multistability (number of solutions of the fixed-point equation $x_0 = f[x_0]$), and to the strength of the nonlinear mixing occurring for a particular solution waveform (this strength can also be evaluated through the degree of the polynomial approximating the nonlinear effect: quadratic when close to an extremum of the nonlinear function, or cubic elsewhere, or even higher degrees for very large amplitude solution waveforms).

For the bandpass linear filter case, it is worth noticing that steady state solution necessarily concerns $x_0 = 0 = (\mathrm{d}y)/(\mathrm{d}t)$. Various amplitude-gain conditions are nevertheless still accessible, through the setting of the offset phase Φ_0.

6.3.4 Delay induced complexity: degrees of freedom, initial conditions, phase space

When the delay in equations (1) or (2) is set to zero (or when the delay can be considered as negligible compared to the other time scale, e. g., the response time γ^{-1}), the dynamical complexity is obviously limited to a scalar nonlinear equation, with a single degree of freedom, eventually with a multistable behavior depending on the number of solutions for the fixed-point equation $x_0 = f[x_0]$. Such a scalar equation has solutions that can be uniquely determined with the definition of a single initial condition, $x_i = x(t = 0)$, leading to a one-dimensional phase space. Its most complex asymptotic state is a fixed point, eventually a limit cycle in the case of a bandpass filter (thus also requiring an additional initial condition value, y_i; The phase space is then 2D).

The presence of a delay (one moreover usually considers large delay situations, where $\gamma\tau_D \gg 1$) changes dramatically the situation in terms of degrees of freedom, in terms of size of the initial conditions, and thus in terms of the actual phase space dimension. This statement is valid even for scalar equations. Indeed, with the presence of a delay, the initial conditions needed to uniquely determine a solution for the delay dynamics, and thus the corresponding phase space required to represent any trajectory, become infinite dimensional: The initial conditions consist in the definition of a functional $\{x_i(t) \mid t \in [-\tau_D, 0]\}$, which consists of an infinite number of consecutive values x_i spanning over a time interval having the length of the delay. The delay is thus an essential ingredient allowing for the development of high-dimensional motions. It provides the possibility for obtaining various waveforms which can expand within a huge phase space, and which are initially determined by functionals that are allowed to adopt an infinite number of possible shapes. This essential property matches nicely the concept of high-dimensional neural networks, that is used in brain inspired computing. A general concept is indeed to expand complex information content into the even more complex phase space of a neural network, so that relevant but initially hidden features of the information can be properly extracted through this expansion. The "Read-Out" operation on this neural network expansion, is an operation to be learned, it often consists in the finding of a specific location in the neural network phase space. This phase space location typically reveals, or isolates, the sought information features through the motion developped within the neural network. The network motion itself is triggered after the input information have been properly injected into the network ("Write-In" operation).

Beyond the magic jump into infinity thanks to the delay, one has to cope more realistically with a few constrains and limitations. The phase space is theoretically infinite, however the actual possible trajectories in delay dynamics, do not necessarily cover the entire available space. One very first obvious constrain lays in the necessarily continuous waveforms that are obtained from a delay differential dynamics (it is not a discrete time map, in which the dynamical variable can switch discontinuously from one value to another, through the iteration function). Differently speaking, the fastest amplitude change is limited by the response time γ^{-1} of the differential process. In the Fourier domain, this corresponds to the high cut-off frequency of the low pass filtering: High frequencies are filtered out, and this forces smooth waveforms only for the dynamics, with fastest oscillations that are limited to oscillation frequencies of the order of γ (in rad/s). This has, e. g., direct consequences on the actual (fractal) dimension exhibited by a chaotic waveform. Such dimensions are indeed finite, and ruled by $\beta\gamma\tau_D$ [15] in Ikeda dynamics, which can be commented as follows: $\gamma\tau_D$ is roughly the number of fast γ^{-1}-oscillations one can fit within a time delay interval; The factor β has then a kind of dimension-amplification effect, through the nonlinear mixing operated by the extrema of the unity amplitude $\cos^2$-function. This upper bound for a reachable dimensionality defines a smallest temporal grain which can be effective only, within a motion developed by the delay dynamics. To further comment the reasons for the dimensionality limitation, one could also refer to the Shannon sampling theory, qualitatively stating that the information content of a delay dynamics motion is limited to the number of a γ^{-1} time scale (the equivalent of a sampling period) that one can fit into a time delay interval. When one wants to properly inject ("Write-In") information to be processed by a delay dynamics, thus expanding this information in the delay system phase space, there is an optimum way in terms of temporal density of information that can match the actually available dimensionality in the delay Reservoir. A too slow sampling rate cannot benefit from all the actual dimensions of the delay dynamics; With a too fast injection rate, one would over-estimate the actually available dimensions, and the corresponding information would be too strongly filtered out by the delay dynamics. Empirical studies [9] for this "Write-In" sampling period have found an optimal value around $(5\gamma)^{-1}$. One could notice here that this quantity is related to the width of the impulse response function $h(\cdot)$ introduced in equations (3) and (4).

6.4 Emulating a dynamical network with a delay system

In the previous section, we have qualitatively introduced the physical and theoretical representation of delay dynamics, revealing their intrinsic high-dimensional character. This specific character was connected to other dynamical systems exhibiting also high-dimensional features, such as a network of neurons. Neural networks have however a very different architecture, since they form a spatially extended network of interconnected dynamical nodes, the neurons. Beyond the very qualitative fact that high-dimensionality is interesting for processing information in a neural network, we would like to point out in this section a closer analogy between delay dynamics and a network of coupled dynamical nodes, e. g., neural networks. This will then be used later to more clearly explain the way delay dynamics can be used as an experimentally efficient and tractable *Reservoir* [16] for processing information, in a way which was initially introduced under the naming *Echo State Networks* (ESN) [17] and *Liquid State Machines* [18].

6.4.1 Space-Time representation of a delay system

Space-Time representation was proposed for delay dynamics in the early 1990s [19]. It allowed to provide a visual representation of complex motions emerging in delay dynamics, allowing to reveal hidden global temporal structures attached to deterministic origins, the latter being however related to very different time scales (delay, and characteristic response times). This early work on Space-Time motions for delay dynamics was concerned by experimental results obtained from a CO_2 laser with a delayed intra-cavity loss modulation (the model being actually close to the Ikeda dynamics).

Since delay dynamics are physically temporal motion only, the idea guiding the principle for their Space-Time representation, is to decompose the time variable s into a combination of two significantly different time scales: one playing the role of a virtual space variable and another one keeping its temporal meaning. These two very different (nearly "separated") time scales need to be chosen such that they properly reflect the underlying physics of a delay dynamics, which is intrinsically a multiple time scale dynamics. According to what was introduced in the previous sections on delay dynamics modeling, a straightforward choice is to highlight:

- The delay time τ_D as a slow "discrete" time scale (τ_D is even converted into unity after normalization, thus pointing onto an integer number counting the successive iterations after each round trip within the ring or feedback architecture);
- And the response time γ^{-1} as a fast "continuous" time scale (changed into ε when normalized).

The fast time fluctuations are thus considered as filling the virtual "space" consisting of a time delay interval. The evolution of the amplitude along this virtual space is ruled by the iteration from one time delay interval to the next. Mathematically written, the normalized time variable s reads as follows after being decomposed into the two time scales:

$$s(n,\sigma) = n(1+\nu) + \sigma \quad \text{with:} \qquad (5)$$
$$n \in \mathbb{N}, \quad \sigma \in [0;1+\nu] \quad \text{and} \quad \nu = O(\varepsilon).$$

One can notice here that there is a small (normalized) time shift ν in the choice for the virtual space span. This virtual space domain thus does not match exactly the unity normalized time delay, but it deviates with a small quantity of the order of $\varepsilon = (\gamma\tau_D)^{-1}$. This deviation finds its origin in the fact that the maximum absolute time correlation from one round trip to the next does not amount exactly to the time delay. This small shift can be interpreted as an additional small contribution to the effective delay, brought by the group delay of the linear filter (thus explaining why this is a $O(\varepsilon)$-small quantity). The fact that both delay and response time appear together in the space span, is also an indication that the two time scales indeed have a coupled influence on the whole behavior generated by the delay dynamics.

The introduction of these new space and time variables for a delay system, results then in a convenient graphical representation for any solution waveform $x(s) = x(\sigma, n)$, in a 3D-space (Figure 6.3): A color-encoded amplitude is attributed to the dynamical variable x, which is considered as depending on both a continuous space variable σ (horizontally) and a discrete time variable n (vertically). The Space-Time plane should actually be viewed (it is however less convenient) as a weakly twisted cylinder, due to physical continuity condition. Indeed, the end of a waveform at $x(1+\nu, n)$ is physically continuously connected to the beginning of the next delay-waveform starting at $x(0, n+1)$.

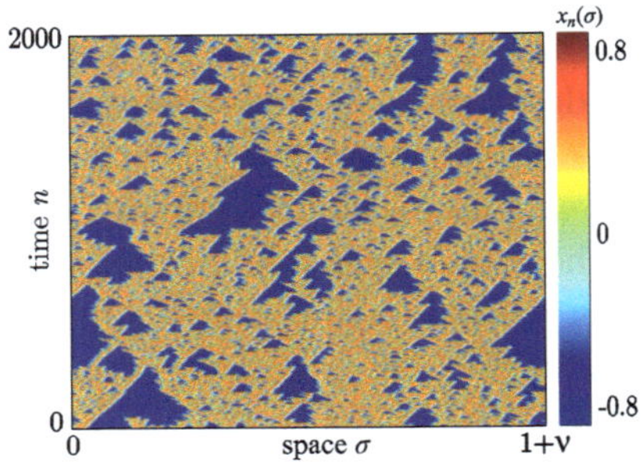

Figure 6.3: Complex chaotic pattern for a delay dynamics, which is revealing specific structures when represented in an appropriate Space-Time domain (σ, n).

The determination of the numerical value for ν is an important issue to be solved, when nice patterns have to be revealed. Indeed, a small positive or negative deviation from the correct value of ν, results in a very sensitive left or right tilt along the discrete time axis n. The ν-mismatch can rapidly cumulate after the successive delay round trips. The correct value for ν is found to depend also on the type of solution (actually determined by all amplitude and time parameters, as well as by the initial conditions), and not on the temporal parameters of the model only $(\varepsilon, \delta, \ldots)$. Specific algorithms for the detection of pattern orientation are typically used in practice, in order to properly adjust the value of ν.

6.4.2 An illustration: chimera states in delay dynamics

Chimera states are corresponding to particular behaviors occurring in a network of coupled oscillators, in which the global network motion is self-organized into clusters of oscillators. Each cluster is adopting incongruent motions, in a stable coexisting fashion. The oscillators motions between clusters are incongruent, while the motions within a cluster are congruent over all oscillators of the considered cluster. The original discovery by Kuramoto in 2002 [20] was considering a network of identical oscillators, each oscillator being similarly coupled to many of its neighbors, according to a coupling function depending on the distance between the oscillator and the considered neighbor oscillator. One typical model constructed to explore "chimera states," which was named as such in 2004 [21], consisted in a network of continuously distributed Kuramoto oscillators:

$$\frac{\partial \phi}{\partial t} = \omega_0 + \int G(x - \xi) \cdot \sin[\alpha + \phi(t, x) - \phi(t, x - \xi)] \, d\xi, \tag{6}$$

where ω_0 is the natural angular frequency of each oscillator, $G(\cdot)$ is a coupling strength function between oscillators depending on their relative distance, and α is an offset phase for the nonlinear coupling. The nonlinear coupling is a sine function depending on the oscillators relative phase. One should notice that chimera states can be obtained only for specific α-values within a small interval, i. e., for a specific operating point of the nonlinear coupling function.

Since the whole network is assumed to be homogeneous by construction, two typical asymptotic network motions are a priori expected, in order to respect the intrinsic symmetry of the network: Either all oscillators are synchronized and coherent, or all oscillators behave in a fully incoherent and desynchronized way. The numerical observation of chimera states was thus corresponding to an unexpected spontaneous symmetry breaking situation, where different incongruent behaviors are coexisting and are gathered into clusters. The phenomenon has attracted lots of attention (for a review see [22]) and continues to do so. The two first experimental observations have been discovered in 2012 [23, 24] in the field of optics and chemistry, followed then by several other experimental discoveries. Among these additional experimental observations of chimera states, an unexpected candidate appeared: delay dynamics, first implemented electronically [13], and slightly later also implemented in a tunable laser diode dynamics performing an Ikeda-like bandpass oscillator [25]. The search for chimera states in delay dynamics was actually motivated by the curiosity to understand how effective their Space-Time representation could be, typically in exhibiting motions that are considered to be specific to spatiotemporal phenomena.

From the more theoretical point of view, manipulating the convolution integral model of delay dynamics in equation (4), even allowed to derive a closely connected mathematical writing of delay dynamics, compared to the Kuramoto oscillator network. The idea was to rewrite equation (4) considering the specific properties of a delay dynamics in the framework of its Space-Time analogy. Doing so, one tried to highlight the importance of the time delay interval $[s-1; s+\nu]$ along which the virtual spatial domain spans, and to consider the discrete time iteration from one round trip to the next. The following mathematical description of a delay system was obtained, whose formulation can be compared with a chimera toy model as the one given in equation (6):

$$x(n+1,\sigma) = I[x(n,\sigma)] + \int_{\sigma-1}^{\sigma+\nu} h(\sigma+\nu-\xi)\cdot f_{\mathrm{NL}}[x(n,\xi)]\,\mathrm{d}\xi, \tag{7}$$

where $I[x(n,\sigma)]$ is a complex, but negligible integral term.[1]

To point out a comparison between the models in equations (6) and (7), one can simply identify the different terms in both equations. They both describe an update rule for the evolution of each dynamical system. They both involve integral terms cumulating contributions over their respective spatial domains (i. e. the virtual space interval $[\sigma-1; \sigma+\nu]$ for the delay dynamics). Both integrals exhibit two characteristic

---

1 The integral term $I[x(n,\sigma)]$ cumulates over the integration domain $]-\infty; n-1]$ the bounded fluctuations of $f_{\mathrm{NL}}[x(s-1)]$, weighted by the impulse response $h(s)$; Since $h(s)$ exhibits typically an exponential decay in s/ε, and since $s \gg 1 \gg \varepsilon$ over the concerned integration domain, the integral term is a very small quantity.

factors. The second factor appears as a nonlinear coupling from any oscillators in the whole space ($\xi \in [\sigma - 1; \sigma + \nu]$ for the delay dynamics), at a fixed time. The first factor ($h(\cdot)$ for the delay dynamics, and $G(\cdot)$ for the network of coupled Kuramoto oscillators) does not depend on time, but on space only; it acts as a weighting factor for the nonlinear coupling. In the delay dynamics case, this weight depends on the virtual distance $\sigma + \nu - \xi$ between the coupled oscillators. One can notice that since $h(\cdot)$ has localized nonzero amplitude when its argument is close to zero, this reveals that essentially the virtual oscillator positions ξ close to $\sigma + \nu$, have a significant contribution to the dynamics. The impulse response $h(t)$ appears thus as a distance dependent weighting function for the coupling between oscillators, by analogy to the function $G(x - \xi)$ in equation (6).

In order to observe chimera motion in delay systems, it was found that two particular delay dynamics settings are important to pay attention to. First, the linear filter needs to be bandpass, which was known to stabilize the period-1 limit cycle [26]. This period-1 waveform is indeed found to be the carrier waveform for chimera states in delay dynamics. A low pass configuration would not allow for such a stable carrier pattern, as it was observed in [27], where systematic coarsening of any complex initial condition functional was reported. Beyond this stability argument for the carrying pattern, one could also discuss the effect of a bandpass filter compared to a low pass one, in terms of the impulse response width. This width is indeed broader for the bandpass case, which would mean that the coupling function allows more influence from farther dynamical nodes in the network. Chimera states are indeed known to be favored when the coupling weight is active over longer distances.

Finally, another important feature for the occurrence of chimera in delay dynamics has been noticed, it concerns the shape of the nonlinear coupling function $f_{\mathrm{NL}}[x]$. This shape requires asymmetric maxima and minima, but this condition can not be fulfilled by the standard cos-function of the original Ikeda equation. For the tunable laser setup [25], this has motivated the use of an Airy function provided by a Pérot–Fabry resonator, instead of a two-wave interference device (e. g., birefringent filter).

Under such conditions for an Ikeda-like delay dynamic, it was possible to obtain both numerically and experimentally nicely controlled chimera patterns, that were nearly magically revealed thanks to the adequate Space-Time representation. Chimera patterns manifest themselves essentially as an alternation of plateaus and chaotic sequences. This is clearly seen in the time domain, from which it is however difficult to extract the sustained repetition of this alternation. Space-Time representations can then easily reveal the regularity of their support waveform with $(1+\nu)$-periodic carrier (see Figure 6.4 for an example).

Depending on the operating point in (ε, δ)-parameter space, a complex but deterministic organization of these chimera patterns was found, concerning the number of possible chimera "heads" within the virtual space. In the framework of the reported analytical analogy with Kuramoto oscillator network, this parameter space is directly

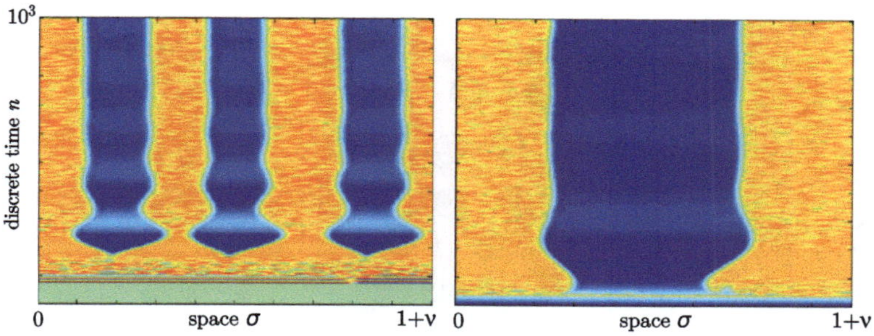

Figure 6.4: Two emerging chimera states (3-headed and single-headed) from different initial conditions, but same operating parameters, as observed in the Space-Time domain (σ, n).

related to the properties of the coupling weight (function $h(\cdot)$) and its span in the neighborhood of each individual dynamical node of the network.

Last but not least, recent results [28] also showed the possibility to obtain two-dimensional virtual space through the use of a second delay. This second delay was chosen significantly larger than the first one, thus emulating another virtual spatial dimension. This additional spatial dimension was still able to host coherent recurrences over both virtual spatial dimensions. Chaotic islands in a quiet sea (2D-plateaus), or its symmetric version, were successfully obtained.

Finally, an important remark has to be highlighted about the controllability, and the excellent matching accuracy of the observed chimera states in delay dynamics experiments when compared to numerical results. The homogeneity of the node dynamics and of their coupling over the whole network forms an important assumption for the interest of chimera states. With inhomogeneities, symmetry breaking would indeed not be that surprising. Such a theoretical assumption is easy to force in numerics, but it is in general difficult to fulfill in real world experiments. In delay dynamics, however, the virtual nature of the emulated dynamical nodes makes delay dynamics highly appropriate to this homogeneity condition. Indeed, the direct consequence of the virtual nature of the network, is that a single physical node is practically shared among the whole virtual network. This construction forces any dynamical node, as well as its coupling environment, to be exactly the same everywhere in the virtual network.

6.4.3 From autonomous to nonautonomous delay dynamics

Most of the previous sections have presented some basic concepts and properties about nonlinear delay dynamical systems. They are considered as complex autonomous dynamics capable to give rise to various motions, depending whether on their intrinsic parameter settings (time and amplitude parameter values), or on the shape of the functions involved in the feedback loop (nonlinear transformation, or linear filter). When an application is concerned through the practical use of a delay

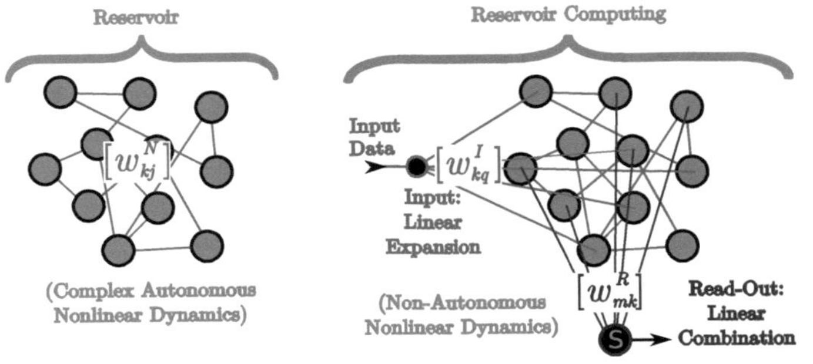

Figure 6.5: A complex dynamical network of coupled nonlinear nodes, autonomous, and non-autonomous for Reservoir Computing application.

system, the dynamics has to be considered in interaction with the external world, i. e., the dynamics becomes necessarily nonautonomous (see Figure 6.5). For chaos encrypted transmission system, the information to be encoded into the chaotic waveform can be typically injected directly into the delay dynamics [29, 30], thus strongly perturbing the chaotic motion itself. This is especially true when the relative amplitude of the information signal is comparable to the stand-alone autonomous chaotic waveform. For optoelectronic delay oscillators generating a high spectral purity microwave signal for Radar applications, the external noise perturbing the periodic oscillation is the small signal to be carefully considered in order to analyze the effective spectral properties of the oscillator [31]. Finally, the nonautonomous operation of a delay dynamics for Reservoir Computing processing [9] is even more pronounced, since the most important motion is the large amplitude transient generated by the Reservoir dynamics (the nonlinear delay system in our case). This transient is the dynamical response of the Reservoir to a large amplitude external drive. The driving signal corresponds to the encoded information to be processed, and it has to be injected onto the various dynamical nodes forming the Reservoir (a dynamical network), as illustrated in the right picture of Figure 6.5. The last section will now enter into the main objective of this chapter, the use of Ikeda dynamics for Reservoir Computing applications. The Reservoir operation of a delay dynamics will be particularly emphasized through the emulation of such dynamics as a virtual network of dynamical nodes. Despite being an emulation of a real dynamical network only, it is expected to be capable for playing the role of the usual neural network that is naturally and mostly considered in brain inspired computing approaches.

6.5 Ikeda-based photonic Reservoir

Reservoir Computing (initially named as Echo State Network or ESN [32] and Liquid State Machine or LSM [18]), is originally thought as a modified recurrent neural network computing concept. It is thus obviously implemented in the framework of a dynamical system constructed as a standard network of coupled dynamical nodes, as usually considered in the neural network computing community. We will briefly recall one popular model in Reservoir Computing, the Echo State Network. We will then propose a transposition of this model within the framework of the Space-Time analogy previously introduced for delay dynamics. Finally, a few examples of physical implementations for photonic Reservoir Computing, inspired by Ikeda dynamics, will be described.

6.5.1 Standard ESN for Reservoir Computing

In a simplified view, an ESN-based Reservoir Computer can be described as follows. The input information vector $\mathbf{u}(n) \in \mathbb{R}^Q$ to be processed, has to be distributed onto each node of the Reservoir. The Reservoir is a recurrent neural network formed by nodes, each following the perceptron model. The input information distribution is performed according to a so-called input connectivity matrix $W^I \in \mathbb{R}^K \times \mathbb{R}^Q$ (this is sometimes referred to the input layer, or "Write-In" layer). The Reservoir itself is ruled by a discrete time map dynamics of the network state vector formed by its K node amplitudes, $\mathbf{x}(n) \in \mathbb{R}^K$. The network mapping is performed according to its internal connectivity matrix $W^N \in \mathbb{R}^K \times \mathbb{R}^K$, which is ruling the iterative law of each node receiving signals from both the other nodes (autonomous dynamical contribution), and the injected inputs (nonautonomous contribution). The output result $\mathbf{y}(n) \in \mathbb{R}^M$ to be computed by the Reservoir Computer, is obtained through the "Read-Out," or output layer, through a linear combination of the internal Reservoir states. This is performed through the matrix $W^R \in \mathbb{R}^M \times \mathbb{R}^K$. The latter matrix is actually the one to be learned during the training stage. This training usually consists of a simple and fast linear regression method, making use of known pairs of data set {*input data, network response, target output*}, in the case of supervised learning. More details are provided in other chapters of this book, which are more focusing on the basic concepts of Reservoir Computing.

Considering individual signals processed in the ESN, the mathematical operations can be written as follows:

– Input amplitude affecting node k in the Reservoir:

$$u_k^{\text{in}}(n) = \sum_{q=1}^{Q} w_{kq}^I \, u_q(n) \tag{8}$$

$\mathbf{u}^{\text{in}}$ is thus a vector with K coordinates, one for each node in the network.

– Reservoir update rule for the ESN:

$$x_k(n) = f_{\text{NL}}\left[\sum_{j=1}^{K} w_{kj}^N \, x_j(n-1) + \rho \cdot u_k^{\text{in}}(n)\right], \tag{9}$$

where ρ is a scaling factor weighting the nonlinear contribution of the injected information into the Reservoir.

– Learned output (linear combination of the ESN internal state):

$$y_m(n) = \sum_{k=1}^{K} w_{mk}^R \, x_k(n). \tag{10}$$

The learning, as already mentioned, is intended to determine the coefficients of the Read-Out matrix. In principle, these coefficients apply to both input data and Reservoir responses. For sake of simplicity, and according to successful empirical trials, one can restrict the action of the Read-Out to the Reservoir response only. In the case of a classification problem, one makes use of a training set of L input vectors $\{\mathbf{u}^l(n) \mid l \in [1,\dots,L], n \in [1,\dots,N_l]\}$ and their corresponding target answers $\tilde{\mathbf{y}}^l(n)$. Each input leads to a Reservoir phase space trajectory $\{\mathbf{x}^l(n) \mid n \in [1,\dots,N_l]\}$. A concatenated $K\times N_l$ matrix A^l formed by the N_l horizontally stacked vectors $\mathbf{x}^l(n)$ (for all $n = 1,\dots,N_l$) is typically constructed for each input information sequence $\{\mathbf{u}^l(n)\}$. It is associated to a similar $M \times N_l$ target matrix $\tilde{B}^l$ obtained from the concatenation of target vectors sequence $\tilde{\mathbf{y}}^l(n)$ of the same duration. Each of these matrices are further concatenated horizontally for all L input vectors of the training set, thus leading to a single $K\times(\sum N_l)$ Reservoir state learning matrix A, and a single $M \times (\sum N_l)$ target matrix $\tilde{B}$. A correct Read-Out matrix for the training set is expected to verify $W^R \cdot A = \tilde{B}$. When inverting this ill-posed linear problem, one can try a ridge regression with a regression parameter λ, resulting in the following calculation for obtaining an optimal Read-Out matrix:

$$W_{\text{opt}}^R = \tilde{B} A^{\mathrm{T}} \left(A A^{\mathrm{T}} - \lambda I\right)^{-1}, \tag{11}$$

where the matrix inversion is calculated numerically, e. g., from a standard Moore–Penrose algorithm. To test the performance of the learned matrix, one proceeds then with untrained input data $\mathbf{u}^u(n)$ as follows:

– Inject these data into the Reservoir according to the same encoding as in equations (8) and (9),
– Record the Reservoir nonlinear transient response $\mathbf{x}^u(n)$,
– Format the response into a matrix A^u,
– Calculate the resulting output B^u using W_{opt}^R,
– Compare with the right answer $\tilde{B}^u$,
– And extract an error measure.

Error rate is obtained from processing and evaluation of several untrained data.

6.5.2 Transposing the ESN model to delay dynamics

The transposition of a Reservoir Computing approach from the ESN model to an Ikeda dynamics one is supported by the previously introduced Space-Time analogy of a delay dynamics. It will make use of the writing in equation (7) and will extend it to the case of a nonautonomous operation (external information injected into the dynamical system). The analogy is unfortunately not perfectly matching the ESN framework, since in delay dynamics, the time n only is discretized. The space σ is unfortunately continuous in delay systems, whereas it is discrete in the ESN model as indicated by the spatial position integer $k = 1, \dots, K$.

One straightforward solution is then to sample the continuous space of a delay dynamics with a sampling period μ. This sampling period will correspond to a spacing between two adjacent nodes, and the quantity $\mu^{-1}(1+\nu)$ will correspond to the actual number of virtual nodes (taking the integer part) in the emulated neural network over the round trip time interval. In this new framework, the normalized time variable s is rewritten as depending now on two discrete coordinates, space k and time n:

$$\begin{aligned} & s(n,\sigma) = s(n,\sigma_k), \quad \text{where} \quad \sigma_k = (k-1)\cdot\mu \\ & s(n,k) = n\cdot(1+\nu) + (k-1)\cdot\mu, \\ & \text{with:} \quad n \in \mathbb{N}, \quad k = 1,\dots,K \quad \text{and} \quad K\mu = 1+\nu. \end{aligned} \tag{12}$$

The former continuous space dependency for delay dynamics can now be transformed into a discretized version: $x(n,\sigma) = x(n,(k-1)\mu) = x_k(n)$, thus proposing a one-to-one notation correspondence with the ESN node amplitude in equation (9).

It should however be noticed that this fully discretized vision for a delay dynamics is more justified by a notation convenience in order to connect delay dynamics to ESN, than supported by a rigorous mathematical approach. Indeed, physical delay differential dynamics stay a continuous time dynamical system, due to its intrinsic differential nature. The transient waveform it generates from any input, keeps continuous, and the discretized vision is mostly a representation convenience.

Let us now rewrite the Reservoir Computing processing model in equations (8) to (10), for the case where delay dynamics is used instead of an ESN.

– Input amplitude affecting node k in the Reservoir:

$$u^{\text{in}}(n,\sigma) = \sum_{k=1}^{K}\left[\sum_{q=1}^{Q} w_{kq}^{I}\, u_q(n)\right] p_\mu(\sigma-\sigma_k), \tag{13}$$

  where $p_\mu(\cdot)$ is a temporal window of width μ, starting at zero. This function is well known in conventional sampling theory, as a zero-order sample and hold process. Such a formulation allows to translate mathematically the actual piecewise constant input waveform, in which each constant amplitude at the virtual space position k is addressing node k, according to the same input connectivity matrix W^I as the one defined for equation (8). One could notice here that addressing temporal position with the desired amplitude is a well-known technique in communication systems, called Time Division Multiplexing (TDM).

- Reservoir update rule for the delay dynamics

$$x_k(n) \simeq \int_{\sigma_k-1}^{\sigma_k+\nu} h(\sigma_k + \nu - \xi) \cdot f_{\mathrm{NL}}[x(n-1,\xi) + \rho \cdot u^{\mathrm{in}}(n,\xi)] \, \mathrm{d}\xi. \tag{14}$$

  This is a formal writing of the delay Reservoir dynamics, obviously derived from equation (7) for the case of a nonautonomous operation according to RC concepts. In effective simulations, one practically performs the numerical integration of the delay differential model, in which the injected information $\rho \cdot u^{\mathrm{in}}$ is added to the argument of the nonlinear delayed feedback function $f_{\mathrm{NL}}[\cdot]$.
- Learned output (linear combination of the delay dynamics state vector coordinates, whose discrete values are obtained by sampling the response of the delay dynamics):

$$y_m(n) = \sum_{k=1}^{K} w_{mk}^R \, x_k(n). \tag{15}$$

  The Read-Out is then exactly the same expression as for the ESN. However, one has to remember here that the Reservoir response is a continuous waveform. Thus the sampling at the output does need necessarily to be synchronized with the one used at the input for addressing each virtual node through a TDM principle. Small desynchronizations, or time shifts, are constituting a priori degrees of freedom that one can explore at the Read-Out layer. It was indeed found under some specific computing tasks (classification for speech recognition) that small sampling frequency deviations between "Write-In" and "Read-Out" can lead to significantly improved RC performances [33].

The reported possible analogy between delay dynamics emulating a virtual network, and an ESN, was definitely a very convenient way to harness Reservoir Computing with physical implementations. Through this, it was possible to demonstrate that analogue hardware can efficiently implement the concept of RC. This transposition of ESN to analogue physical hardware has however up to now well identified limitations. Some limitations are related to the fully discrete nature of an ESN, whereas delay dynamics are intrinsically continuous and analogue solutions. There are thus necessarily lots of horizons and directions for deeper investigating more advanced and better matched hardware configurations. A lot of processing steps with RC physical hardware, among which are delay dynamics, are still performed off-line in conventional computers or on-line in digital boards like FPGA: Information expansion into the network, training, Read-Out, and so on. Digital approaches are obviously straightforward for various implementations of these steps. They however suffer, through their necessarily finite and discretized representation of the information, from an intrinsic simplification of the actual complexity developed by natural neural networks like the biological brain. The most important challenge in exploring analogue hardware mimicking the way the brain is processing information, is probably the opportunity to understand its fundamental mechanisms, and to find a way on how to open the black box of the fundamental processing and computing mechanisms of biological neural networks.

In the last section, we will present a few experimental implementations of Reservoir Computing with photonic Ikeda-like dynamics, more precisely the ones performed at the FEMTO-ST institute. They belong to the very first contributions to the field of photonic RC, together with other research groups (many of them being co-authors of this book). They correspond necessarily to partial achievements for the RC hardware implementation. Lots of progress is still to be done, which makes the topic highly challenging from many scientific areas, from physics of complex systems, nonlinear dynamics, to brain processing concepts, through mathematics [34, 35] and information theory.

6.5.3 Examples of Ikeda-based photonic RC implementations

Photonic RC is by far not limited to Ikeda models. As already mentioned, the very first delay-Reservoir [9] was proposed in electronics, building a circuit reproducing the Mackey–Glass model, which is nevertheless very close to an Ikeda dynamics, since the nonlinearity only is changed into a single maximum function instead of the multiple extrema provided by a sine two-wave interference function. Photonic Reservoir, to mention only a few, was also proposed with external cavity laser dynamics [36], network of integrated SOA [37], fiber ring laser with SOA [38], network of photonic crystal structures [39], network of coupled ring resonators [40], network of integrated coupled delay lines [41], etc. Ikeda-like dynamics have the great advantage to have a system-based approach, with an excellent experimental control accuracy and flexibility. They are usually based on commercially available devices, which one needs to arrange in the desired way in order to explore the properties closely matching a simple dynamical model. We will start by presenting the most popular and broadly used Ikeda dynamics, the one involving an electrooptically tunable interferometer [3, 4, 6, 42, 43, 44, 45, 46].

6.5.3.1 Intensity

The electrooptic intensity modulation was the first experimental approach used to demonstrate efficient photonic Reservoir Computing [7, 8]. The ease of handling and controlling of this setup is mostly explaining why it was adopted and why it led so rapidly to a successful demonstration. A schematic of the setup, and a picture of its realization, is represented in Figure 6.6. A 4-km monomode Telecom fiber spool (20 μs delay) was used with a Lithium Niobate ($LiNbO_3$) integrated optic Mach–Zehnder modulator (rf half-wave voltage V_π of 4.2 V). The latter was seeded by a standard

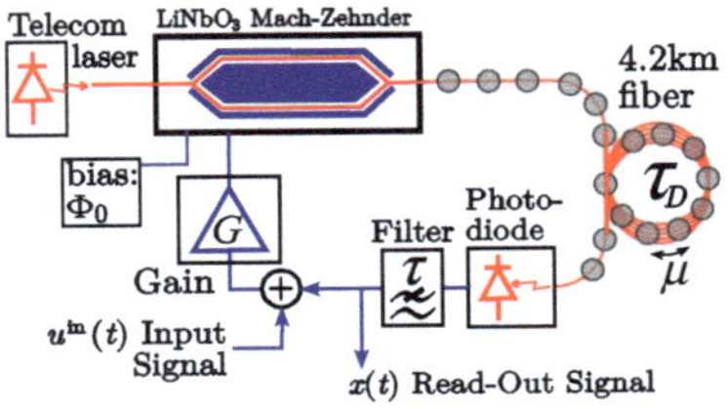

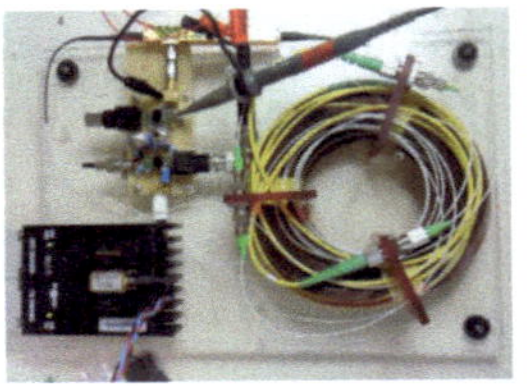

Figure 6.6: Photonic Reservoir computer based on an integrated optics Mach–Zehnder modulator, and a 4 km fiber delay line along which virtual nodes are distributed.

1.5 μm and 20 mW Telecom laser diode. The fiber output was simply detected by a photodiode, whose electrical signal was then processed with standard home-made electronics, involving simple but flexible low frequency operational amplifier circuits. The circuits were aimed to provide the following functions:

- Low-pass filter @ ca. 1 MHz;
- Extract the delay-reservoir response signal $x(t)$ for monitoring and recording it in the large memory of a digital oscilloscope;
- Add the external input information signal to be processed; This signal was generated by an arbitrary waveform generator (AWG) in which the encoded TDM signal $u^{\text{in}}(t)$ was properly programmed, with the adequate sampling rate for the correct node spacing of the experiment, and with the appropriate amplitude allowing for the optimum voltage span (amounting slightly above V_π for this input signal peak-to-peak amplitude);
- Amplify and drive the 50 Ω matched input of the Mach–Zehnder modulator, with the appropriate voltage span (up to 12 V, for sufficient nonlinear operation along the sine nonlinear transformation).

The operating point of the Mach–Zehnder (parameter Φ_0 in equation (2)) was set through the control of a DC voltage applied to the separated bias electrode of the Mach–Zehnder. This was maybe the most critical point to pay attention to, since bias in $LiNbO_3$ integrated devices can have potentially significant drifts, depending on the fluctuations of the environmental conditions. When operating with long sequences of data to be processed, "silences" (typically zero level input information $u^{\text{in}}(t)$ at the end of a sequence) were carefully monitored, in order to ensure that the static level of the photodiode signal was maintained to a constant value (i. e., proportional to $\cos^2 \Phi_0$) during these quiescent phases over the whole processing time. The best RC processing for parameter Φ_0 was empirically obtained (confirmed by numerics) for a value resulting in an average operating point of the Mach–Zehnder modulator (i. e., the $\cos^2$-function) close to an extremum (but not exactly at the extremum). This was interpreted as the requirement for the nonlinear function to provide a rich equivalent polynomial behavior, linear, essentially quadratic, and partly cubic depending on the drive amplitude applied to the Mach–Zehnder.

The parameter β in equation (2) was controlled through the laser diode operating optical power, since the normalized gain parameter β is directly proportional to this optical power. The first instability threshold of the autonomous delay oscillator (most often the Hopf instability) is known to occur when $|\beta \sin(2\Phi_0)|$ is close to unity. Depending on the chosen Φ_0 parameter (defining the local profile of the nonlinear feedback transformation), one typically increases the optical power from zero, up to the birth of the typical two-delay periodic oscillation (when operating along a negative feedback slope of the nonlinear function). The optimal β value was empirically found (both numerically and experimentally) to be 0.7 times lower than the Hopf threshold.

According to the time parameters of the setup, we adopted a scaling of the TDM data injection such that the emulated network had $K = 400$ virtual nodes, corresponding to the empirically found optimal value for the data injection sampling, $\mu \simeq \varepsilon/5$ (see equation (12)).

The reference benchmark test explored through the here reported Ikeda-based Reservoirs, is the speech recognition of a data set extracted from the TI-46 database (500 spoken digits from zero to nine, uttered 10 times by 5 different female speakers). For this particular test (notice that other benchmark tests have been successfully explored, such as time series prediction), the Mach–Zehnder Reservoir enabled to obtain a word error rate (WER) very close to zero, and sometimes reaching zero for the best processing trials. We used an arbitrary splitting of the 500 spoken digits partition into 20 subsequences of 25 spoken digits. Practically, 475 (19 subsets) were used for training with equation (11) performed off-line in a computer, and the remaining 25 digits were used for testing (off-line as well, using the W^R_{opt} found after training and applying it to the recorded Reservoir responses of the 25 test digit sequences). We used cross-validation in the selection of the 19 sequences for training and one for testing, such that each digit was used once in the testing set.

It is worth mentioning that the practiced speech recognition test was chosen for comparison issues, and not for state of the art performance motivation. Currently, available commercial speech recognition softwares are far more advanced and performing much better on much more complex and difficult database. Modern software systems are however making use of incomparably more computational power with respect to the very first and very simple photonic Reservoir Computers, and they are also benefiting from decades of accumulated digital processing research experience on speech recognition software. It is anyway surprising that so simple systems as delay-based Reservoirs, compared to the current high abstraction level speech recognition solutions, are able to obtain very interesting performances.

One should however fairly try to estimate the contribution of the hardware Reservoir compared to software processing. A simple way to do this, is to remove the hardware in our experiment, and to process directly the TDM sequence $u^{\text{in}}(t)$. This sequence is the result of the randomly spanned original information -actually the cochleagram, or the time-frequency representation, obtained from the original acoustic speech waveform, according to the input connectivity matrix W^I (a sparse and randomly defined matrix). One could try then to directly apply the Read-Out procedure onto this temporal input waveform u^{in}, the one actually generated in the experiment by the AWG. A WER of about 8 to 10 % is obtained if we apply the same training and testing procedure. The contribution of the Reservoir is thus to lower down to zero this Reservoir-free WER, a result that is however not necessarily straightforward, even for some purely software techniques.

6.5.3.2 Wavelength

The wavelength Ikeda dynamics is a representative one, since it is the first having been developed in our group, leading to one of the first experimental demonstrations of optical chaos communication [47, 5], patented one year earlier. This setup was recently re-used for the discovery of chimera states in photonic delay dynamics [25]. It is finally also worth mentioning that the original wavelength-chaos concept was recently adapted to the visible range, in a scientific work targeted toward a broad audience dissemination (International Year of Light 2015), demonstrating a “chaotic rainbow” [48].

The basic physics idea in this setup is to modulate a two wave interference in a different way than through its optical path difference (the case of an electro-optically tunable medium inside one interferometer arm). The phase difference involved in such a two-wave interferometer reads $2\pi\Delta/\lambda$, where Δ is the optical path difference (the quantity usually modulated through a Pockels electrooptic effect), and λ is the wavelength of the monochromatic coherent light involved in the interference phenomenon. If one uses a strongly imbalance static interferometer (big Δ), the interference condition can also be modulated through a small wavelength modulation $\lambda = \lambda_0 + \delta\lambda$. The latter is easily obtained from a tunable semiconductor laser (a double section-gain and phase sections – distributed Bragg reflector Pérot–Fabry laser diode), even with a 1 nm continuous wavelength tunability around $\lambda_0 \simeq 1.55\,\mu\text{m}$ ($\Delta \simeq 1$ cm was used through the two polarization directions of a calcite birefringent crystal, placed between two crossed polarizers for generating the interference).

The main advantage of the setup is the extremely high achievable nonlinearity, since up to 14 extrema can be obtained. Another advantage is the flexibility in changing the shape of the nonlinear transformation, since any static interferometer with a sufficiently large optical path difference can be used. For chimera states exploration, the required asymmetric extrema function was performed by a Pérot–Fabry interferometer designed with a 5 mm glass slab with moderate reflection coatings (70 %). The drawback is related to the low tuning speed of the wavelength (typical maximum bandwidth up to 20 kHz or 100 kHz). The direct consequence is that significant delays (compared to the response time of the delay feedback loop oscillation) are difficult to achieve with optical fibers (100 km would be needed). Electronic delay lines are thus needed, with time delays of the order of ms. Flexibility for the fine and easy tuning of

Figure 6.7: Photography of a multiple delay optoelectronic Reservoir Computer, involving a FPGA board (courtesy of R. Martinenghi).

this digitally controlled delay line, is however an interesting advantage. When using a Field Programmable Gate Array (FPGA), multiple programmable delays can be implemented relatively easily. The latter possibility was particularly used in [49] to evaluate the possible improvements to be obtained when involving a random and sparse connectivity matrix W^R. Indeed, the first delay-based Reservoir, due to their single delay architecture, were concerned by very regular internal connectivity matrices W^R, since essentially its diagonal W^R is nonzero, and a few lines below it (the number of these few lines depending on the actual width of the impulse response $h(t)$).

In [49] (see Figure 6.7 for a setup photography), 15 and 150 parallel delay lines were implemented through the FPGA, both situations corresponding for W^R to 10 % sparsity and full connectivity, respectively. The weights for each delay line were randomly defined, resulting in a random filter in the feedback path, instead of the all-pass filter corresponding to a single delay of unity weight. The full connectivity configuration, as previously found numerically for ESN, did not improve the performance of the speech recognition task. However, the sparse and random connectivity allowed to obtain comparable results with less than half the number of nodes, thus indicating a possible optimization of the network size through this randomly weighted multiple delay architecture.

6.5.3.3 Phase

The optical phase variable is well known in telecommunication to provide better signal to noise ratio to transmission systems, especially at very high bit rate. A specific phase modulation delay dynamics [50] was initially designed to further improve the speed of optical chaos communication [30]. Phase modulation techniques in optical communication systems are typically associated at the receiver side to differential phase shift keying (DPSK) demodulators, which are consisting typically in an imbalance fiber based interferometer. When phase modulation is performed over time scales that are shorter than the DPSK time imbalancing $\delta\tau_D$, the interference condition can be dynamically and continuously scanned. The corresponding demodulation transfer function can be strongly nonlinear as soon as the phase modulation spans over more than a π phase shift. Despite the fact the general recipe for an Ikeda delay dynamics is followed, the electrooptic phase delay dynamics requires some changes in the model equation, essentially due to the DPSK device, now providing a nonlocality in time for the nonlinear transformation. The significant time imbalancing indeed introduces an additional small, but nonnegligible, time delay, which is about 400 ps, thus comparable or sometimes even greater than the response time of typical optical telecommunication electronics (from 10 ps to a few 100 ps). The temporally nonlocal nonlinear transformation now reads:

$$f_{\mathrm{NL}}[\phi(t), \phi(t-\delta\tau_D)] = \beta\cos^2\{\Phi_0 + [\phi(t) - \phi(t-\delta\tau_D)]/2\}. \tag{16}$$

One could notice here that, even in the nonlinear transformation, the static solutions ($\phi(t)$ = constant) are necessarily suppressed in the oscillation loop, because of the phase difference. The nonlinear function itself is thus a Fourier filter, having a high-pass profile at low frequencies, and a periodic bandpass profile for high frequencies.

As expected, the setup adapted to the Reservoir Computing configuration (see Figure 6.8, [33]) was able to process information at an unprecedented speed, up to one million words per second in the case of speech recognition. That processing speed was moreover not limited by the devices, however by the maximum speed of the ultrafast AWG which could not deliver waveforms above 24 GS/s. The sampling step $\mu\tau$ was set to 56 ps, corresponding to 17 GS/s in the AWG. The large delay in the feedback

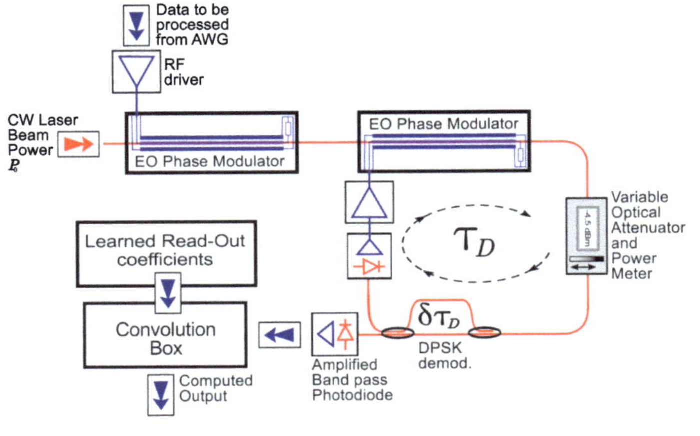

Figure 6.8: Schematic setup of the electrooptic phase delay dynamics.

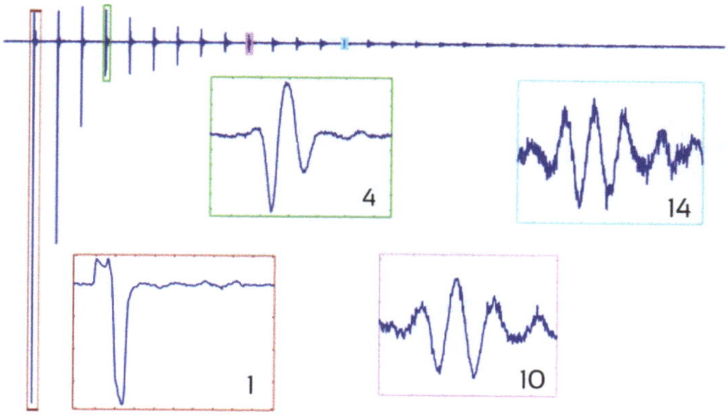

Figure 6.9: Impulse response of the phase delay dynamics, and its many echoes sustained through the set of sufficiently high value of the feedback gain β. The first large echoes reveal strong non-linear distortion of the original single pulse seeded by the AWG. The small echoes at the end are linearly attenuated essentially.

loop was measured to 63.33 ns, a minimum value that was imposed by the different fiber and rf-cable patchcords needed to connect the different devices, and cumulating a few meters. This set of temporal parameters was resulting in a total number of 1113 nodes.

The other optimal operating parameter values are: Peak-to-peak input information u^{in}, $1.2V_\pi$; $\Phi_0 \simeq 2\pi/5$, confirming the optimal best average operating nonlinear transformation close to an extremum of the interference (notice that an active control of this phase interference condition was used); The feedback gain $\beta \simeq 0.7$ has been set to a value that would correspond to 1.4 in the original Ikeda setup, since in the phase delay setup, the first Hopf threshold occurs at $\beta = 0.5$ instead of 1. Figure 6.9 is an experimental illustration on the effect of β in the feedback, which is equivalent to the setting of a "feedback memory," also referred to as Echoes as expressed by the name ESN. This experimental characterization of the Reservoir displays many echoes the Reservoir is capable of generating, for the appropriately chosen value of β. When β is decreased, the echoes are vanishing more rapidly, and the Reservoir does not have enough memory. When β is increased, too close from the Hopf instability, the echoes are sustained and the Reservoir exhibits a too long memory: This is also not desirable for an efficient processing. This optimal tuning for β is what is sometimes referred as to "the edge of chaos" in the RC literature (actually the edge of the first instability in our case).

Figure 6.10 is a picture of the whole electrooptic phase setup arranged as a Reservoir, and connected to the different instruments needed for both "Write-In" (AWG) and "Read-Out" (digital oscilloscope).

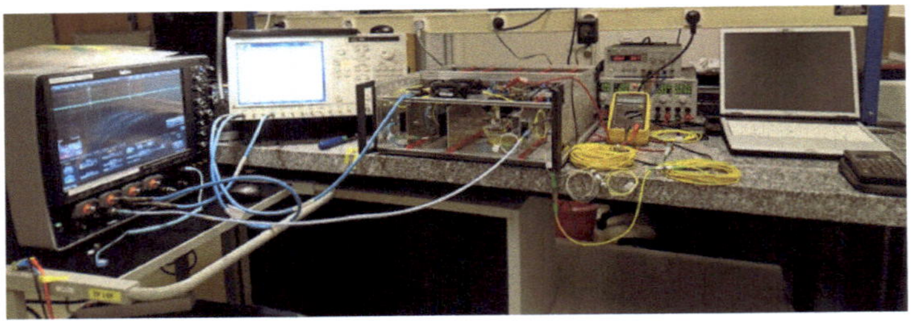

Figure 6.10: Rack-mounted electrooptic broadband phase delay Reservoir, and its processing environment (AWG, broadband oscilloscope).

6.6 Conclusion

Ikeda delay dynamics represents an important toy model for the physics investigations of photonic Reservoirs. It has allowed to achieve several key results, consolidating and supporting the continuation of the photonic hardware investigations aimed at proposing the future generation of brain-inspired computers. Fundamental concepts explaining the efficiency of Reservoir Computing are progressing, together with practical applications that are currently being addressed.

The topic of photonic neuromorphic processors has fundamentally appeared as a cross-disciplinary research direction, involving of course the original machine learning and brain cognitive research communities, but also now physics, nonlinear dynamics, mathematics, and probably more and more importantly information and communication theory. The progress in the understanding will continue to involve more and more signal processing concepts, as we already now know, e. g., from the importance of filtering and convolution operations in the different steps identified in various artificial intelligence approaches. Pattern formation will be also of great importance, maybe to capture how unsupervised learning can produce unexpected Read-Out filters or matrices, that could emerge from a complex dynamical system seeded by specific initial information, such as the way one has discovered chimera states.

The living biological brain needs of course to continue to provide insights, but one needs also to take at the same time the necessary distance to find the best physics and technology compromise for its hardware implementation solutions, thus keeping computing efficiency compatible with technological feasibility.

Bibliography

[1] K. Ikeda. "Multiple-valued stationary state and its instability of the transmitted light by a ring cavity system". In: *Optics Communications* 30.3 (Aug. 1979), pp. 257–261.

[2] M. C. Mackey and L. Glass. "Oscillation and chaos in physiological control systems". In: *Science* 197 (1977), pp. 287–288.

[3] H. M. Gibbs et al. "Observation of chaos in optical bistability". In: *Physical Review Letters* 46.7 (Feb. 1981), pp. 474–477.

[4] A. Neyer and E. Voges. "Dynamics of electrooptic bistable devices with delayed feedback". In: *IEEE Journal of Quantum Electronics* 18.12 (Dec. 1982), pp. 2009–2015.

[5] J.-P. Goedgebuer, L. Larger, and H. Porte. "Optical cryptosystem based on synchronization of hyperchaos generated by a delayed feedback tunable laser diode". In: *Physical Review Letters* 80.10 (June 1998), pp. 2249–2252.

[6] X. S. Yao and L. Maleki. "High frequency optical subcarrier generator". In: *Electronics Letters* 30.18 (Sept. 1994), pp. 1525–1526.

[7] L. Larger et al. "Photonic information processing beyond Turing: an optoelectronic implementation of reservoir computing". In: *Optics Express* 20.3 (Jan. 2012), pp. 3241–3249.

[8] Y. Paquot et al. "Optoelectronic reservoir computing". In: *Scientific Reports* 2 (Feb. 2012), p. 287.

[9] L. Appeltant et al. "Information processing using a single dynamical node as complex system". In: *Nature Communications (London)* 2.468 (Sept. 2011), pp. 1–6.

[10] L. Larger et al. "From flow to map in an experimental high-dimensional electrooptic nonlinear delay oscillator". In: *Physical Review Letters* 95 (July 2005), p. 043903.

[11] L. Larger. "Complexity in electro-optic delay dynamics: modeling, design, and applications". In: *Philosophical Transactions of the Royal Society of London. A* 371 (Aug. 2013), p. 20120464.

[12] M. Peil et al. "Routes to chaos and multiple time scale dynamics in broadband bandpass nonlinear delay electro-optic oscillators". In: *Physical Review E* 79 (Feb. 2009), p. 026208.

[13] L. Larger, B. Penkovskyi, and Y. L. Maistrenko. "Virtual chimera states for delayed-feedback systems". In: *Physical Review Letters* 111 (Aug. 2013), p. 054103.

[14] M. C. Soriano et al. "Complex photonics: dynamics and applications of delay-coupled semiconductors lasers". In: *Reviews of Modern Physics* 85.1 (Mar. 2013), pp. 421–470.

[15] M. Le Berre et al. "Conjecture on the dimension of chaotic attractors of delayed-feedback dynamical systems". In: *Physical Review A* 35.9 (May 1987), pp. 4020–4022.

[16] D. Verstraeten et al. "Isolated word recognition with the Liquid State Machine: a case study". In: *Information Processing Letters* 30.6 (Sept. 2005), pp. 521–528.

[17] H. Jaeger and H. Haas. "Harnessing nonlinearity: predicting chaotic systems and saving energy in wireless communication". In: *Science* 304 (Apr. 2004), pp. 78–80.

[18] W. Maass, T. Natschlager, and H. Markram. "Real-time computing without stable states: a new framework for neural computation based on perturbations". In: *Neural Computation* 14.11 (2002), pp. 2531–2560.

[19] F. T. Arecchi et al. "Two-dimensional representation of a delayed dynamical system". In: *Physical Review A* 45.7 (Apr. 1992), R4225–R4228.

[20] Y. Kuramoto and D. Battogtokh. "Coexistence of coherence and incoherence in nonlocally coupled phase oscillators". In: *Nonlinear Phenomena in Complex Systems* 5.4 (Dec. 2002), pp. 380–385.

[21] D. M. Abrams and S. H. Strogatz. "Chimera states for coupled oscillators". In: *Physical Review Letters* 93.17 (Oct. 2004), p. 174102.

[22] M. J. Panaggio and D. M. Abrams. "Chimera states: coexistence of coherence and incoherence in networks of coupled oscillators". In: *Nonlinearity* 28 (Feb. 2015), R67–R87. DOI: 10.1088/0951-7715/28/3/R67.

[23] A. M. Hagerstrom et al. "Experimental observation of chimeras in coupled-map lattices". In: *Nature Physics (London)* 8 (Sept. 2012), pp. 658–661.

[24] M. R. Tinsley, S. Nkomo, and K. Showalter. "Chimera and phase-cluster states in populations of coupled chemical oscillators". In: *Nature Physics (London)* 8 (2012), p. 662.

[25] L. Larger, B. Penkovsky, and Y. L. Maistrenko. "Laser chimeras as a paradigm for multistable patterns in complex systems". In: *Nature Communications (London)* 6 (July 2015), p. 7752. DOI: 10.1038/ncomms8752.

[26] L. Weicker et al. "Slow-fast dynamics for time-delay problems: theory and experiments". In: *Philosophical Transactions of the Royal Society of London. A* 371 (Feb. 2013), p. 20120459.

[27] G. Giacomelli et al. "Coarsening in a bistable system with long-delayed feedback". In: *Europhysics Letters* 99 (Sept. 2012), p. 58005.

[28] D. Brunner et al. "Spatio-temporal complexity in dual-delayed nonlinear feedback systems: chimeras and dissipative solitons". In: *Chaos* 28 (Oct. 2018), p. 103106.

[29] A. Argyris et al. "Chaos-based communications at high bit rates using commercial fiber-optic links". In: *Nature (London)* 438 (Nov. 2005), pp. 343–346.

[30] R. Lavrov, M. Jacquot, and L. Larger. "Nonlocal nonlinear electrooptic phase dynamics demonstrating 10 Gb/s chaos communications". In: *IEEE Journal of Quantum Electronics* 46.10 (Oct. 2010), pp. 1430–1435.

[31] Y. K. Chembo et al. "Determination of phase noise spectra in optoelectronic microwave oscillators: a Langevin approach". In: *IEEE Journal of Quantum Electronics* 45.2 (Feb. 2009), pp. 178–186.

[32] H. Jaeger. "The "echo state" approach to analysing and training recurrent neural networks". In: *GMD Report* 148 (2001).

[33] L. Larger et al. "High-speed photonic reservoir computing using a time-delay-based architecture: million words per second classification". In: *Physical Review X* 7 (Feb. 2017), p. 011015.

[34] L. Grigoryeva et al. "Stochastic nonlinear time series forecasting using time-delay reservoir computers: performance and universality". In: *Neural Networks* 55C (May 2014), pp. 59–71.

[35] L. Grigoryeva and J.P. Ortega. "Echo state networks are universal". In: *Neural Networks* 108 (Dec. 2018), pp. 495–508.

[36] D. Brunner et al. "Parallel photonic information processing at gigabyte per second data rates using transient states". In: *Nature Communications* 4 (2013), p. 1364.

[37] K. Vandoorne et al. "Parallel reservoir computing using optical amplifiers". In: *IEEE Transactions on Neural Networks* 22.9 (Sept. 2011), pp. 1469–1481.

[38] F. Duport et al. "All-optical reservoir computing". In: *Optics Express* 20.20 (Sept. 2012), pp. 22783–22795.

[39] M. A. A. Fiers et al. "Nanophotonic reservoir computing with photonic crystal cavities to generate periodic patterns". In: *IEEE Transactions on Neural Networks and Learning Systems* 25.2 (Jan. 2014), pp. 344–355.

[40] C. Mesaritakis, V. Papataxiarhis, and D. Syvridis. "Micro ring resonators as building blocks for an all-optical high-speed reservoir-computing bit-patternrecognition system". In: *Journal of the Optical Society of America. B, Optical Physics* 30.11 (Oct. 2013), pp. 3048–3055.

[41] K. Vandoorne et al. "Experimental demonstration of reservoir computing on a silicon photonics chip". In: *Nature Communications (London)* 5 (Mar. 2014), p. 3541.

[42] J.-P. Goedgebuer et al. "Optical communication with synchronized hyperchaos generated electrooptically". In: *IEEE Journal of Quantum Electronics* 38.9 (Sept. 2002), pp. 1178–1183.

[43] M. W. Lee, L. Larger, and J.-P. Goedgebuer. "Encryption system using chaotic delays between lightwaves". In: *IEEE Journal of Quantum Electronics* 39.7 (July 2003), pp. 931–935.

[44] N. Gastaud et al. "Electro-optical chaos for encoded multi-10 Gb/s optical transmissions". In: *Electronics Letters* 40.14 (July 2004), pp. 898–899.

[45] L. Illing and J. Gauthier Daniel. "Hopf bifurcations in time-delay systems with band-limited feedback". In: *Physica D* 210 (Aug. 2005), pp. 180–202.

[46] A. B. Cohen et al. “Using synchronization for prediction of high-dimensional chaotic dynamics”. In: *Physical Review Letters* 101 (Oct. 2008), p. 154102.
[47] G. D. VanWiggeren and R. Roy. “Communicating with chaotic lasers”. In: *Science* 279.3 (Feb. 1998), pp. 1198–1200.
[48] Y. K. Chembo et al. “Ikeda-like chaos on a dynamically filtered supercontinuum light source”. In: *Physical Review A* 94 (Aug. 2016), p. 023847.
[49] R. Martinenghi et al. “Photonic nonlinear transient computing with multiple-delay wavelength dynamics”. In: *Physical Review Letters* 108 (Jan. 2012), p. 244101.
[50] R. Lavrov et al. “Electro-optic delay oscillator with non-local non linearity: optical phase dynamics, chaos, and synchronization”. In: *Physical Review E* 80 (Aug. 2009), p. 026207.

Guy Van der Sande and Miguel C. Soriano

7 Semiconductor lasers as reservoir substrates

7.1 Introduction

Semiconductor lasers are the most common type of lasers produced, with a wide range of applications that include optical storage systems, communication systems (ranging from short-distance data communication systems to long-haul fiber-optic networks), as pump sources, for material processing, and in many other applications [1]. A recent application, and the topic of this chapter, is the use of semiconductor lasers as substrates for photonic reservoir computers. Here, we present an overview of recent publications and discuss the main properties of photonic reservoir computers based on semiconductor lasers subject to optical feedback. The optical feedback serves as a way to create the recurrence required for reservoir computing in the context of delay-based reservoir computers. For more information on the formal definition of delay-based reservoir computers, we refer the reader to Chapter 5.

7.2 Laser basics and semiconductor types

A laser device comprises the amplifying medium, where the stimulated emission takes place, and a resonant cavity, which provides an adequate feedback mechanism and serves as a frequency selective element. An example of an amplifying medium is a semiconductor p–n junction where population inversion is achieved through an externally applied current flow. In the simple case of a two-level system, population inversion is reached when the higher-lying electronic level has a higher population than the lower-lying level.

The laser cavity provides the selective feedback through multiple transits of the photons before leaving the cavity, as shown in Figure 7.1. Frequencies for which constructive round trip interference occurs are sustained and the others are suppressed. Two parameters define the temporal characteristics of the laser cavity: the laser cavity round trip time defines the frequencies of operation and the inverse of the photon lifetime describes the rate at which photons are lost from the cavity.

Even when population inversion is achieved, the net rate of stimulated emission is not necessarily enough to overcome the losses present in the laser cavity. These losses mainly are the transmission losses at the laser facets, light scattering and light absorption. The device only starts lasing when the injection current exceeds a value called the threshold current at which the gain overcomes all the different losses.

https://doi.org/10.1515/9783110583496-007

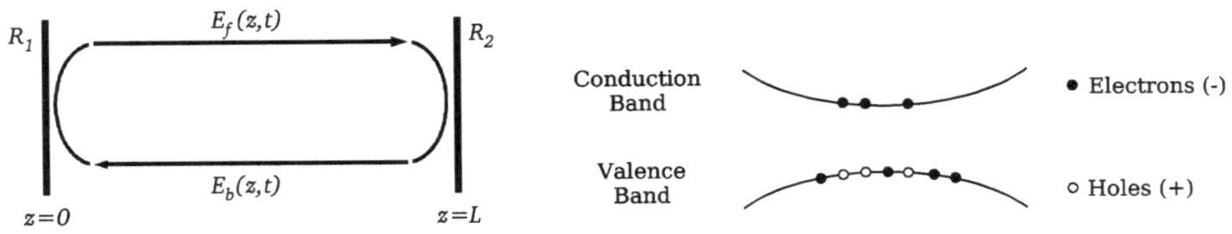

Figure 7.1: (Left) Schematic arrangement of an optical resonator. (Right) Electrons and holes in an arbitrary distribution between bands.

In semiconductor materials, the charge carriers (electrons) are distributed between the conduction and the valence band, as shown in Figure 7.1. The energy difference between both bands is called the band gap. Stimulated emission can occur when the energy of the incoming light exceeds the energy band gap of the semiconductor material. When an electron is excited from the valence band to the conduction band, it leaves behind what is called a "hole" (the absence of a charge carrier) in the valence band. The annihilation of an electron and a hole leads to spontaneous emission, or stimulated emission when an extra photon induced the annihilation. By doping a semiconductor material, one can alter the natural distribution of electrons and holes. A p-doped semiconductor has an excess of holes and an n-doped semiconductor an excess of electrons, originating from acceptors and donors respectively. In the junction of a p-doped and a n-doped semiconductor, stimulated emission dominates over absorption if there is population inversion. The semiconductor medium is said to be transparent when the rates of stimulated emission and absorption are equal. An incoming beam of photons experiences a gain in a p–n junction with population inversion, meaning that the number of outgoing photons is larger than the number of incoming photons. For this reason, this region is called the active layer of the laser.

Semiconductors materials have a complex energy band structure. In the context of semiconductor lasers and other optoelectronic devices, direct bandgap semiconductor materials are preferred since the valence band maximum and the conduction band minimum both occur at the same wave vector $\vec{k}$. A wide range of III–V material systems is used for the production of semiconductor lasers. Table 7.1 lists a number of these materials and their output wavelengths [2].

Semiconductor lasers are typically built with standard multi-layer epitaxial growth techniques and come in several different geometries. The more conventional lasers are known as Edge Emitting semiconductor diode Lasers (EELs). In EELs, light travels in the lateral direction inside an active layer where electrons and holes can recombine. Since these devices provide a high gain per round trip, no special reflecting coatings or structures at the facets of the device are necessary but often used. It is also important to note that the light field in EELs is linearly polarized along two fixed directions, namely along the heterojunction plane (TE polarization) or perpendicular to it (TM polarization), due to the combination of wave guiding and gain. However, in

Table 7.1: Material systems typically used for the production of semiconductor lasers [2].

Material system	Wavelength
AlGaAs / GaAs	680–890 nm
InGaAs / GaAs	950–1100 nm
InGaAsP / InP	1000–1700 nm
AlGaInP / GaAs	600–700 nm
ZnCdSSe	450–550 nm
AlGaInN	200–640 nm
GaInNAs / GaAs	1300–1500 nm
GaN / AlGaN	400–550 nm

standard EELs the TM mode generally experiences a larger loss at the facets than the TE mode. Therefore, light emission in EELs is predominantly TE polarized.

Under normal conditions, the laser operates close to the maximum of the material gain. A number of modes could experience similar amplification, depending on the width of the gain profile compared to the frequency spacing of the optical resonator modes. When a gain medium amplifies a strong laser beam, the gain is saturated, i. e. reduced to some extent. Occasionally, the saturation can be inhomogeneous, i. e., it can be stronger around the wavelength of the laser beam than at other wavelengths. This illustrates that the gain saturation can force a change of lasing mode. In general, the laser chooses the mode that has the highest instantaneous gain. Single-longitudinal mode operation can be guaranteed by, e. g., creating a periodic structure within the laser cavity that increases its frequency selectivity.

7.3 Single-mode semiconductor lasers for reservoir computing

Semiconductor lasers exhibit nonlinear interactions between the laser field and the semiconductor medium, resulting in complex behavior when subjected to feedback, electrical modulation, or optical injection [3]. In this section, we show how to implement the computational concept of reservoir computing (RC) in photonics with single-mode semiconductor lasers subject to optical feedback. The reservoir and information injection is realized all-optically, allowing for high-speed information processing. This can be achieved by, e. g., utilizing the analogue transient dynamics generated by the semiconductor laser coupled to a fibre-optic feedback loop. Following the delay-based RC concept [4], we use a single semiconductor laser as the nonlinear node in which nonlinear transient states are generated in the context of previous input responses. Consequently, the system is capable of processing temporal sequences of information.

7.3.1 Modeling and numerical results

We start by presenting the rate equations for a single mode semiconductor laser diode [5]. These rate equations describe the time evolution of the electric field $|E(t)|$ and the number of charge carriers $N(t)$. The electric field is normalized in such a way that $|E(t)|^2$ corresponds to the number of photons $P(t)$. The rate equations then read

$$\frac{\mathrm{d}E(t)}{\mathrm{d}t} = \left(\frac{1+i\alpha}{2}\right)\left[G_N(N(t)-N_o) - \frac{1}{\tau_p}\right]E(t) + F_E(t), \tag{1}$$

$$\frac{\mathrm{d}N(t)}{\mathrm{d}t} = \frac{I}{e} - \frac{N(t)}{\tau_c} - G_N(N(t)-N_o)|E|^2. \tag{2}$$

Here, α is the linewidth enhancement factor, G_N is the differential gain, τ_p is the photon lifetime, N_o is the carrier number at transparency, I is the pump current, e is the elementary charge, and τ_c denotes the electron lifetime. Spontaneous emission effects are included in the model by adding a Langevin noise term F_E to the field equation. This spontaneous emission noise is implemented as a complex Gaussian white noise term $F_E = F_1 + iF_2$ in the field equation with zero mean $\langle F_E(t)\rangle = 0$, where the real and imaginary parts are independent random processes and the following holds: $\langle F_j(t)F_j(t')\rangle = \beta/\tau_c N(t)\delta(t-t')$. Here, β is the spontaneous emission factor that describes the fraction of spontaneously photons emitted into the respective lasing mode.

For information processing in the context of reservoir computing, the external input can be added as an optical injection term and the recurrences are included via optical feedback. Following these assumptions, equation (1) can be rewritten as

$$\frac{\mathrm{d}E(t)}{\mathrm{d}t} = \left(\frac{1+i\alpha}{2}\right)\left[G_N(N(t)-N_o) - \frac{1}{\tau_p}\right]E(t) + \kappa E(t-\tau_{\mathrm{ec}}) + E_{\mathrm{inj}}(t)e^{i\Delta\omega t}, \tag{3}$$

where κ denotes the feedback strength, τ_{ec} is the delay time, and $\Delta\omega$ is the detuning between the response semiconductor laser and the optical injection. The optical feedback is modeled following the Lang–Kobayashi rate equations [6] and the external signal is injected via $E_{\mathrm{inj}}(t)$. For practical reasons and to compare to the experimental implementation, the external modulation of the injected light is performed via a Mach–Zehnder electrooptic modulator. The addition of the input signal $S(t)$ is then modeled via an injected power P_{inj} modulated with a sine-square around a mean value $\bar{P}_{\mathrm{inj}}$ yielding the signal:

$$P_{\mathrm{inj}}(t) = \bar{P}_{\mathrm{inj}}\left[1/4 + 3/2\sin^2\left(\frac{\pi}{4}S(t) + \Phi_0\right)\right], \tag{4}$$

which means the injected power is modulated between ±75 % around the average injected power $\bar{P}_{\mathrm{inj}}$. For the optical injection, we consider symmetric modulation with $S(t)$ normalized between ±1 and $\Phi_0 = \frac{\pi}{4}$. The chosen values for the laser parameters as used during the numerical simulations are shown in Table 7.2 [7].

Table 7.2: Laser parameter values used in the numerical simulations.

Parameter	Value
α	3.0
τ_p	5 ps
β	10^{-6}
κ	$10\ \text{ns}^{-1}$
τ_{ec}	80 ns
$\Delta\omega$	0.0
τ_c	1 ns
G_N	$10^{-5}\ \text{ns}^{-1}$
N_o	$1.8 * 10^8$
I_{thr}	32.0 mA

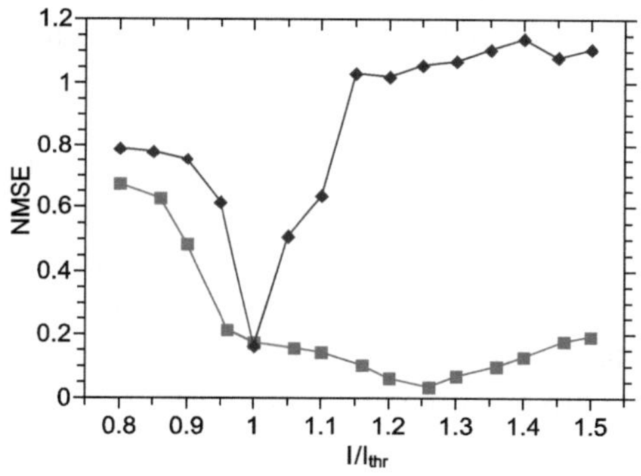

Figure 7.2: Normalized mean squared error (NMSE) for the Santa Fe time series prediction task versus the bias current I for $\bar{P}_{inj} = 436\,\mu\text{W}$ (red squares), and $\bar{P}_{inj} = 11\,\mu\text{W}$ (blue diamonds), respectively. The feedback rate is set to $\kappa = 10\ \text{ns}^{-1}$. The other parameters were chosen as in Table 7.2. Note that the lines only serve as a guide to the eyes. Figure reprinted with permission from Ref. [7], IEEE.

For reservoir computing purposes, the external input $S(t)$ is constructed as described in Chapter 5 with a random input connectivity mask that repeats every τ_{ec}. In the numerical analysis, the reservoir consists of $N = 400$ virtual nodes, resulting in a virtual node spacing of $\Theta = 200$ ps (τ_{ec}/N).

As an example of the performance of a semiconductor laser system as reservoir, we tackle a time series prediction task. Here, we evaluate the performance of this scheme in predicting the respective next point of a chaotic time series. We specifically employ data from the Santa Fe time series competition, data set A [8]. For the evaluation of the prediction error, we take 4000 data points of this data set, created by a far-Infrared laser operating in a chaotic regime [9]. We used 75 % of the points for training and 25 % for testing. To characterize the performance of the system for this task, we compute the normalized mean square error (NMSE) of the prediction, defined as the normalized difference between the predicted and its target value. We study the dependence of the NMSE with the laser bias current for a fixed feedback rate. This prediction task requires the system to have memory, i. e., optical feedback is crucial for this task.

In Figure 7.2, we show the NMSE as a function of the laser bias current for two different values of the optical injection power. We first discuss the one corresponding to a large average power of the injected light, compared to the power of the laser subject to feedback, $\bar{P}_{\text{inj}} = 436\,\mu$W. In this case, the NMSE for the Santa Fe time series prediction task is below 0.2 for a wide range of currents (see squares in Figure 7.2), with a minimum NMSE at $I = 1.25I_{\text{th}}$ of 0.036. Figure 7.2 also presents the results for a smaller average power of the optical input injection, $\bar{P}_{\text{inj}} = 11\,\mu$W. In this case, we find that low prediction errors are restricted to laser bias currents close to the solitary lasing threshold (see diamonds in Figure 7.2), with a minimum NMSE value of 0.164 at $I = I_{\text{th}}$. For larger bias currents, the prediction error rises significantly for the case of low injection power, due to the onset of delayed feedback instabilities. For the case of a high injection power, however, the prediction error keeps reducing from $I = I_{\text{th}}$ to $I = 1.25I_{\text{th}}$ and then it rises slowly. In both cases, the prediction errors strongly increase for injection currents below $I = I_{\text{th}}$. Overall, competitive prediction errors can be achieved for optimized parameters. It is interesting to note, though, that a larger average optical injection power allows for a wider range of bias currents providing good performance.

Remarkably, when decreasing the node distance to a smaller value similar results are obtained even when this value is far below the relaxation oscillation period [10]. In Figure 7.3, it is shown that the NMSE does not significantly change with the node distance when the system operates above the threshold current. This good performance indicates that the reservoir remains in the transient state for all the values of the node distance explored in Figure 7.3. This is due to the optical injection of the data signals. Optical injection is an integral part of the setup and an optical signal with constant bias power is always present even in the case of no information being injected in the semiconductor laser ($S(t) = 0$ in equation (4)). It turns out that an optically injected laser can react at speeds related to the laser locking phenomenon, which can be much faster than the relaxation oscillations. This feature is related to the phase dynamics. The node distance θ can therefore be freely chosen between the fastest time scale (of the optical injection) and the relaxation oscillation period without significantly degrading the performance when the system operates above the solitary threshold [10]. These numerical results are obtained for realistic levels of spontaneous emission

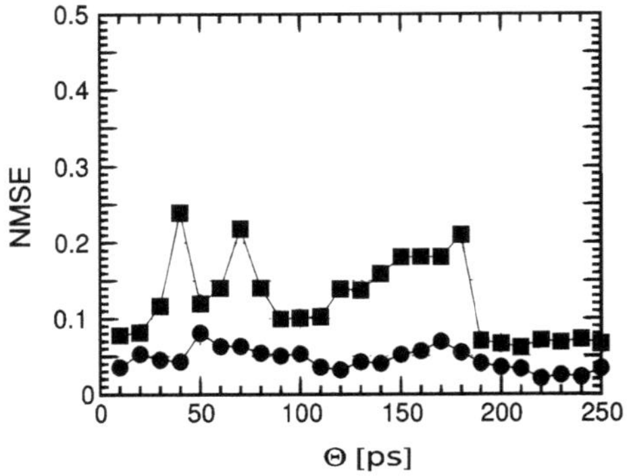

Figure 7.3: Normalized mean squared error (NMSE) for the Santa Fe time series prediction task versus the node distance current θ for injection current above threshold (circles; $I = 1.1I_{\text{th}}$) and below threshold (squares; $I = 0.9I_{\text{th}}$). The feedback rate is set to $\kappa = 10\,\text{ns}^{-1}$ and the number of virtual nodes is set to $N = 200$. The other parameters were chosen as in Table 7.2. Note that the lines only serve as a guide to the eyes. Figure reprinted with permission from Ref. [10], OSA.

noise, while a limited signal to noise ratio in the detection has not been considered. In practice, the overall contribution of the different noise sources will likely impose a more restrictive limit on the minimum node distance.

However, below threshold, for $I < I_{th}$, the RC performance does depend on the node distance. In Figure 7.3, NMSE< 0.1 is obtained only for specific θ values. As phase dynamics exists even below the threshold, good performance can be obtained below the threshold current provided that the node distance is suitably chosen.

Taking $N = 200$, this broad range of node distances leads to an overall delay length between 2 ns and 50 ns. It is worth noting that small θ values, and thus small delay lengths, are useful for compact on-chip implementations. Furthermore, a short delay also allows to increase the processing speed as the data is fed at the delay period.

7.3.2 First experimental implementation with a single-mode semiconductor laser

The scheme of the first experimental implementation of semiconductor laser-based RC is shown in Figure 7.4 [11]. A standard edge emitting laser diode with an emission wavelength of $\lambda = 1542$ nm is employed as the nonlinear node. The stand-alone single-longitudinal mode laser has a longitudinal mode splitting of ~ 150 GHz, with a side-mode suppression exceeding 40 dB. Using free-space optics, the emission is collected in a standard single mode fiber. A fiber loop provides delayed optical feedback (delay $\tau_D = 77.6$ ns), with the loop comprising an optical circulator, an optical attenuator, a polarization controller, and two fiber splitters utilized for signal detection and optical injection. An optical attenuator and a polarization controller facilitate the control of the optical feedback conditions. In this experiment, the reservoir consists of $N = 388$ virtual nodes, resulting in a virtual node spacing of $\Theta = 200$ ps. This experimental arrangement follows the delay-based RC protocol introduced in [4] and discussed in Chapter 5.

For information processing, the modulated light of a tunable laser is injected into the single-mode semiconductor laser. The external input signal is encoded via injection intensity modulation using a Mach–Zehnder modulator. While the laser diode current and the Mach–Zehnder modulator have a modulation bandwidth going up to or exceeding 10 GHz, the information is injected at a rate of 5 GSamples/s. In these experiments, the arbitrary waveform generator, used for generating the input information, is the bandwidth limiting factor [11].

The performance of this experimental RC system, evaluated using the Santa Fe time series prediction task, is depicted in Figure 7.5 for the case of optical data injection. Particular to time series prediction tasks is a high sensitivity to noise. Hence an injection power of 7.5 μW is chosen as the constant bias optical power in the absence of information, allowing for a better signal to noise ratio. In addition to serving as a data injection source, the external laser therefore additionally acts as an injection locking

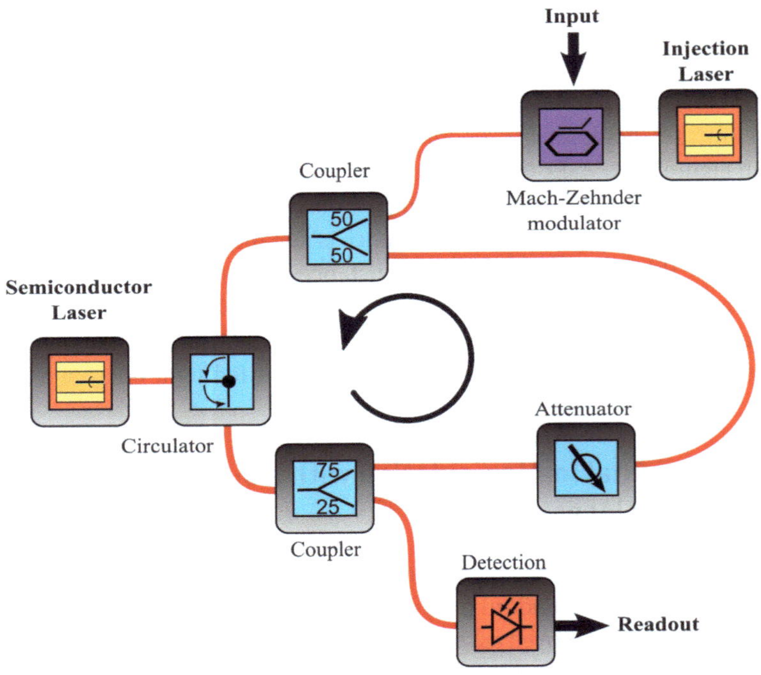

Figure 7.4: Scheme of the all-optical reservoir computer based on a semiconductor laser subject to delayed optical feedback. The experimental setup comprises the laser diode, a tunable laser source to optically inject the information, a Mach–Zehnder modulator, an optical attenuator, a circulator, couplers, and a fast photo diode (PD) for signal detection.

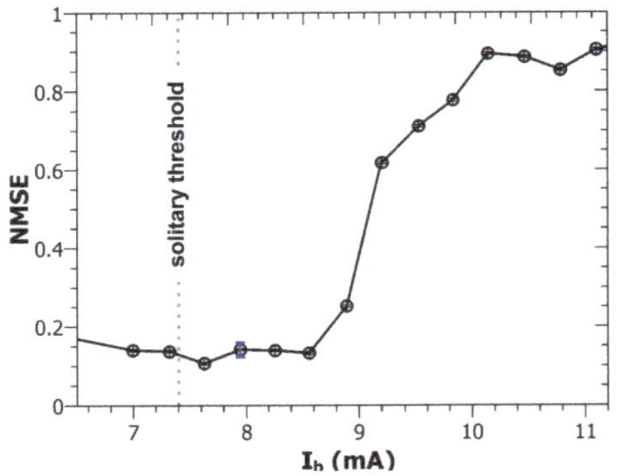

Figure 7.5: Prediction error in the Santa Fe time series prediction task. The prediction error increases dramatically for $I_b > 8.9$ mA, a regime in which the steady-state dynamics becomes unstable. The blue error bar represents the standard deviation for different training/testing partitions of the data. Figure adapted with permission from Ref. [7], IEEE.

source, increasing the performance significantly by reducing the noise in the steady-state dynamics. The NMSE for the prediction task, depending on the bias current, is shown in Figure 7.5. The best performance is obtained for I_b close to the solitary laser threshold, with a prediction error of 0.106 ($I_b = 7.62$ mA, feedback attenuation 10 dB) at a prediction rate of 1.3×10^7 points per second (dictated by the inverse of the delay time). The performance significantly degrades for large bias currents ($I_b > 8.9$ mA).

7.3.3 Further experimental implementations with single-mode semiconductor lasers

After the pioneering work of Brunner et al. [11], several experimental implementations have focused on understanding the fundamental properties of semiconductor laser-based RC for nonlinear prediction tasks. On the one hand, it has been shown that the conditions to achieve good prediction performance are linked to the injection locking, consistency, and memory properties of the system [12]. On the other hand, Kuriki et al. [13] illustrated the influence of the input mask on the performance of the system.

A semiconductor laser subject to optical feedback and optical injection exhibits a wide range of dynamical phenomena [14]. As shown in [12], the ability of the semiconductor laser-based RC system to process information is tightly linked to its underlying dynamical properties in the absence of external input. In particular, the lowest prediction error for a nonlinear prediction task occur at the injection locking boundary. Injection locking refers to the state in which the optical frequency of the response laser locks to the optical frequency of the injection laser, which happens for certain combinations of optical frequency detuning between the lasers and optical injection strength [15]. These results can be interpreted as follows: the injection locking boundary provides an optimum compromise between the diversity of the nonlinear responses in the system and the reproducibility of these responses for similar input signals [12]. This reproducibility of the responses can be quantified by the consistency correlation measure [16]. In Figure 7.6, we show how the best computational performance for a prediction task (NMSE) is found for the system parameters that lead to an optimum compromise between a sufficient memory capacity and a large consistency correlation.

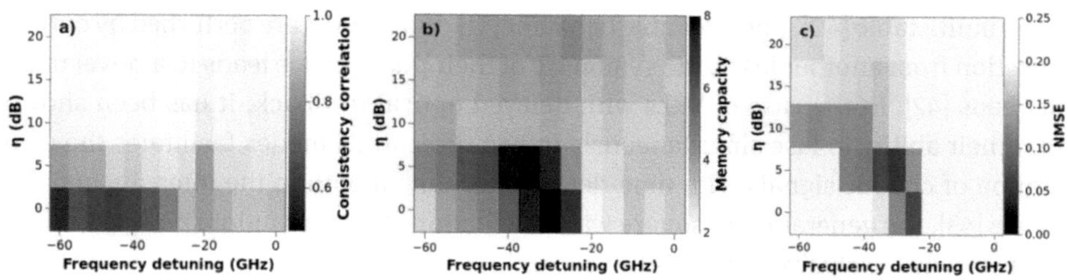

Figure 7.6: Experimental results for the (a) consistency correlation, (b) memory capacity, and (c) normalized mean square error (NMSE) for the prediction of a Mackey–Glass chaotic time series as a function of the feedback attenuation (η) and the frequency detuning for a fixed bias current $I_b = 0.99 I_{th}$. Figure adapted with permission from Ref. [12], OSA.

Semiconductor laser-based RC systems belong to the fastest hardware implementations of this machine learning concept, operating at multi-GHz speeds. Due to the intrinsic high bandwidth of the reservoir system, the experimental apparatus used to drive the system with the external input, the reservoir itself and the detection need to have at least comparable analog bandwidths. In this context, Kuriki et al. [13] shows that an input mask matched in bandwidth to the response laser presents an improved prediction performance. The bandwidth-adapted mask is superior to binary, multi-level or uniform random masks. Together, the results in [12] and [13] highlight the importance of consistency as one of the key properties of nonlinear photonic systems to process information.

7.4 Other photonic systems as reservoir substrates

7.4.1 Semiconductor ring lasers for reservoir computing

Semiconductor ring lasers (SRLs) are currently the focus of a rapidly thriving research activity due to their unique feature of directional bistability [17] and the fact that they do not require cleaved facets or gratings for optical feedback. Hence, SRLs are particularly suited for a monolithic integration [18]. SRLs have been suggested to fulfill several practical applications [19–25]. All optical flip-flops based on a single or two coupled microring lasers have been fabricated. These devices can be switched between counter-propagating modes by injection of a signal counter-propagating to the lasing mode [22, 26]. In addition, switching schemes based on injection only on one side of the SRL have been suggested as well [27, 28]. Monolithic SRLs exhibiting unidirectional operation are also highly desirable in applications because of their wavelength stability [19, 20, 29, 30, 21]. The bistability of the SRLs opens the possibility of using them in systems for all-optical switching, gating, wavelength-conversion functions, and optical memories [20, 22, 31–38]. Moreover, SRLs have been recognized to be ideal optical prototypes of nonlinear Z2-symmetric systems [39] exhibiting, in the solitary case, multistable [40], and excitable behavior [41]. When SRLs are perturbed by optical injection from another laser, the symmetry of their phase space leads to a novel route to chaos [42]. In the case of SRLs with delayed optical feedback, it has been shown that their ability to lase simultaneously in two-directional modes facilitates the generation of chaotic signals with time-delay concealment both in the intensity and the phase [43], the generation of square wave oscillations [44], or random bits generation using bitwise Exclusive-OR operations [45].

A general rate-equation approach for SRLs has been suggested by Sorel et al. in [19]. The model consists of two mean-field equations for the counter-propagating modes in the SRL, and a third rate equation for the carriers. The model accounts for self- and cross-gain saturation effects and includes backscattering contributions originating at the coupling to an output-waveguide. In the same work, they have experimentally observed bidirectional and unidirectional regimes of continuous-wave mode operation. Moreover, a bidirectional regime where the two counter-propagating modes experience harmonic alternate oscillations has been observed as well. These different features are adequately described by the rate-equation model for SRLs in [19]. Although this general rate-equation approach explains certain experimentally observed features, problems involving e. g. wavelength changes fall outside the scope of such rate-equation models, and a traveling wave model is more suited to tackle such questions [46, 47].

We consider a SRL operating in a single-longitudinal, single-transverse mode. In the limit of small outcoupling from the ring cavity, the total electric field oscillating in the ring can be written as the sum of two counter-propagating waves, clockwise (CW), and counter-clockwise (CCW):

$$E(z,t) = E_{\mathrm{CW}}(t)\exp[i(\omega_0 t - k_0 z)] + E_{\mathrm{CCW}}(t)\exp[i(\omega_0 t + k_0 z)] + \text{c.c.}. \tag{5}$$

Here, k_0 is the longitudinal wavenumber and ω_0 is the optical frequency of the mode. In the slowly varying envelope approximation, the amplitudes of the clockwise E_{CW} and counter-clockwise propagating modes E_{CCW} vary on time scales which are orders of magnitude slower that ω_0. The rate-equation model is formulated mathematically in terms of two rate equations for the slowly varying amplitudes $E_{\mathrm{CW,CCW}}$ and one rate equation for the carrier number N. The equations read [19]:

$$\dot{E}_{\mathrm{cw}} = \kappa(1+i\alpha)[g_{\mathrm{cw}}N - 1]E_{\mathrm{cw}} - (k - \Delta k/2)e^{i(\phi_k - \Delta\phi_k/2)}E_{\mathrm{ccw}}, \tag{6}$$

$$\dot{E}_{\mathrm{ccw}} = \kappa(1+i\alpha)[g_{\mathrm{ccw}}N - 1]E_{\mathrm{ccw}} - (k + \Delta k/2)e^{i(\phi_k + \Delta\phi_k/2)}E_{\mathrm{cw}}, \tag{7}$$

$$\dot{N} = \gamma[\mu - N - g_{\mathrm{cw}}N|E_{\mathrm{cw}}|^2 - g_{\mathrm{ccw}}N|E_{\mathrm{ccw}}|^2] \tag{8}$$

where dot represents differentiation with respect to time t, $g_{\mathrm{cw}} = 1 - s|E_{\mathrm{cw}}|^2 - c|E_{\mathrm{ccw}}|^2$, $g_{\mathrm{ccw}} = 1 - s|E_{\mathrm{ccw}}|^2 - c|E_{\mathrm{cw}}|^2$, κ is the field decay rate, γ is the carrier decay rate, α is the linewidth enhancement factor and μ is the renormalized injection current with $\mu \approx 0$ at transparency and $\mu \approx 1$ at lasing threshold. The two counter-propagating modes are considered to saturate both their own and each other gain due to, e. g., spectral hole burning effects. Self- and cross-saturation effects are added phenomenologically and are modeled by s and c, respectively. For a realistic device, cross-saturation is stronger than self-saturation. Reflection of the counter-propagating modes occurs at the point where light is coupled out of the ring cavity into a coupling waveguide and can also occur at the end facets of the coupling waveguide. These localized reflections result in a linear coupling between the two fields characterized by an amplitude k and a phase shift ϕ_k. Moreover, due to unavoidable imperfections in the SRL introduced during the fabrication process, the SRL will have a certain asymmetry in the linear coupling between both counter-propagating modes. This asymmetry is introduced in equations (6)–(8) as Δk and $\Delta\phi_k$, representing the difference in backscat-

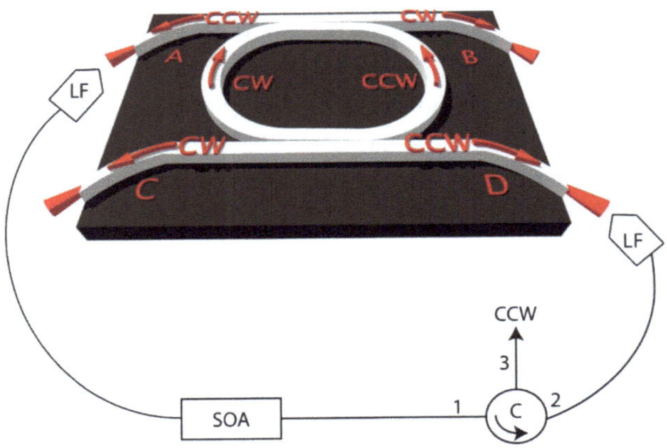

Figure 7.7: Schematic of a SRL with one feedback loop. In this example, the CW mode is subject to cross-feedback from the CCW mode. Red symbols A, B, C, and D are output ports, LF: lensed fibers, C: circulator, SOA: semiconductor optical amplifier.

tering strength and phase, respectively. This asymmetry is necessary to describe phenomena such as excitability in SRLs [41].

SRLs can be used as the nonlinear node in a delay based reservoir computer, just as standard semiconductor lasers. However, the fact that single-mode SRLs can emit in two counter-propagating directional modes with nearly the same frequency allows for a larger flexibility in the reservoir computing set-up. First of all, one has the choice to send different data signals into the SRL separate to each directional mode, or to send the same data signal to both. In other words, one can opt to use the higher modal dimension of the SRL to process two tasks in parallel. If not, and both CW and CCW are used for the same data signal, one can expect that the number of virtual nodes spread out over the delay length can now be distributed also over two modes. As a result, more virtual nodes can be used for the same delay length and reservoir computation can be sped up by a factor of two. Secondly, the optical delay line can be coupled in multiple ways to the same SRL device (see Figure 7.7 for a possible configuration). One can therefore realize either self-feedback in the two-modes (i. e., the CW (CCW) mode is coupled back in to the CW (CCW) mode after a certain delay) or cross-feedback (i. e., the CW (CCW) mode is coupled back in to the CCW (CW) mode after a certain delay). Of course, it is possible to implement delayed feedback in only one mode. However, this would defeat the purpose of the dimensional increase offered by the two-directional modes. Hence, we do not consider this case.

Nguimdo et al. have investigated these scenarios by expanding the SRL rate equation model with the appropriate Lang–Kobayashi feedback terms and optical injection [48]. Backscattering was left symmetric. The rate equation, e. g., the optical field of the CW mode then reads:

$$\dot{E}_{\mathrm{cw}} = \kappa(1 + i\alpha)[g_{\mathrm{cw}}N - 1]E_{\mathrm{cw}} - ke^{i\phi_k}E_{\mathrm{ccw}} + \eta_{\mathrm{cw}}F_{\mathrm{cw}}(t) + k_1E_1(t). \tag{9}$$

The rate equation for the optical field of the CCW mode can be written in an analogous manner. The feedback terms are $F_{\mathrm{cw}}(t)$ and $F_{\mathrm{ccw}}(t)$, which can be explicitly defined

Table 7.3: SRL parameter values used in the numerical simulations.

Parameter	Value
α	3.5
s	0.005
c	0.01
κ	100 ns^{-1}
γ	0.2 ns^{-1}
k	0.44 ns^{-1}
ϕ_k	1.5
$\theta_{cw,ccw}$	0
$\eta_{cw,ccw}$	10 ns^{-1}
$k_{1,2}$	10 ns^{-1}
N	100
Θ	20 ps
$T_{cw,ccw}$	2 ns

depending on the feedback configuration. For the cross-feedback configuration,

$$F_{\mathrm{cw}}(t) = E_{\mathrm{ccw}}(t - T_{\mathrm{ccw}})e^{-i\theta_{\mathrm{ccw}}}, \tag{10}$$

$$F_{\mathrm{ccw}}(t) = E_{\mathrm{cw}}(t - T_{\mathrm{cw}})e^{-i\theta_{\mathrm{cw}}}, \tag{11}$$

where T_{cw} and T_{ccw} are delay times and θ_{cw} and θ_{ccw} are the constant feedback phases. The self-feedback configuration can be defined in a similar way. The last term in equation (9) is the injected field $E_1(t)$, containing the data of the corresponding tasks to be processed, with k_1 being the injection strength. Realistic levels of the spontaneous emission noise are always taken into account for the numerical simulations. Data signals are injected through optical injection using a Mach–Zenhder modulator as described before in equation (4). The preprocessing of the data signals occurs according to the masking procedure of delay-based RC described in [4]. The length of one input sample was matched to the delay length. It is worth noting that the preprocessing and the post-processing of the signals in the two modes are independent. The mask, the number of virtual nodes N and the node distance Θ can therefore differ from one mode to another if desired. The values of the SRL parameter used in the numerical simulations can be found in Table 7.3. The information processing tasks can be different in the two modes (e. g., CW for task 1 and CCW for task 2).

Figure 7.8(a) shows the numerically obtained prediction error for the Santa Fe time series as a function of the pump current μ for self- and cross-feedback configurations. The results point out that there is a broad range of the pump current for which SRL-based RC can successfully predict the next step in the Santa Fe time series. The smallest prediction error $\approx 3\,\%$ is obtained around $\mu \approx 1.5$ for the self-feedback configuration while for the cross-feedback configuration, NMSE $\approx 4\,\%$ is obtained around $\mu \approx 1.3$. The optimum performance of the system for Santa Fe time series prediction is similar

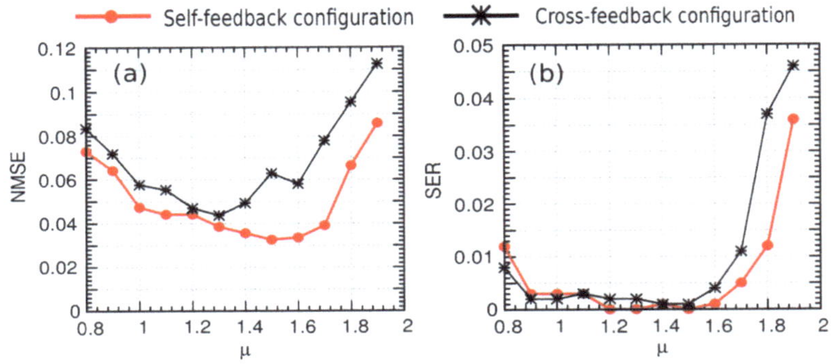

Figure 7.8: Performance of a SRL based reservoir computer on a single task versus the pump current μ. The task is (a) Santa Fe time series prediction and (b) nonlinear channel equalization. The task is processed only in the CW mode. The NMSE and SER values are the average of 10 realizations with different randomly generated masks. Figure reprinted with permission from Ref. [48], IEEE.

to that in other RC schemes based on semiconductor lasers with optical feedback. It can be noted here that the error increases very slowly with pump current for, both, the self- and cross-feedback configuration. This can be attributed to the small node distance that is considered. In Figure 7.8(b), we show the performance for a nonlinear channel equalization task for different pump currents. Details about the channel equalization task can be found in [49]. Very good performance is obtained for this task in both feedback configurations, both below and above the pump current threshold. In addition, the range of pump currents that yield good performance agrees well with that found for the Santa Fe time series prediction. In particular, the smallest symbol error rate (SER) for the nonlinear channel equalization task is $< 0.1\,\%$ and $\approx 0.1\,\%$ for self- and cross-feedback configurations, respectively.

Having two directional modes that can be both addressed by on optical data signal independently, SRLs are promising candidates for parallel processing of two tasks. For this purpose, we consider in the Santa Fe data set the first 4000 points for task 1 to be processed in the CW mode and the last 4000 points for task 2 to be processed in the CCW mode. The input data is rescaled so that $-\pi \leq S_{1,2}(t) \leq \pi$. Figure 7.9 displays the performance of the system for a simultaneous prediction of the future sample in each Santa Fe time series for different values of the pump current considering a SRL with (a) self-feedback and (b) cross-feedback. For the self-feedback configuration, it turns out that there is a broad range of pump currents for which the NMSE is smaller than or equal to 10 % for both tasks. This means the system succeeds in simultaneously predicting the future sample in each Santa Fe time series in this range of the pump current. In particular, the prediction errors for the two tasks are very similar and reach a minimum of $\approx$ 4 % for task 1 and $\approx$ 6 % for task 2 around $\mu \approx 1.3$ [see Figure 7.9(a)]. Compared to the performance in Figure 7.8(a), it is clear that a parallel simultaneous prediction of the next sample in the two chaotic time series does not sig-

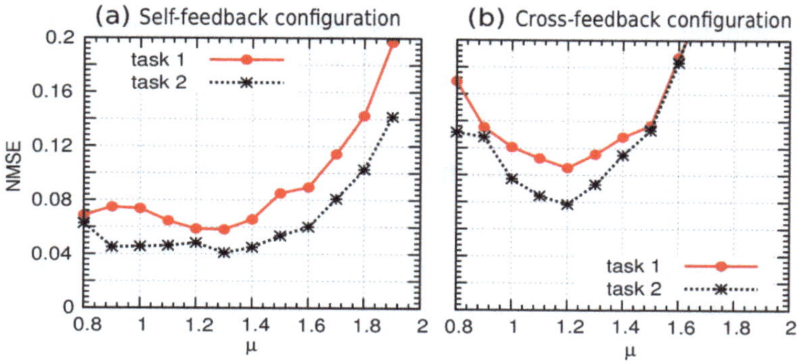

Figure 7.9: Performance of a SRL based reservoir computer on two parallel tasks versus the pump current μ. The task is the Santa Fe time series prediction performance is expressed as NMSE. NMSE obtained for task 1 (•,red) and task 2 ($\star$, black) in the case of double self- feedback (a) and double cross-feedback (b). Figure reprinted with permission from Ref. [48], IEEE.

nificantly degrade the performance in a self-feedback configuration. This shows that, despite the two modes being coupled, the interaction effects are not significant. As displayed in Figure 7.9(b), the prediction errors for a simultaneous one-step prediction in the two chaotic time series are worse for the cross-feedback configuration. In addition, there is only a very narrow range of pump current values for which a NMSE 10 % can be simultaneously obtained for both tasks. The minimum error is $\approx$ 8 % for task 1 and $\approx$ 10 % for task 2 in this case. This is twice as large as the minimum prediction error found when only one task is processed considering the same configuration [see Figure 7.8(a)]. We find an increase in the prediction error because the cross-feedback configuration introduces additional coupling between the two counter-propagating modes. The amount of information transferred from the CW mode to the CCW mode and vice versa is therefore larger when compared to the self-feedback configuration, hampering the computational performance.

7.4.2 Erbium-doped microchip lasers

In [50], it was experimentally and numerically shown that diode-pumped erbium-doped microchip lasers subject to optical feedback can also be used to implement RC systems for prediction tasks. Using a Santa Fe time series as a bench-mark, the authors found similar performances as with semiconductor lasers. Besides investigating RC in a different material type of laser, which requires to be optically pumped instead of electrically, the authors consider an injection sample separation of the input data (i. e., the inverse of the processing speeds) close or corresponding to the time delay. Also, in [50], it was explored whether data can be directly coupled to the feedback light beam (i. e., modulating the feedback path) instead of using an additional laser for optically injecting the electrical data into the reservoir.

7.4.3 Semiconductor optical amplifiers

The first experimental realization of all-optical RC was based on the nonlinear response of a semiconductor optical amplifier (SOA) placed in a ring optical cavity [51]. In this case, the external input was injected as a modulated optical field and the output layer was implemented off-line after detection. The reservoir speed was at the time limited by the rate of the input signal generator, yielding to an input processing rate of 7.8 μs/symbol. As a result of a thorough comparison between numerical and experimental results, the authors highlight that the noise present in this analog system degrades the performance on tasks such as the memory capacities, the nonlinear channel equalization, and isolated spoken digit recognition [51]. This observation holds for most photonic implementations of RC to date.

7.5 Conclusion

The development of high-speed implementations of photonic reservoir computers is an active field of research that is still in progress. As shown in this chapter, a semiconductor laser subject to optical feedback fulfills the requirements for a high-speed implementation of the reservoir. Current trends for the advancement of photonic reservoir computers include the possibility to integrate most of the photonic components and to develop an all-optical implementation of the full system. Photonic integration aims at targeting a robust implementation of the reservoir, while full system implementations must also include the input and output layers. Besides the technological challenges to implement the full photonic system based on a semiconductor laser, it is still unclear what is the precise influence of the laser nonlinearity in the computational performance. In this context, a precise experimental characterization of the amplitude and phase response of the semiconductor laser may shed some light on the role of the nonlinearity for the case of the optically driven semiconductor laser.

Recent works suggest certain modifications to the original scheme that could still improve the performance of laser-based RC. These modifications include the use of two delay loops [52] in order to extend the fading memory of the system, or the combination of responses for different laser parameters in order to enhance the computational power of the system. From the application point of view, it is crucial to design photonic reservoir computers than can be compatible with existing technologies. Altogether, there remain several challenges ahead, both at the fundamental and technological level in order to establish photonic reservoir computers as contenders to current technologies for real-world applications. In this context, laser-based RC can potentially have an advantage in terms of power efficiency and of processing speed, eliminating the electronic bottleneck in photonic applications. In photonics, a potential application that can benefit from these properties is the recovery of optical communication signals using a driven semiconductor laser and the RC paradigm [53].

Bibliography

[1] G. P. Agrawal. *Fiber-optic communication systems*. Vol. 222. John Wiley & Sons, 2010.

[2] T. E. Sale. *Vertical cavity surface emitting lasers*. Research Studies Press, 1995. ISBN: 0471957402.

[3] M. C. Soriano et al. "Complex photonics: dynamics and applications of delay-coupled semiconductors lasers". In: *Reviews of Modern Physics* 85.1 (2013), pp. 421–470.

[4] L. Appeltant et al. "Information processing using a single dynamical node as complex system." In: *Nature Communications* 2 (2011), p. 468. DOI: 10.1038/ncomms1476.

[5] K. Petermann. "Laser diode modulation and noise". In: *Advances in optoelectronics (ADOP)*, Dordrecht: Kluwer and Tokyo: KTK Scientific Publishers (1991).

[6] R. Lang and K. Kobayashi. "External optical feedback effects on semiconductor injection laser properties". In: *IEEE Journal of Quantum Electronics* 16.3 (1980), pp. 347–355.

[7] K. Hicke et al. "Information processing using transient dynamics of semiconductor lasers subject to delayed feedback". In: *IEEE Journal of Selected Topics in Quantum Electronics* 19.4 (2013), p. 1501610.

[8] A. S. Weigend and N. A. Gershenfeld. *Time series prediction: forecasting the future and understanding the past*. Vol. 80. Addison-Wesley, 1993. URL: ftp://ftp.santafe.edu/pub/Time-Series/Competition.

[9] U. Hubner, N. B. Abraham, and C. O. Weiss. "Dimensions and entropies of chaotic intensity pulsations in a single-mode far-infrared NH3 laser". In: *Physical Review A* 40.11 (1989), pp. 6354–6365. DOI: 10.1103/PhysRevA.40.6354.

[10] R. M. Nguimdo et al. "Fast photonic information processing using semiconductor lasers with delayed optical feedback: role of phase dynamics". In: *Optics Express* 22.7 (2014), pp. 8672–8686.

[11] D. Brunner et al. "Parallel photonic information processing at gigabyte per second data rates using transient states". In: *Nature Communications* 4 (2013), p. 1364.

[12] J. Bueno et al. "Conditions for reservoir computing performance using semiconductor lasers with delayed optical feedback". In: *Optics Express* 25.3 (2017), pp. 2401–2412. DOI: 10.1364/OE.25.002401.

[13] Y. Kuriki et al. "Impact of input mask signals on delay-based photonic reservoir computing with semiconductor lasers". In: *Optics Express* 26.5 (2018), pp. 5777–5788.

[14] G. H. M. Van Tartwijk and D. Lenstra. "Semiconductor lasers with optical injection and feedback". In: *Quantum and Semiclassical Optics: Journal of the European Optical Society Part B* 7.2 (1995), p. 87.

[15] S. Wieczorek et al. "The dynamical complexity of optically injected semiconductor lasers". In: *Physics Reports* 416.1–2 (2005), pp. 1–128.

[16] A. Uchida, R. Mcallister, and R. Roy. "Consistency of nonlinear system response to complex drive signals". In: *Physical Review Letters* 93 (2004), p. 244102. DOI: 10.1103/PhysRevLett.93.244102.

[17] M. Sorel et al. "Operating regimes of GaAs-AlGaAs semiconductor ring lasers: experiment and model". In: *IEEE Journal of Quantum Electronics* 39.10 (2003), pp. 1187–1195.

[18] P. J. R. Laybourn, T. Krauss and J. S. Roberts. "CW operation of semiconductor ring lasers". In: *Electronics Letters* 26.25 (1990), pp. 2095–2097. DOI: 10.1049/el:19901349.

[19] M. Sorel et al. "Alternate oscillations in semiconductor ring lasers". In: *Optics Letters* 27 (2002), pp. 1992–1994.

[20] J. J. Liang et al. "Unidirectional operation of waveguide diode ring lasers". In: *Applied Physics Letters* 70.10 (1997), pp. 1192–1194. DOI: 10.1063/1.118527.

[21] V. R. Almeida and M. Lipson. "Optical bistability on a silicon chip". In: *Optics Letters* 29 (2004), pp. 2387–2389.

[22] M. T. Hill, H. J. S. Dorren, T. de Vries, X. J. M. Leijtens, J. H. den Besten, B. Smalbrugge, Y. S. Oei, H. Binsma, G. D. Khoe and M. K. Smit. "A fast low-power optical memory based on coupled micro-ring lasers". In: *Nature* 432 (2004), pp. 206–209.

[23] B. Li et al. "All-optical response of semiconductor ring laser to dual-optical injections". In: *IEEE Photonics Technology Letters* 20 (2008), pp. 770–772.

[24] Z. Wang et al. "Storing 2 bits of information in a novel single semiconductor microring laser memory cell". In: *IEEE Photonics Technology Letters* 20 (2008), pp. 1228–1230.

[25] S. Furst and M. Sorel. "Cavity-enhanced four-wave mixing in semiconductor ring lasers". In: *IEEE Photonics Technology Letters* 20 (2008), pp. 366–368.

[26] T. Perez et al. "Bistability and all-optical switching in semiconductor ring lasers". In: *Optics Express* 15 (2007), pp. 12941–12948.

[27] L. Gelens et al. "Optical injection in semiconductor ring lasers: backfire dynamics". In: *Optics Express* 16 (2008), pp. 16968–16974.

[28] L. Gelens et al. "Phase-space approach to directional switching in semiconductor ring lasers". In: *Physical Review E* 79 (2009), p. 16213.

[29] S. Zhang et al. "Ring-laser optical flip-flop memory with single active element". In: *IEEE Journal of Selected Topics in Quantum Electronics* 10 (2004), pp. 1093–1100.

[30] V. R. Almeida et al. "All-optical switching on a silicon chip". In: *Optics Letters* 29 (2004), pp. 2867–2869.

[31] M. Hentschel and T. Y. Kwon. "Designing and understanding directional emission from spiral microlasers". In: *Optics Letters* 34 (2009), pp. 163–165.

[32] C. M. Kim et al. "Continuous wave operation of a spiral-shaped microcavity laser". In: *Applied Physics Letters* 92 (2008), p. 131110.

[33] J. Y. Lee, X. S. Luo, and A. W. Poon. "Reciprocal transmissions and asymmetric modal distributions in waveguide-coupled spiral-shaped microdisk resonators". In: *Optics Express* 15 (2007), pp. 14650–14666.

[34] J. Wiersig, S. W. Kim, and M. Hentschel. "Asymmetric scattering and nonorthogonal mode patterns in optical microspirals". In: *Physical Review A* 78 (2008), p. 53809.

[35] M. Waldow et al. "25 ps all-optical switching in oxygen implanted silicon-on-insulator microring resonator". In: *Optics Express* 16 (2008), pp. 7693–7702.

[36] L. Shang, L. Y. Liu, and L. Xu. "Single-frequency coupled asymmetric microcavity laser". In: *Optics Letters* 33 (2008), pp. 1150–1152.

[37] A. R. Bahrampour et al. "All-optical set-reset flip-flop based on frequency bistability in semiconductor microring lasers". In: *Optics Communications* 282 (2009), pp. 2451–2456.

[38] S. J. Chang et al. "A compact and low power consumption optical switch based on microrings". In: *IEEE Photonics Technology Letters* 20 (2008), pp. 1021–1023.

[39] S. Beri et al. "Topological insight into the non-Arrhenius mode hopping of semiconductor ring lasers". In: *Physical Review Letters* 101 (2008), p. 93903.

[40] L. Gelens et al. “Exploring multistability in semiconductor ring lasers: theory and experiment”. In: *Physical Review Letters* 102.19 (2009), p. 193904.

[41] S. Beri et al. “Excitability in optical systems close to Z2-symmetry”. In: *Physics Letters A* 374.5 (2010), pp. 739–743. DOI: 10.1016/j.physleta.2009.11.070.

[42] W. Coomans, G. Van der Sande, and L. Gelens. “Oscillations and multistability in two semiconductor ring lasers coupled by a single waveguide”. In: *Physical Review A* 88 (2013), p. 33813.

[43] R. M. Nguimdo et al. “Loss of time-delay signature in chaotic semiconductor ring lasers”. In: *Optics Letters* 37.13 (2012), pp. 2541–2543.

[44] L. Mashal et al. “Square-wave oscillations in semiconductor ring lasers with delayed optical feedback”. In: *Optics Express* 20 (2012), pp. 22503–22516.

[45] R. M. Nguimdo et al. “Fast random bits generation based on a single chaotic semiconductor ring laser”. In: *Optics Express* 20 (2012), pp. 28603–28613.

[46] R. Ciegis, M. Radziunas, and M. Lichtner. “Numerical algorithms for simulation of multisection lasers by using traveling wave model”. In: *Mathematical Modelling and Analysis* 13 (2009), pp. 327–348.

[47] J. Javaloyes and S. Balle. “Emission directionality of semiconductor-ring lasers: a travelling-wave description”. In: *IEEE Journal of Quantum Electronics* 45 (2009), pp. 431–438.

[48] R. M. Nguimdo et al. “Simultaneous computation of two independent tasks using reservoir computing based on a single photonic nonlinear node with optical feedback”. In: *IEEE Transactions on Neural Networks and Learning Systems* 26.12 (2015), pp. 3301–3307. DOI: 10.1109/TNNLS.2015.2404346.

[49] H. Jaeger and H. Haas. “Harnessing nonlinearity: predicting chaotic systems and saving energy in wireless communication”. In: *Science* 304 (2004), pp. 78–80.

[50] R. M. Nguimdo et al. “Prediction performance of reservoir computing systems based on a diode-pumped erbium-doped microchip laser subject to optical feedback”. In: *Optics Letters* 42.3 (2017), pp. 375–378.

[51] F. Duport et al. “All-optical reservoir computing”. In: *Optics Express* 20 (2012), pp. 22783–22795.

[52] Y. Hou et al. “Prediction performance of reservoir computing system based on a semiconductor laser subject to double optical feedback and optical injection”. In: *Optics Express* 26.8 (2018), pp. 10211–10219.

[53] A. Argyris, J. Bueno, and I. Fischer. “Photonic machine learning implementation for signal recovery in optical communications”. In: *Scientific Reports* 8 (2018), p. 8487. DOI: 10.1038/s41598-018-26927-y.

Piotr Antonik, Serge Massar, and François Duport

8 Advanced reservoir computers: analogue autonomous systems and real time control

8.1 Introduction

Ideally brain-inspired information processing systems should be autonomous devices that receive an analog input, produce an analog (or, possibly, digital) output, and in addition should be able to learn by themselves to produce the desired output. Such devices could then be used as black boxes that could be connected to other devices, or possibly to themselves. They would thus be the elementary brick used to build more and more complex analogue information processing systems.

However, implementing such autonomous black box devices is highly complex. For this reason, in initial experiments, only a (small) part of the device is implemented experimentally, and the remaining pieces are implemented digitally. The complex engineering task of building fully autonomous systems is left for later. This step-by-step approach is very natural in science. The development of experimental reservoir computing illustrates very well this piecewise implementation. A good example of this approach is one of the pioneering papers [1] which used an experimental optoelectronic nonlinearity but implemented all the rest of the system digitally.

In the case of reservoir computing, a further important simplification was introduced in [2]: namely the architecture based on a single nonlinear node and a delay line described in Chapter 5. In addition to its conceptual interest, this architecture also considerably simplified experimental implementations because it replaces a parallel system by a sequential system which is easier to build and debug. For this reason, this architecture has become an ideal place to start studying reservoir computing using novel substrates, with the additional simplification that one can—if necessary in a first instance—use a digital delay line. Experiments that used this approach include e. g. the electronic system of [2], the first optoelectronic reservoir computers [3, 4], the first all optical reservoir computer [5], the first reservoir computer based on a laser with delayed feedback [6], the first reservoir computer using a nanoscale spintronic oscillator [7], and the first reservoir computer using a nonlinear mechanical oscillator [8].

A few experiments did not use the architecture based on delay dynamical systems, but they also needed to considerably simplify the system; see for instance [1] mentioned above, and the first on-chip reservoir computer [9] (see Chapter 3), which had to resort to sequential measurements of the neurons.

All these experiments are important milestones in the development of experimental reservoir computing. But as mentioned above, they only implemented part of a reservoir computer, and are far from an autonomous system. A few steps have been

https://doi.org/10.1515/9783110583496-008

taken toward the idea of a fully-analogue reservoir computer. In Section 8.3, we describe how one can implement a more complete system, that comes close to a fully autonomous one, in which the input layer, a photonic reservoir computer based on a delay dynamical system, and an output layer are implemented experimentally. (Note, however, that the output layer in the experiment we describe is not strictly speaking fully autonomous, as it produces a continuous signal that needs to be sampled at specific times to produce the desired output.) These results are based on [10].

The most difficult part of the system [10] seems to be the output layer. Indeed, an analogue output is a linear combination, with positive and negative weights of the reservoir states. Such linear combinations are easy to implement digitally. But analogue implementations are difficult, as a small error on the output weights, or any additional noise that occurs in the output layer, can have a large effect on the output signal. For this reason, we believe that analogue output layers will be a key difficulty to overcome in the development of experimental reservoir computers.

In Section 8.4, we consider a photonic reservoir computer with which we can interact in real time. Indeed, in any real world application, reservoir computers will have to interact with their environment in real time. This for instance could be due to the fact that the task the reservoir computer must accomplish changes, or because the reservoir dynamics are slowly evolving. We explore these possibilities by coupling the reservoir computer to a Field-Programmable Gate Array (FPGA) equipped with an Analogue to Digital Converter (ADC) and a Digital to Analogue Converter (DAC). However, it must be acknowledged that in these experiments we simplified the reservoir computer by implementing the output layer (which as mentioned is challenging to implement experimentally) digitally on the FPGA. A natural next step would be to repeat these experiments using a reservoir with an analogue output layer.

As an application of the above concept, we consider online training, in which the output weights of the reservoir computer are updated in real time. This allows the reservoir computer to cope with tasks that change with time. This experiment is based on [11]. This method also potentially allows one to train analogue output layers without having to model the output layer in detail, as demonstrated numerically in [12]. It is interesting to compare the results we present here with the more recent work of [13] in which online training was used on an analogue output layer, but with an update rate for the reservoir (i. e., the rate at which successive inputs are processed) of approximately 5 Hz, compared to above 130 kHz for [11].

Finally, we show (in Section 8.5) how real time control of the reservoir computer allows the implementation of output feedback, in which the past output is used to drive the system. This allows novel functionalities, such as pattern generation and emulation of chaotic systems which are not possible with the simple reservoir architecture. Output feedback for emulation of chaotic systems was introduced in [14]. Interest in this question has recently grown again, with several theoretical contributions [15–18]. The experiment we describe here is based on [19].

The main conclusion of this Chapter, see Section 8.6, is that developing autonomous reservoir computers with real-time control is both essential for applications and enables new functionalities. It is thus a key challenge in the field.

8.2 A simple optoelectronic reservoir computer

The experiments presented in this chapter are based on the optoelectronic implementation of reservoir computing first introduced in [3, 4] and illustrated in Figure 8.1; see also Chapter 6. For completeness, we recall here its operation principles. We first recall the basic equations describing a reservoir computer [14, 20]. The system is driven by an external signal $u(n)$, contains N internal variables $x_i(n)$ $(i = 0, \ldots, N-1)$, and produces an output $y(n)$, where n is discretized time. It is described by the following equations:

$$s_i(n) = M_i u(n) + b_i \qquad \text{input layer} \qquad (1)$$

$$x_i(n) = f\Big(\sum_j W_{ij}^{\text{res}} x_j(n-1) + s_i(n)\Big) \qquad \text{reservoir layer} \qquad (2)$$

$$y(n) = \sum_i W_i x_i(n) \qquad \text{output layer} \qquad (3)$$

where $f(\cdot)$ is a nonlinear function, M_i is the input mask (also noted W_i^{in}), b_i the input bias, W_{ij}^{res} the reservoir interconnection matrix, W_i the output weights (also noted W_i^{out}). In most implementations, M_i, b_i, W_{ij}^{res} can be chosen at random from some distribution, except for global scaling which is adjusted for best performance. The output weights W_i are chosen to optimize performance.

In the case of reservoir computers based on delay dynamical systems, described in [22] and Chapter 5, equation (2) is replaced by

$$\begin{aligned} x_i(n) &= f(\alpha x_{i-k}(n-1) + \beta s_i(n)) \quad &\text{for } i = k, \ldots, N \\ x_i(n) &= f(\alpha x_{N+i-k}(n-2) + \beta s_i(n)) \quad &\text{for } i = 0, \ldots, k-1 \end{aligned} \qquad (4)$$

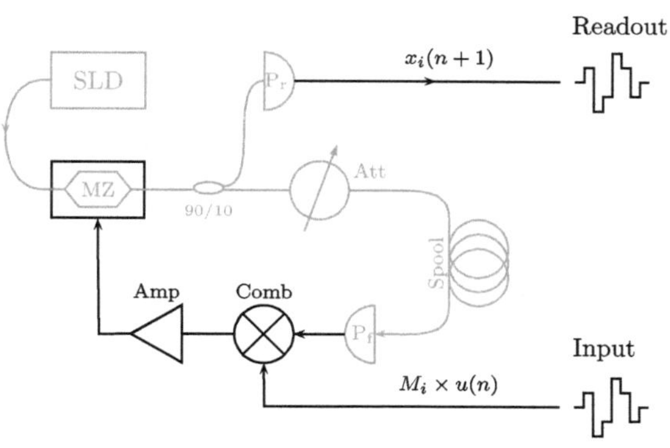

Figure 8.1: Schematic representation of the photonic reservoir layer. It contains a light source (SLD or DFB laser), a Mach–Zehnder intensity modulator (MZ), a 90/10 beam splitter, an optical attenuator (Att), a fiber spool (Spool), two photodiodes (P_r and P_f), a resistive combiner (Comb), and an amplifier (Amp). Optical and electronic components are shown in grey and black, respectively. Adapted with permission from [21].

where α and β are global scaling parameters of the interconnection and input matrices W_{ij}^{res} and M_i, respectively, called *feedback gain* and *input gain*; $k = 0$ (synchronized regime, in which case f must include a low-pass filter, see [2, 3]), or $k > 0$ (unsynchronized regime).

Figure 8.1 depicts an optoelectronic setup that implements equation (4). The reservoir layer consists of a delay line and a single nonlinear node. Similar systems have been studied previously in the general context of nonlinear dynamics; see, e. g., [23–25]. This reservoir layer is essentially identical to the optoelectronic reservoir used in [3, 4, 26–30].

The delay line consists of a spool of optical fiber (roughly 1.7 km of SMF28e). The internal variables x_i are time-multiplexed along the delay line. They are represented by the light intensity that travels along the delay line within fixed temporal windows. At the end of the fiber, the optical feedback signal is converted to a voltage by the feedback photodiode $\mathrm{P_f}$. The resulting signal is then amplified to drive a Mach–Zehnder (MZ) light intensity modulator. The light source can vary between different experiments. We typically use either a DFB laser (Covega -SFL-1550p-NI- with a wavelength around 1550 nm) or a superluminiscent diode (SLD, Thorlabs SLD1550P-A40), also emitting at the standard telecommunication wavelength of 1550 nm.

The sine response of this M–Z modulator is used as the nonlinearity of the reservoir (nonlinear function f in equation (2) and equation (4)). During the experiments, the bias point of the M–Z modulator is regularly tuned to ensure a proper sine response. In other words, if no signal is applied to the RF port of the M–Z modulator, its transparency is at 50 %. In some works [3], the bias point of the M–Z is considered as a tuneable parameter which allows one to modify the nonlinear function f. Here, the bias point, and hence f is kept fixed. At the output of the M–Z modulator, 10 % of the light intensity is picked up by the readout photodiode $\mathrm{P_f}$, the remaining 90 % is attenuated by a tuneable optical attenuator before going into the optical delay line. This optical attenuator allows adjusting the feedback gain of the cavity (α coefficient in equation (4)). A typical value for the round-trip time of the cavity is $T \approx 8.4\,\mu\mathrm{s}$. If we omit the constant part of the signal (that is in any case filtered out by the amplifier), the dynamics of the system without input can be approximated by

$$x(t) = \sin(\alpha x(t - T)). \tag{5}$$

In order to carry out computation, we drive the cavity with a desynchronized signal as in [4]. To this end, we define the duration θ of each internal variable by the relation

$$T = (N + k)\theta, \tag{6}$$

where we recall that T is the round-trip time, and k denotes the degree of desynchronization. We convert the discrete time input $u(n)$ into a continuous signal $u(t)$ by a sample and hold procedure for duration $T' = N\theta$. Thus, in continuous time, the input of the reservoir computer is represented by a step signal given by

$$u(t) = u(n) \quad \text{for } t \in \left](n-1)T', nT'\right]. \tag{7}$$

The value of the internal variable $x_i(n)$ is given by the average value of $x(t)$ in the interval $t \in](i-1)\theta + (n-1)T', i\theta + (n-1)T']$. The input signal $s_i(n)$ is set to the input $u(n)$ mulitiplied by the input mask m_i, see equation 1, with the bias b_i set to 0. In continuous time, the input mask is represented by a periodic function $m(t)$ of period T'.

The continuous-time dynamics of the driven system can thus be described by

$$x(t) = \sin(\alpha x(t-T) + \beta m(t)u(t)). \tag{8}$$

In discrete time, the variable $x_i(n)$ is connected to either $x_{i-k}(n-1)$ if $i > k$ or to $x_{N+i-k}(n-2)$ if $i \leq k$. The corresponding dynamics in discrete time is thus given by equation (4).

8.3 Experimental implementation of analogue input and output layers

Despite the increasing interest in the reservoir computing paradigm, its potential in terms of processing easiness and speed has not yet been fully exploited. In particular, all previous experiments, presented throughout this book, required either digital preprocessing of the inputs, or digital post-processing of the outputs, or both (i. e., at least either the input layer or the output layer were digitally implemented). This is indeed a major limitation if one intends to use physical reservoir computers as versatile and efficient stand-alone solutions. Moreover, besides the advantages of speed and versatility, a fully analogue device would allow for the feedback of the output of the reservoir into the reservoir itself, enabling new training techniques [31] as well as the exploitation of reservoir computers for new kinds of tasks, such as pattern generation [14, 32] (see also Section 8.5).

Here, we report the work of [10]. Note that some steps toward a fully analogue reservoir had already been taken. A first analogue output layer had been reported in [26], but with less good performance. And in an unpublished manuscript [30], it was shown how to implement an analogue input layer. In fact an analogue input layer is comparatively simpler to implement, as it consists of multiplying the input signal with random weights. The exact values of these weights are not very important, as they can be chosen at random up to some global scaling. Optimization of the input weights has been considered in [27, 29, 33, 34].

The first report of a reservoir computer with an analogue output layer was given in [26]. This solution was tested on a single task, and the results obtained were not as good as those obtained using a digital output. The difficulty of constructing an analogue output layer is due to the nature of the computation that needs to be carried out. Indeed the output of the reservoir computer is a linear combination, with positive and negative weights, of many internal states, which requires a very high computation accuracy. While this accuracy is obtained naturally with a digital computer, it is rather challenging to reach it with an analogue integrating device.

In this section, we present the first fully analogue reservoir computer, originally reported in [10]. This implementation takes as input an analogue optical signal, and produces as output an analogue electrical signal proportional to the output requested by the task. It thereby proves that the concept of reservoir computing can be entirely implemented by means of analogue signals handled by analogue components. This opens up the route to new promising developments based on high-bandwidth standalone reservoirs as well as feedback loop reservoirs.

8.3.1 Experimental setup

The experimental setup is depicted in Figure 8.2. It consists of input, reservoir, and output layers. The operation principles of the optoelectronic reservoir have been presented in Section 8.2. The following two sections are devoted to the input and readout layers.

8.3.1.1 Analogue input layer

In a reservoir computer based on a delay dynamical system with a single nonlinear node, the input mask m_i plays a crucial role as it breaks the symmetry of the system, giving each internal variable $x_i(n)$ a different dependence on the inputs $u(n)$. For this reason the optimization of the input mask has been the subject of several studies [27, 29, 33, 34]. In the present implementation, the input mask $m(t)$ is introduced independently of the input and is intrinsically continuous, which greatly simplifies its hardware implementation.

The optical input signal is generated as follows. A superluminescent light emitting diode (SLED Denselight DL-CS5254A) is modulated using a Mach–Zehnder (M–Z) modulator (Photline MXAN-LN-10) to generate an optical signal proportional to the input $u(t)$ of the reservoir computer. However, a M–Z modulator exhibits a sine response to the applied voltage. The following paragraph explains how we precompensate the signals which drive the M–Z modulators inside the optical input layer and inside the readout layer (see Figure 8.2) so as to obtain a linear response.

Let us consider the light intensity I_{in} at the input of a M–Z modulator, with the insertion loss ρ and half wave voltage V_π. The light intensity I_{out} at the output of the

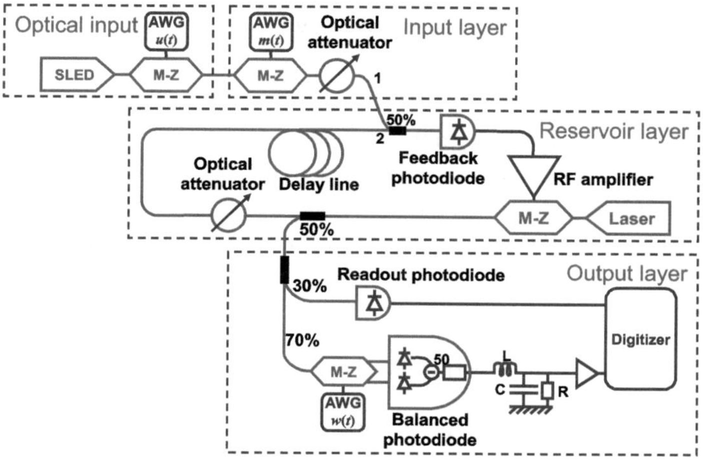

Figure 8.2: Schematic of the experimental reservoir computer with analogue input and output layers. The optical input produces the signal that must be processed. The input layer multiplies the input signal by the input mask. The reservoir layer is a delay dynamical system in which a M–Z modulator acts as nonlinearity. The output layer produces an analogue electric signal proportional to the desired output. Electric components are in green, optical components in red and purple, with purple used for polarization maintaining fiber components (used to avoid the use of polarization controllers) and red for nonpolarization maintaining fiber components. AWG denotes Arbitrary Waveform Generator; RF amplifier denotes Radio Frequency amplifier; R, L, C denote resistor, inductor and capacitor, respectively. Reprinted from [10].

modulator as a function of the driving voltage $v(t)$ is given by

$$I_{\text{out}}(t) = \rho I_{\text{in}} \frac{1}{2}\left(1 + \sin\left(\frac{\pi}{V_\pi} v(t)\right)\right). \tag{9}$$

The goal of the precompensation is to obtain at the output of the M–Z modulator a light intensity proportional to the input light intensity multiplied by the signal $f(t)$. This is obtained by taking $v(t)$ to be equal to

$$v(t) = \frac{V_\pi}{2} \frac{2}{\pi} \arcsin(f(t)), \tag{10}$$

where we assume that the signal $f(t)$ belongs to the interval $[-1, 1]$. Therefore, the signal

$$g(t) = \frac{2}{\pi} \arcsin(f(t)) \tag{11}$$

should be loaded in the Arbitrary Waveform Generator (AWG) and generated with an amplitude of $V_\pi/2$.

This precompensated input signal is generated with a sample rate close to 200 MS/s and a resolution of 16 bits (NI PXI-5422). The intensity profile of the optical signal sent to the reservoir computer is thus given by

$$I_{\text{in}}(t) = I_0 u(t), \tag{12}$$

where the input has been scaled to belong to the interval $u(t) \in [0, 1]$.

The multiplication by the input mask is achieved with the same sample rate and resolution (200 MS/s and 16 bits) by another AWG (Tabor WW2074) that drives an additional M–Z modulator (Photline MXAN-LN-10). The optical signal after multiplication by the input mask has intensity

$$\begin{aligned} I(t) &= m(t) I_{\text{in}}(t) \\ &= m(t) u(t) I_0, \end{aligned} \tag{13}$$

where the mask is scaled to belong to the interval $m(t) \in [0, 1]$, and for simplicity we have not written the insertion losses of the M–Z modulator.

A tuneable optical attenuator (Agilent 81571A) is used to adjust the strength of the input signal (β coefficient in equation (4)). The use of an incoherent light source (SLED) avoids interference between the signal injected into the cavity and the signal already present in the cavity (which is coherent since it comes from a laser). Therefore, at the output of the 50 % fiber coupler, the feedback photodiode produces an electrical signal proportional to $\alpha x(t - T) + \beta(t) u(t)$ (compare with equation (8)).

Concerning the choice of input mask $m(t)$, we use sinusoidal signals, as in [30]. The simplest mask signal of this type would be a single sine at frequency p/T' with p integer

$$m_p(t) = \frac{1}{2}\left(1 + \sin\left(-\frac{\pi}{2}\cos\left(2\pi\frac{p}{T'}t\right)\right)\right). \tag{14}$$

However, the performances depend strongly on the value of p. For a good choice of p, the results are close to those we can obtain with a random input mask. However, this is true only when the output is post-processed digitally. When the results are obtained with the experimental analogue readout layer they are significantly less good than those resulting from a random mask.

We found that the performance is significantly improved when we use an input mask m_{pq} containing two frequencies p/T' and q/T'

$$m_{pq}(t) = \frac{1}{2}\left(1 + \sin\left(-\frac{\pi}{4}\cos\left(2\pi\frac{p}{T'}t\right) - \frac{\pi}{4}\cos\left(2\pi\frac{q}{T'}\right)\right)\right). \tag{15}$$

The phase of the cosines in equation (15) is chosen in order to ensure that the mask vanishes at times $t = nT'$ when the input $u(t)$ has discontinuities. The signal sent into the cavity is thus a smooth function without any discontinuity and the synchronization between the input signal and the mask is drastically simplified.

A trace of the masked input signal is given in Figure 8.3. Note that it was necessary to scan the values of p and q to get good results.

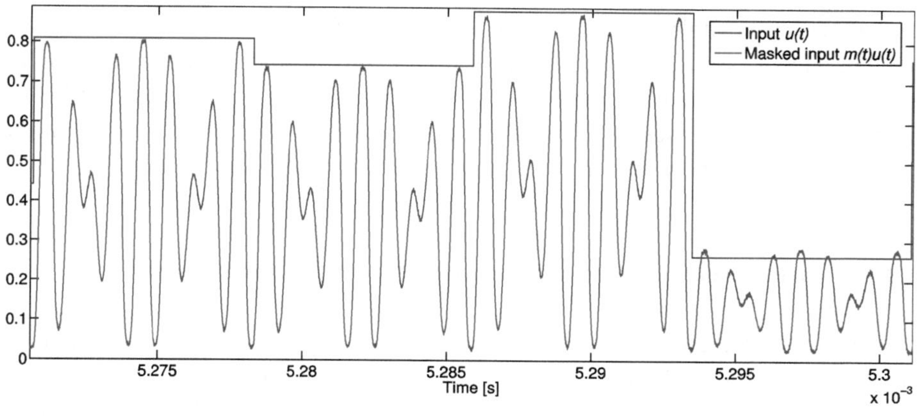

Figure 8.3: Signals injected into the reservoir layer. The blue curve is the optical input $I_{in}(t) = I_0 u(t)$. The green curve is a record of the masked input signal $m_{pq}(t)I_{in}(t)$ with $p = 7$ and $q = 9$, as measured by the photodiode and the digitizer. The vertical axis is scaled so that its maximum range is $[0, 1]$, i. e., $I_0 = 1$. Reprinted from [10].

8.3.1.2 Analogue output layer—general principle

The readout layer is in charge of producing the output $y(n)$ of the reservoir. It consists of two parts, the first measures the internal states $x(t)$ of the reservoir. The second produces the output itself.

As shown in Figure 8.2, 30 % of the light intensity sent to the reservoir layer is detected by the readout photodiode (TTI TIA525). The resulting signal is recorded by a digitizer (NI PXI-5124) at 200 MS/s with a resolution of 12 bits and a bandwidth of 150 MHz. This signal is used during the training phase to compute the values of the internal variables $x_i(n)$ and of the readout coefficients W_i (following the method described below). The remaining 70 % of light intensity is modulated by a dual output M–Z modulator (Photline MXDO-LN-10 with 10 GHz of bandwidth) using a signal produced by an AWG (Tabor WW2074). The two outputs of this modulator are complementary and detected by a balanced photodiode (TTI TIA527 with a cut-off frequency of 125 MHz and output impedance 50 Ω). The bias point of this modulator is regularly tuned to have a sine response. In other words, if no signal is applied on the RF port of the M–Z modulator, both outputs have a transparency of 50 % and the signal at the output of the balanced photodiode is null. If a positive (negative) voltage drives the M–Z modulator, the signal at the output of the balance photodiode is positive (negative). The reason for constructing the readout layer in this way is that the internal variables are given by the optical intensity inside the reservoir, hence their values are positive. But for processing information with the reservoir computer, positive and negative readout coefficients W_i are required. Using a dual output M–Z modulator coupled to a balanced photodiode enables us to modulate the internal variables by coefficients that are either positive or negative. The signal from the balanced photodiode is filtered by a low-pass RLC filter whose role is to carry out an analogue summation of the weighted internal variables. The output of the low-pass filter is then amplified before being recorded by the second channel of the digitizer (NI PXI-5124). The value of the resulting signal at every instant $t = nT'$ is the output of the reservoir $y(n)$.

8.3.1.3 Analogue output layer—computation of the readout coefficients

The balanced M–Z modulator in the output layer is driven by a signal produced by an AWG. Using the method described below, one computes a continuous time weight function $w(t)$. The signal produced by the AWG is precompensated so that the signal at the output of the balanced photodiode is proportional to $w(t)x(t)$.

In the case of a dual output M–Z modulator, the light intensities I_{out1} and I_{out2} at the two outputs of the modulator are similarly given by

$$I_{\text{out1}}(t) = \rho I_{\text{in}}\frac{1}{2}\left(1 + \sin\left(\frac{\pi}{V_\pi}\frac{V_\pi}{2}g(t)\right)\right) = \rho I_{\text{in}}\frac{1}{2}(1 + f(t)) \tag{16a}$$

$$I_{\text{out2}}(t) = \rho I_{\text{in}}\frac{1}{2}\left(1 - \sin\left(\frac{\pi}{V_\pi}\frac{V_\pi}{2}g(t)\right)\right) = \rho I_{\text{in}}\frac{1}{2}(1 - f(t)). \tag{16b}$$

Hence, when detecting these two outputs with a balanced photodiode, the resulting signal should be proportional to the light intensity at the input of the modulator multiplied by the signal $f(t)$.

Figure 8.4 gives the response after the balanced photodiode with and without precompensation when a ramp of 47 weights W_i ranging from –1 to 1 is use for the analogue readout mask.

Denoting by $h(t)$ the impulse response of the RLC filter followed by the amplifier, the signal $y_c(t)$ detected by the second channel of the digitizer can be expressed as

$$\begin{aligned} y_c(t) &= (w(t)x(t)) * h(t) \\ &= \int w(\tau)x(\tau)h(t-\tau)\,d\tau. \end{aligned} \tag{17}$$

Since we use a real (causal) filter, the integration in equation (17) is done over the interval $\tau \in]-\infty, t]$. The continuous time weight function $w(t)$ is a stepwise function of period T' defined by

$$w(t) = W_i \quad \text{for } nT' + (i-1)\theta \le t \le nT' + i\theta, \tag{18}$$

with $1 \le i \le N$ and $n \in \mathbb{Z}$, where we recall that θ is the duration of each internal variable. The output of the reservoir computer $y(n)$ is a function of discrete time. It is

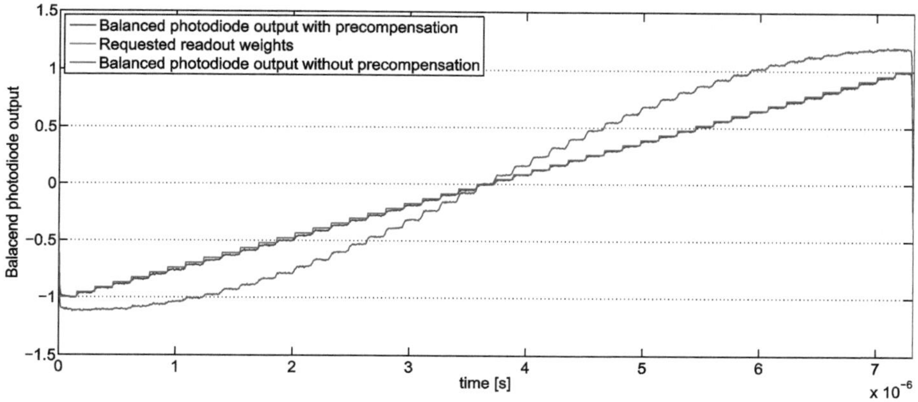

Figure 8.4: Precompensation of the dual output M–Z modulator. The signal $f(t)$ used for this test is a stepwise function of 47 values from −1 to 1 over the period T' (red curve). The green record is the output of the balanced photodiode when no precompensation is applied on $f(t)$. The blue curve is normalized output of the balanced photodiode when the precompensation is used. Reprinted from [10].

equal to the continuous output $y_c(t)$ at time nT': $y(n) = y_c(nT')$. It can be expressed as

$$\begin{aligned} y(n) &= y_c(nT') \\ &= \sum_{r=0}^{n-1} \sum_{i=0}^{N} W_i \int_{rT'+(i-1)\theta}^{rT'+i\theta} x(\tau)h(nT' - \tau)\, d\tau. \end{aligned} \tag{19}$$

In order to calculate the readout coefficients for the analogue readout layer, new internal variables $x_i(n)$ are defined by

$$x_i(n) = \sum_{r=0}^{n-1} \int_{rT'+(i-1)\theta}^{rT'+i\theta} x(\tau)h(nT' - \tau)\, d\tau. \tag{20}$$

In practice, the impulse response of the readout layer has a finite length. Let l be an integer such that the impulse response is shorter than lT' (i. e., $h(t) \approx 0$ for $t > lT'$), the sum over r in equations (19) and (20) can be limited to values of r from $n - 1 - l$ to $n - 1$. Note that since the impulse response lasts longer than T', the current output $y(n)$ contains contributions from the light intensity $x(t)$ up to l input periods in the past, which is a small difference with respect to the traditional reservoir computer; see equation (3). In our experiment, for the channel equalization task we use $l = 10$, and for NARMA10 and the radar signal forecasting, $l = 15$.

At the beginning of the experiment, we record the step response (response to the Heaviside function) of the analogue readout layer by applying a voltage step on the dual output M–Z modulator. The derivative of the recorded signal is the impulse response $h(t)$ of the analogue readout layer.

Note that a key point to obtain good results is to optimize the extinction of the signal when a readout weight equal to zero is applied. Indeed, the balanced M–Z modulator and the balanced photodiode have some imperfections, namely different insertion losses and on/off ratios for the two outputs of the modulator, and the difference in responsivity of each photodiode. For these reasons, a null voltage on the RF port of the M–Z modulator does not give, at the output of the readout layer, full extinction of the optical input. This effect, if not taken into account, degrades the performance of the output layer. To compensate for it, we measure the small offset needed to obtain a full extinction of the signal at the output of the reservoir layer. This offset is taken into account when we precompensate the readout mask.

During the training phase, we record the output $x(t)$ of the reservoir using the readout photodiode (first channel of the digitizer). This record is then combined with the impulse response $h(t)$ of the analogue readout layer to compute the new internal variables $x_i(n)$ (see equation (20)). From these internal variables, we compute the readout weights W_i using Tikhonov (ridge) regularization [35]. The corresponding stepwise periodic signal $w(t)$ is normalized with the highest absolute value of W_i, so as to fit the maximum modulation capabilities of the analogue readout layer. The corresponding gain (the highest absolute value of W_i) is applied on the recorded signal after acquisition of $y_c(t)$ and finally an offset correction is applied.

Note that the AWG that produces the output signal $w(t)$ has a finite resolution and, therefore, exhibits quantification noise which degrades the quality of the output $y_c(t)$. This effect is minimized if the amplitudes of the W_i are all comparable. This can be enforced by increasing the ridge regularization parameter. In the present experiments, we found it useful to take a ridge regularization parameter 10 times larger than when we use a digital output layer.

The performance of the analogue output layer is obviously dependent on the impulse response $h(t)$, and different tasks work better with different impulse responses. In practice, we first tested numerically different choices of R, L, and C, and then implemented experimentally those that provide good results. Typical values used are R in the range 1.6 kΩ–10 kΩ, C in the range 760 pF to 1.2 nF, with $L = 1.8$ mH.

8.3.2 Results

The fully analogue reservoir computer was tested on three tasks commonly considered in the reservoir computing community, namely equalization of a nonlinear communication channel, NARMA10, and the forecast of a radar signal. The results are compared to [4], in which a practically identical optoelectronic reservoir computer was used, and in the case of the radar task with the all-optical reservoir [5]. Both [4] and [5] used a similar number of internal variables, but without analogue input and output layers.

In all cases, we used $N = 47$ internal variables and $k = 5$. The two frequencies of the input mask p/T' and q/T', are $7/T'$ and $9/T'$. Either numerically before the experiment or during the experiment itself, the feedback gain (α in equation (4)) and the input gain (β in equation (4)) are scanned in order to find their optimal values. For each set of parameters, several data sets are used in the experiment in order to have sufficient statistics. In our experiment, a feedback gain α equal to 1 is obtained when the optical attenuator inside the loop is set to 9.5 dB. At this attenuation, when no input signal is applied, small oscillations appear in the cavity. This corresponds to a maximum optical power received by the feedback photodiode (i. e., at maximum transparency of the M–Z modulator inside the loop) of 264.4 µW. For comparison, the optical signal received by the feedback photodiode when the input is on, and the optical attenuator in the input layer is set to 0 dB, is 1.46 mW. When the input of the reservoir belongs to the interval $[0, 1]$, the input optical attenuation to obtain a β coefficient of 1 is around 7.4 dB.

It is important to note that we do not carry out any time averaging on the acquired signal $y_c(t)$. For this reason, the output suffers from quantification noise (see discussion below). Moreover, note that because each data set is sent to the experiment twice (once to measure $x(t)$ and compute $w(t)$, once to measure $y_c(t)$), the stability of the experiment is more important than in experiments with digital postprocessing. To ensure stability, we regularly adjust the working points of all M–Z modulators.

8.3.2.1 Nonlinear channel equalization

The aim of this task is to compensate for the distortion of a wireless communication link affected by a small nonlinearity and a memory effect. It was used previously in the reservoir computing literature; see, e. g., [14, 36]. A sequence of symbols $d(n)$, randomly drawn from the set of values $\{-3, -1, 1, 3\}$, passes through a channel model with intersymbols interference (due to multipath travels and/or band-pass filters at the channel ends)

$$\begin{aligned} q(n) = {} & 0.08d(n+2) - 0.12d(n+1) + d(n) \\ & + 0.18d(n-1) - 0.1d(n-2) + 0.091d(n-3) \\ & - 0.05d(n-4) + 0.04d(n-5) + 0.03d(n-6) \\ & + 0.01d(n-7), \end{aligned} \tag{21}$$

followed by a nonlinear transformation

$$u(n) = q(n) + 0.036q^2(n) - 0.011q^3(n) + \text{noise}. \tag{22}$$

The signal to noise ratio (SNR) is scanned from 12–32 dB using a step of 4 dB. The input of the reservoir computer is the noisy and distorted sequence $u(n)$, while the target

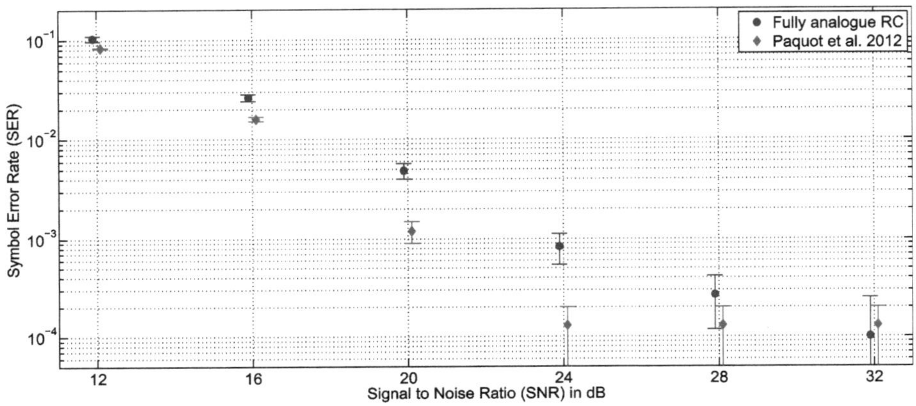

Figure 8.5: Results obtained for the equalization of the nonlinear channel for signal to noise ratios (SNR) ranging from 12 to 32 dB. For each SNR, the symbol error rate (SER) is given with its corresponding error bar over 5 data sets. The blue circles are the results obtained with the full analogue reservoir, and the red diamonds are the results presented in [4] (similar optoelectronic reservoir computer, but without the analogue input and output layers). Reprinted from [10].

output is the original sequence of symbols $d(n)$. For each SNR, the quality of the equalization is given by the symbol error rate (SER). We use 5 different data sets. For each data set, the reservoir is trained over 3000 time steps, and then a second sequence of 6000 time steps is used to test its performances (evaluate the SER). Results are presented with their corresponding standard deviation in Figure 8.5. A slight degradation is observed compared to the results obtained in [4]. The presented results are significantly better than those presented in [26] (for instance, at SNR of 32 dB, a SER of 10^{-4} compared to 10^{-2}). This is due in part to the larger number of internal variables that are used (47 instead of 28), but also to a better characterization of the output layer, and a better choice of the output filter impulse response.

For this task, the feedback optical attenuator is set to 11.25 dB, the input optical attenuator is set to 5 dB, and the analogue output layer had parameters $R = 1.6\,\mathrm{k\Omega}$, $C = 1.2\,\mathrm{nF}$, $L = 1.8\,\mathrm{mH}$. Figures 8.6, 8.7, and 8.8 show the measured impulse and step responses of the analogue readout layer, a sample of the readout signal $y(t)$, and a plot of the readout weights W_i.

8.3.2.2 NARMA10

The aim of this task is to train a reservoir computer to behave like a 10th order Nonlinear Auto Regressive Moving Average system (NARMA10) in which an input $u(n)$, randomly drawn from a uniform distribution over the interval $[0, 0.5]$, is injected. The

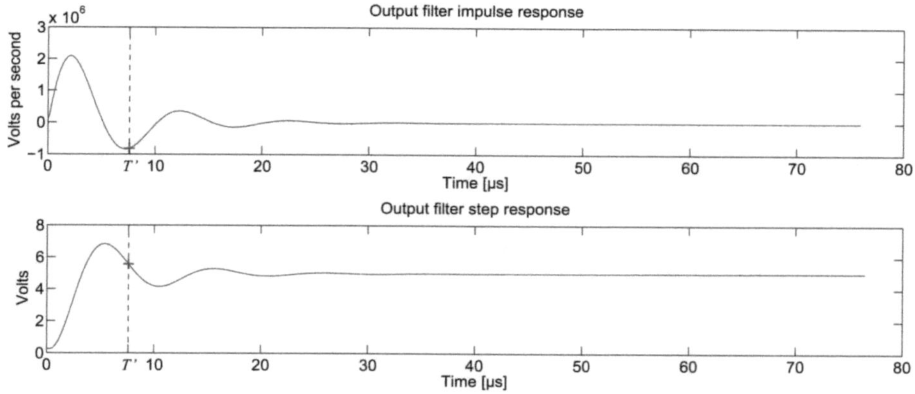

Figure 8.6: Impulse and step responses of the analogue readout layer used for the equalization of a nonlinear channel. The step response is recorded at the beginning of the experiment. Its derivative gives the impulse response of the analogue readout layer. The red cross gives the signal value at $T' = 7.598\,\mu s$. Reprinted from [10].

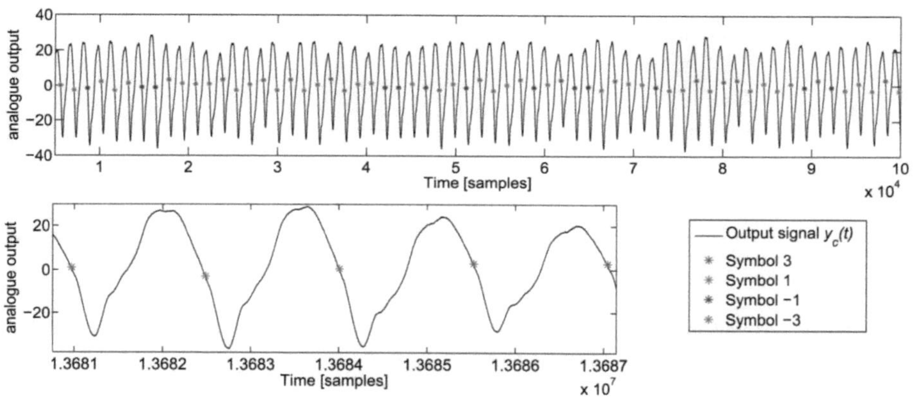

Figure 8.7: Signal at the output of the analogue readout layer for the nonlinear channel equalization task. The time is in number of samples (at 200 MS/s). The black curve is the acquired signal $y_c(t)$ with a final gain correction (multiplication by the maximum absolute value of the readout weights W_i). The stars are the output values $y_c(nT') = y(n)$. The different colours correspond to the different symbol values. Reprinted from [10].

following equation defines the targeted output:

$$\hat{y}(n+1) = 0.3\hat{y}(n) + 0.5\hat{y}(n)\left(\sum_{i=0}^{9} \hat{y}(n-i)\right) + 1.5u(n-9)u(n) + 0.1. \tag{23}$$

For this task, the reservoir is trained over a sequence of 1000 time steps and tested over another sequence of 1000 time steps; this process is repeated 10 times to obtain

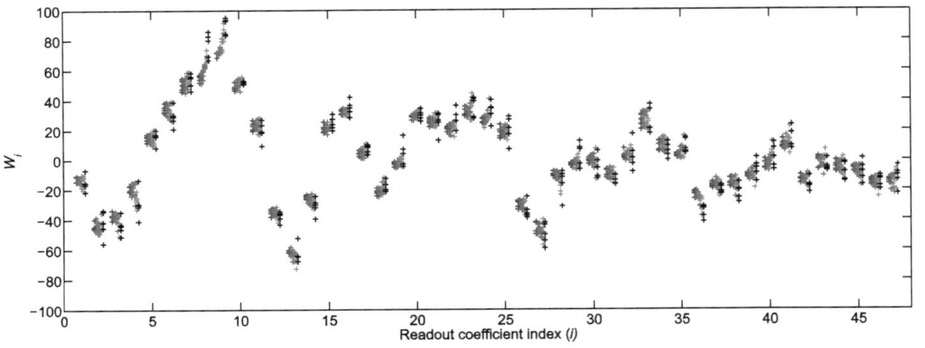

Figure 8.8: Readout coefficients for nonlinear channel equalization. The readout coefficients W_i are given for the six investigated SNRs: 32 dB SNR is in blue, 28 dB in red, 24 dB in green, 20 dB in magenta, 16 dB in cyan, and 12 dB in black. The output signal is taken at the end of the 47th internal variable. Vertical scale is arbitrary. For each investigated SNR, five independent data sets were used. The readout coefficients were computed independently for each data set. For each SNR, we have plotted these five sets of readout coefficients. Thus, for each index i we have plotted 30 values W_i (5 data sets per SNR and 6 SNRs). One sees from the figure that the values W_i for a given index i are all very similar. This is not unexpected since the tasks corresponding to different SNRs are very similar. Reprinted from [10].

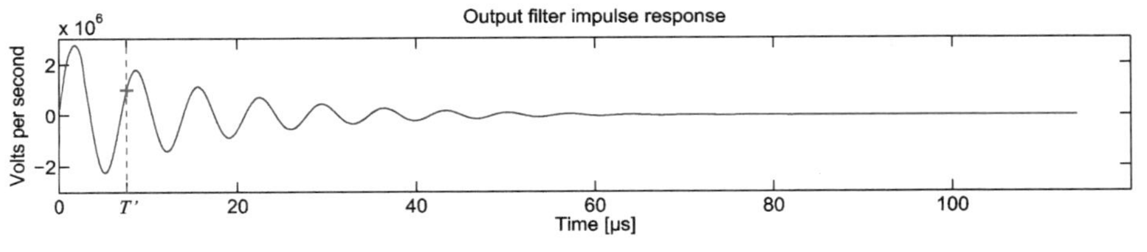

Figure 8.9: Impulse response of the analogue readout layer for NARMA10. The red cross gives the signal value at $T' = 7.598\,\mu s$. Reprinted from [10].

the statistics. The performance on this task is measured using the NMSE. This task is commonly studied in the reservoir computing community; see, e. g., [36, 37].

For this task, the feedback optical attenuator is set to 9.2 dB (i. e., slightly above the threshold for oscillations), the input optical attenuator is set to 9.5 dB, and the analogue output layer had parameters $R = 10\,k\Omega$, $C = 760\,pF$, $L = 1.8\,mH$. The impulse response of the analogue readout layer is given in Figure 8.9.

The test NMSE for the fully-analogue system is 0.230 ± 0.023. For the sake of comparison, note that a reservoir that carries out no computation (i. e., produces a time independent output $y(n) = \text{const}$) has a NMSE = 1, the system reported in [4] provides a NMSE of 0.168 ± 0.015, an ideal linear shift register (no nonlinearity in the reservoir) can reach a NMSE of 0.16, and using a different experimental architecture based on a coherently driven passive cavity a NMSE as low as 0.107 ± 0.012 was reported in [38].

Note that the fully-analogue performance is slightly worse than that of the linear shift register but significantly better than a system that carries out no computation.

8.3.2.3 Radar signal forecasting

This task consists in predicting a radar signal one to ten time steps in the future from a radar signal backscattered from the ocean surface (data collected by the McMaster University IPIX radar). The quality of the forecasting is evaluated by computing the NMSE between the predicted signal and the actual data one to ten time steps in the future. The experiment uses a single recorded radar signal under low sea state conditions, corresponding to an average wave height of 0.8 m (max 1.3 m). The recorded signal has two dimensions, corresponding to the in-phase and in-quadrature outputs (resp., I and Q) of the radar demodulator. Therefore, for each data set, the in-phase and in-quadrature signals are successively processed (predicted) by the experiment. The training and test sequences contain 1000 inputs each. This task has been previously used to evaluate the performance of reservoir computers, see e. g. [36, 39]. The results are presented in Figure 8.10.

For this task, the feedback optical attenuator is set to 9.9 dB, the input optical attenuator varied between 7 and 10 dB, and the analogue output layer had parameters $R = 10\,\mathrm{k\Omega}$, $C = 810\,\mathrm{pF}$, $L = 1.8\,\mathrm{mH}$. The impulse response of the analogue readout layer is given in Figure 8.11.

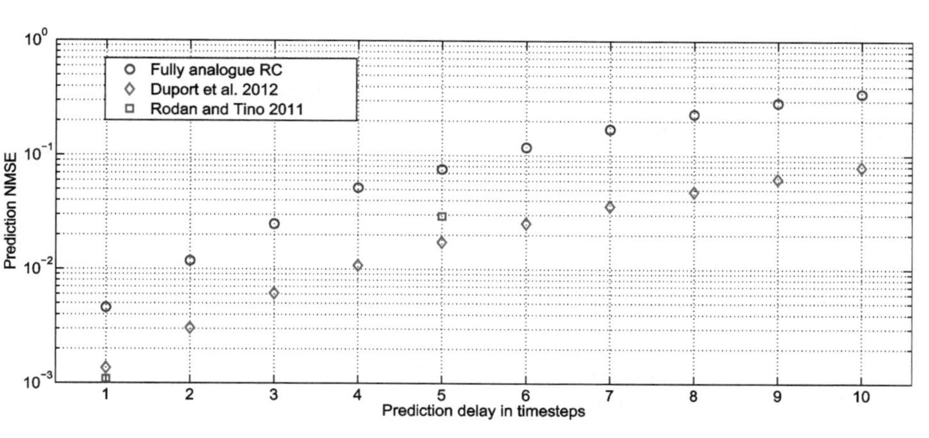

Figure 8.10: Radar signal forecasting error (NMSE) with respect to the number of time steps of the prediction (one to ten time steps in the future). The blue circles are the results obtained with the fully analogue reservoir, the red diamonds are the results published in [5] obtained with an all optical reservoir computer with similar number of internal variables, but without the analogue input and output layers. The green squares are the numerical results of [36]. Reprinted from [10].

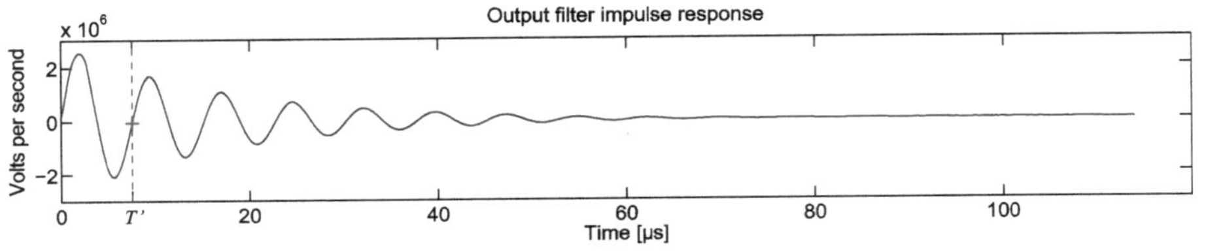

Figure 8.11: Impulse response of the analogue readout layer used for the radar signal forecasting. The red cross gives the signal value at $T' = 7.598$ μs. Reprinted from [10].

8.3.3 Discussion

The novel feature of the experimental reservoir computer presented here is the simultaneous inclusion of analogue input and output layers. The interest of this configuration is that it represents the necessary step toward the development of stand-alone reservoir computers for future complex and high-bandwidth applications. The only role of the external computer in our experiment is to compute the output weight function $w(t)$.

Concerning the analogue input layer, we proposed the use of sinusoidal functions as input mask, as these will be simple to generate in future hardware implementations. Upon using the sum of two sines as input mask, we did not observe significant degradation of performance compared to using the standard random step function.

As for the analogue output layer, whose aim is to produce a linear combination of the internal states that yields the desired output, the key difficulty is the accuracy needed in the summation that involves a large number of adjustable factors (the output weights).

The results presented here are obtained without any temporal averaging of the recorded signal, which makes them sensitive to quantification noise. This is important in our case since the total range of the output signal $y_c(t)$ is much larger than the range of the outputs $y_c(nT') = y(n)$; see Figure 8.7. In the case of the channel equalization task, which essentially constitutes a classification task, quantification noise is not such a problem since a signal that is correctly classified will in general continue to be so if a small amount of noise is added. But in the case of NARMA10 and the radar task, we measure the performance by how close the output is to the desired output using the NMSE. Quantification noise then directly affects the performances. Note that the effects of quantification noise and methods to counteract it have been studied previously in the context of reservoir computing in [27, 28].

Quantification noise also affects the readout mask $w(t)$. For this reason, the ridge regularization parameter was optimized in order to minimize the range of $w(t)$, as discussed in Section 8.3.1.3.

We note that for different tasks, different output filters were used (values of the constants R, L, and C). We do not have a complete explanation of why the optimal filters are different for each task. In a previous work [26], a simpler RC filter was used. This filter typically has a long impulse response, but on the other hand the resulting signal is much smaller, which leads to an increase of the output quantification noise. In the present work, we used a second-order RLC filter that also exhibits a long impulse response, but keeps a larger signal range.

In summary, we presented here the first study of a fully analogue reservoir computer, initially reported in [10]. At stake is the development of future analogue computers dedicated to complex and high-bandwidth signal-processing tasks. Due to the added complexity of this experiment, some degradation of performance is naturally observed compared to previous experiments in which the input and output layers were implemented digitally through digital pre- and post-processing. However, this present experiment can be considered as a proof of principle that suggests the feasibility of fully stand-alone reservoir computers. In this sense our work can also be seen as an important step toward the development of novel applications in which reservoir computers are cascaded or looped on themselves. As emphasized in the above discussion, many technical problems remain to be solved. For instance, some of the difficulties related to the use of fast-electronics may be circumvented by an all-optical output layer.

8.4 Online training

The performance of hardware reservoir computers relies on the training algorithm used to compute the readout weights. Offline learning methods, used up to now in experimental implementations [1–6, 9, 40, 41] provide good results, but become detrimental in real-time applications, as they require large amounts of data to be transferred from the experiment to the post-processing computer. This operation may take longer than the time it takes the reservoir to process the input sequence [4, 5, 41]. Moreover, offline training is only suited for time-independent tasks, which is not always the case in real-life applications. The alternative (and more biologically plausible) approach is to progressively adjust the readout weights using various recursive learning algorithms such as simple gradient descent, recursive least squares or reward-modulated Hebbian learning [42]. Such procedures can be implemented so as to require minimal data storage, with the advantage of being able to deal with a variable task: should any parameters of the task be altered during the training phase, the reservoir computer would still be able to produce good results by properly adjusting the readout weights.

The use of a fast computing unit, such as a FPGA board, is inevitable in the context of online learning, as the system needs to be trained in real time, that is, in parallel with the optoelectronic experiment. Such a system could, in principle, be applied to any kind of signal processing tasks, in particular to those that depend on time. A good example of such a task is the wireless channel equalization.

Indeed wireless communications is by far the fastest growing segment of the communications industry. The increasing demand for higher bandwidths requires pushing the signal amplifiers close to the saturation point which adds significant nonlinear distortions to the channel. These have to be compensated by an equalizer on the receiver side [43]. The main bottleneck lies in the Analogue-to-Digital Converters (ADCs) that have to follow the high bandwidth of the channel with sufficient resolution to sample correctly the distorted signal [44]. Current manufacturing techniques allow producing fast ADCs with low resolution, or slow ones with high resolution, combining both being very costly. This is where analogue equalizers become interesting, as they could equalize the signal before the ADC and significantly reduce the required resolution of the converters, thus potentially cutting costs and power consumption.

The results presented in this section are based on the experiment published in [11], in which the online approach was applied to the optoelectronic reservoir computer, described in Section 8.2, to demonstrate that such an implementation would be well suited for real-time data processing, and in particular for equalization of time varying communication channels. Section 8.4.1 presents the online learning algorithm, Section 8.4.2 outlines the specific features of the experimental setup, and Section 8.4.3 contains the results of this investigation.

8.4.1 Stochastic gradient descent algorithm

Gradient descent is a simple and popular recursive optimization algorithm. It is by far the most common method used to optimize neural networks. The idea is to compute the gradient of the cost function $E(w_i, x_i)$ in order to follow down the slope until the minimum is reached.

The gradient descent algorithm exists in three variants, depending on how much data is used to compute the gradient of the cost function at each iteration: stochastic, batch, and mini-batch [45]. Stochastic, or online, gradient descent updates the weights at each instance of the training set. It is usually fast to compute and can therefore be used online. However, applying multiple updates with high variance can cause heavy fluctuations of the cost function. The batch, or offline gradient descent, computes the average gradient over the entire training set available. This approach performs redundant computations, but avoids the fluctuations of the stochastic version. The mini-batch gradient descent attempts to combine the best of the stochastic and batch methods. As the name suggests, the training is performed over smaller sets of training instances, in order to reduce computational complexity, while maintaining accurate values of the gradient.

Online training of a hardware reservoir computer requires the use of the stochastic gradient descent algorithm. By definition, the updates of the parameters to optimize—

in this context, the readout weights w_i—are given by

$$w_i(n+1) = w_i(n) - \lambda \nabla_{w_i} E(n, w_i, x_i), \tag{24}$$

where $n \in \mathbb{Z}$ is the discrete time and λ is a user-defined coefficient called the *learning rate*. It controls the convergence speed of the parameters and allows to prevent divergence at early stages of the training process. The cost function $E(n, w_i, x_i)$ for an experimental reservoir computer on the channel equalization task is given by

$$E(n, w_i, x_i) = \left(d(n) - \sum_{i=1}^{N} w_i(n) x_i(n) \right)^2, \tag{25}$$

where $d(n)$ is the target output at timestep n. Therefore, the gradients with respect to the readout weights can be obtained as follows:

$$\begin{aligned}
\nabla_{w_i} E(n, w_i, x_i) &= \frac{\partial}{\partial w_i} \left(d(n) - \sum_{i=1}^{N} w_i(n) x_i(n) \right)^2 \\
&= \frac{\partial}{\partial w_i} \left(d(n)^2 - 2d(n) \sum_{i=1}^{N} w_i(n) x_i(n) + \left(\sum_{i=1}^{N} w_i(n) x_i(n) \right)^2 \right) \\
&= 2x_i(n) \left(\sum_{i=1}^{N} w_i(n) x_i(n) - d(n) \right) \\
&= 2x_i(n)(y(n) - d(n))
\end{aligned}$$

where $y(n)$ is the output of the reservoir computer at timestep n. Thus, the update rule for the readout weights (equation (24)) becomes

$$w_i(n+1) = w_i(n) - \lambda x_i(n)(y(n) - d(n)), \tag{26}$$

where the factor 2 has been absorbed by the learning rate λ for simplicity.

The learning rate parameter plays a great role in the performance of the training algorithm, affecting both the accuracy of the optimal solution and the time required to reach it. In the simplest case, the learning rate is set to a constant value. However, choosing the right value is already a challenge. A learning rate that is too small leads to slow convergence. If set too high, it could hinder convergence. Furthermore, this approach yields suboptimal results, in terms of convergence time, for more intricate cost functions. In fact, vanilla gradient descent has troubles progressing through ravines, that is, regions where the gradient is much steeper in one or several directions than the others [45]. Such scenarios are common around local optima and cause the algorithm to oscillate across the slopes with higher gradient and only make slow progress toward the optimum. Several gradient acceleration techniques have been developed to deal with this challenge, such as Nesterov momentum, Adadelta, RMSprop, and Nadam,

Table 8.1: Gradient descent algorithm parameters.

λ_0	λ_{min}	γ	k
0.4	0	0.999	10–50

to name a few. Since their discussion lies beyond the scope of the present chapter, we refer the reader to a comprehensive overview [45].

For simplicity of implementation on the FPGA chip, a learning rate schedule was used in the present work. Such a schedule adjusts the learning rate during training according to a certain fixed law. A popular and quite efficient example of learning rate schedule is annealing, in which the rate is reduced according to a pre-defined function. The evolution of the learning rate λ is given by

$$\lambda(n+1) = \lambda_{\text{min}} + \gamma(\lambda(n) - \lambda_{\text{min}}), \tag{27}$$

where γ is the annealing rate and λ_{min} is the minimal learning rate. The annealing starts at $\lambda(0) = \lambda_0$ and the learning rate is decreased every k timesteps.

In practice, for a stationary channel equalization, the training parameters are summarized in Table 8.1 Setting λ_{min} to zero means that the training process stops after a certain number of iterations, more precisely, when $\lambda(n)$ reaches the numerical precision of the FPGA. For time-dependent tasks, such as drifting and switching channels (that will be discussed Sections 8.4.3.2 and 8.4.3.3), the training needs to be continued to optimize the reservoir for the changing task. One then sets $\lambda_{\text{min}} > 0$, so that the readout weights can be adjusted as long as necessary.

8.4.2 Experimental setup

The experimental setup, depicted in Figure 8.12, contains three distinctive components: the optoelectronic reservoir, the FPGA board, and the computer.

The reservoir part has been discussed in Section 8.2. The input and output readout layers are carried out by the FPGA board. The synchronization of the latter with the reservoir delay loop is crucial for the performance of the experiment. For proper acquisition of reservoir states, the ADC has to output an integer number of samples per round-trip time. The daughter card contains a flexible clock tree that can drive the converters either from the internal clock source, or an external clock signal. The former being limited to the fixed frequencies of the onboard oscillator, the latter option is employed here. The clock signal is generated by a Hewlett Packard 8648A signal generator. With a reservoir of $N = 51$ neurons (one neuron is added to desynchronize the inputs from the reservoir, see Section 8.2) and a round-trip time of 7.94 µs, the sampling frequency is set to 128.4635 MHz, thus producing 20 samples per reservoir state. To get rid of the transients, induced mainly by the finite bandwidths of the ADC and

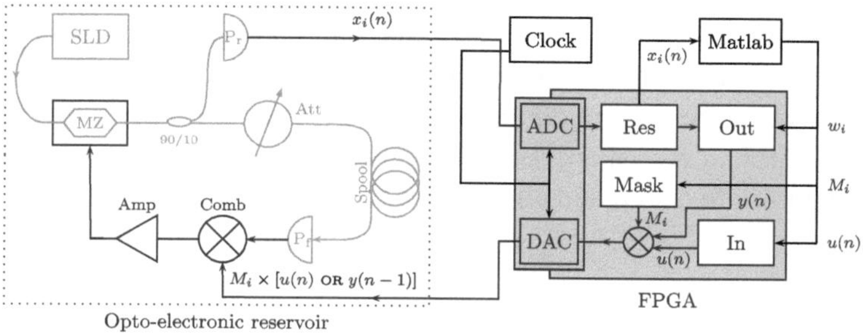

Figure 8.12: Schematic representation of the experimental setup used for online training. The opto-electronic reservoir is delimited with dotted lines and has been introduced in Section 8.2. The FPGA board implements both the input and output layers, generating the input symbols and training the readout weights. The computer controls the devices and records the results. Reprinted with permission from [11].

DAC, the 6 first and 6 last samples are discarded, and the neuron value is averaged over the remaining 8 samples.

The potentials of the electric signals to and from the daughter card need to be adjusted in order to achieve the most efficient interface without damaging the hardware. The DAC output voltage of $2\,V_{p\text{-}p}$ is sufficient for this experiment, as typical voltages of the input signal range between 100 mV and 200 mV. The ADC is also limited to $2\,V_{p\text{-}p}$ input voltage. With settings described in the previous section, the output voltage of the readout photodiode does not exceed $1\,V_{p\text{-}p}$.

Achieving the best performance from the experimental setup requires optimising its parameters, which are: the input gain β, the decay rate k, the channel signal-to-noise ratio and the feedback attenuation, that corresponds to the feedback gain parameter α in equation (4). The first three parameters are set on the FPGA board, while the last one is tuned on the optical attenuator. The input gain β is stored as a 18-bit precision real in $[0, 1[$ and was scanned in the $[0.1, 0.3]$ interval. The decay rate k is an integer, typically scanned from 10 up to 50 in a few wide steps. The noise ratios were set to several predefined values, in order to compare our results with previous reports. The feedback attenuation was scanned finely between 4.5 dB and 6 dB. Lower values would allow cavity oscillations to disturb the reservoir states, while higher values would not provide enough feedback to the reservoir.

The experiment is fully automated and controlled by a Matlab script, running on a computer. It is designed to run the experiment multiple times over a set of predefined values of parameters of interest and select the combination that yields the best results. For statistical purposes, each set of parameters is tested several times with different random input masks.

At launch, connections to the optical attenuator and the FPGA board are established, and the parameters on the devices are set to default values. After generating a set of random input masks, the experiment is run once and the elapsed time is measured. The duration of one run depends on the lengths of train and test sequences and varies from 6 s to 12 s. The script runs through all combinations of scanned parameters. For each combination, the values of the parameters are sent to the devices, and the experiment is run several times with different input masks and the resulting error rates are stored in the Matlab workspace. Once all the combinations are tested, the connections to the devices are closed and all collected data is saved to a file.

8.4.3 Results

Three different problems have been considered in this study. Section 8.4.3.1 proposes a comparison of online and offline training approaches on the same stationary channel equalization task, using very similar experimental setups. Sections 8.4.3.2 and 8.4.3.3 introduce the time variable to demonstrate the benefits of online training. A slowly drifting channel is considered in Section 8.4.3.2, while Section 8.4.3.3 deals with a switching channel.

8.4.3.1 Equalization of a stationary channel

The channel equalization task is described by equations (21) and (22) in Section 8.3.2.1. The noise term in equation (22) is given here by $v(n) = A \cdot r(n)$, where A is the amplitude, and $r(n)$ is drawn from a uniform distribution over the interval $[-1, +1]$ (for ease of implementation on the FPGA chip). Noise amplitude values A are chosen to produce the same signal-to-noise ratios as in [4, 5], where Gaussian noise was used.

Figure 8.13 presents the performance of the online-trained reservoir computer for different Signal-to-Noise Ratios (SNRs) of the wireless channel (black curve). For each SNR, the experiment was repeated 20 times with different random input masks. Average SERs are plotted on the graph, with error bars corresponding to maximal and minimal values obtained with particular masks. The RC performance was tested over one million symbols, and in the case of a noiseless channel the equalizer made zero error over the whole test sequence with most input masks.

The experimental parameters, such as the input gain β and the feedback attenuation α, were optimized independently for each input mask. The equalizer shows moderate dependence on both parameters, with an optimal input gain located within 0.225 ± 0.025 and an optimal feedback attenuation of 5.1 ± 0.3 dB.

For comparison, results reported in [4], obtained with the same optoelectronic reservoir, trained offline are plotted in Figure 8.13 with grey dots. For high noise levels (SNR $\leq$ 20 dB), the results are similar. For low noise levels (SNR $\geq$ 24 dB), the performance of the online-trained implementation is significantly better. Note that the previously reported results are only rough estimations of the equalizer's performance as the input sequence was limited by hardware to 6000 symbols [4]. In the present experiment, the SER is estimated more precisely over one million input symbols.

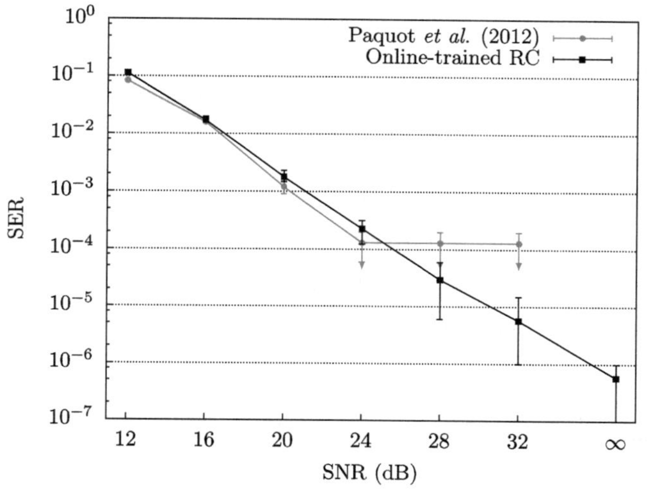

Figure 8.13: Experimental results on the stationary channel equalization task (black curve). Symbol Error Rates (SERs) are plotted against the Symbol-to-Noise Ratio (SNR). The equalizer was tested with 20 different random input masks over one million input symbols, average values are plotted on the graph. For the noiseless channel (SNR $= \infty$), for most choices of input mask, the RC made no errors over the test sequence. Grey dots show the results of the optoelectronic setup with offline training [4]. For low noise levels, the online-trained system produces error rates significantly lower than [4], and for noisy channels the results are similar. Reprinted with permission from [11].

For the lowest noise level (SER = 32 dB), a SER of 1.3×10^{-4} was reported in [4], while the online-trained reservoir yields an error rate of 5.71×10^{-6}. One should remember that common error detection schemes, used in real-life applications, require the SER to be lower than 10^{-3} in order to be efficient. SERs around 10^{-4} have been reported in [4, 5, 41] and a passive-cavity-based setup [38] achieved a 1.66×10^{-5} rate (this value is limited by the use of a 60000-symbol test sequence). However, one should keep in mind that, had it been possible to test [4] on a longer sequence, it is possible that comparable SERs would have been obtained. That is, online learning does not much improve the performance of a reservoir computer on a stationary task, but allows to test it on a longer test sequence and thus to accurately evaluate the error rate. The true strength of online learning resides in the adaptability to a changing environment, as will be shown in the following sections.

8.4.3.2 Equalization of a slowly drifting channel

The model given by equations (21) and (22) describes a stationary communication scheme, that is, the channel remains the same during the transmission. However, in wireless communications, the environment has a great impact on the received signal.

Given its highly variable nature, the properties of the channel may be subject to important changes in real time.

To investigate such scenarios, consider a more general channel model, given by

$$\begin{aligned} q(n) = {} & (0.08 + m)d(n + 2) - (0.12 + m)d(n + 1) \\ & + d(n) + (0.18 + m)d(n - 1) \\ & - (0.1 + m)d(n - 2) + (0.091 + m)d(n - 3) \\ & - (0.05 + m)d(n - 4) + (0.04 + m)d(n - 5) \\ & + (0.03 + m)d(n - 6) + (0.01 + m)d(n - 7), \end{aligned} \tag{28}$$

$$u(n) = p_1 q(n) + p_2 q^2(n) + p_3 q^3(n), \tag{29}$$

where the parameters p_i and m can be either stationary or time-dependent. Their default values are given by $m = 0$, $p_1 = 1$, $p_2 = 0.036$, and $p_3 = -0.011$ (see equations (21) and (22)).

To confront the online-trained reservoir computer with a nonstationary task, we performed a series of experiments with a "drifting" channel model, where parameters p_i or m_i were varying in real time during the signal transmission. These variations occurred at slow rates, much slower than the time required to train the reservoir computer. A simple real-life example of such a situation is a receiver moving away from the transmitter, causing the channel to drift more or less slowly, depending on the relative speed of the receiver. We studied two variation patterns: a monotonic increase (or decrease) and slow oscillations between two fixed values.

A drifting channel is a good example of a situation where training the reservoir online yields better results than offline. Numerical simulations have reported that training a reservoir computer offline on a nonstationary channel results in an error rate ten times worse than with online training [46]. The present work demonstrates that an online-trained experimental reservoir computer performs well even on a drifting channel if λ_{min} is set to a small nonzero value (see Section 8.4.1).

Figure 8.14(a) shows the experimental results for the case of monotonically decreasing p_1 from 1 to 0.652. The grey curve presents the resulting SER with $\lambda_{min} = 0$, that is, with training process stopped after 45000 input symbols. The black curve depicts the error rate obtained with $\lambda_{min} = 0.01$, so that the readout weight can be gradually adjusted as the channel drifts. Note that while in the first experiment the SER grows up to 0.329, it remains much lower in the second case. The increasing error rate in the latter case is due to the decrease of p_1 resulting in a more complex channel. This shows that the nonstationary version of the training algorithm allows a drifting channel to be equalized with a significantly lower error rate.

Figure 8.14(b) depicts error rates obtained with p_1 linearly oscillating between 1 and 0.688. With $\lambda_{min} = 0$ (grey curve) the error rate is as low as 1×10^{-4} when p_1 is around 1, and grows very high elsewhere. With $\lambda_{min} = 0.01$ (black curve), the obtained SER is always much lower, even at $p_1 = 0.688$, it stays at SER $= 5.0 \times 10^{-3}$.

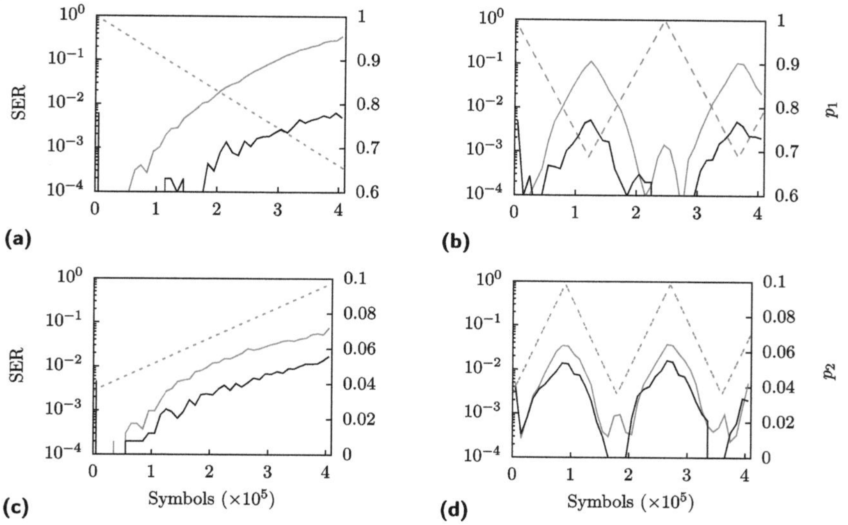

Figure 8.14: Symbol error rates (left axis, log scale), averaged over 10000 symbols, produced by the experimental setup with a drifting channel. Each panel presents data obtained from one experimental run with a fixed input mask and optimal parameters α, β and k. The grey curves show the results produced with $\lambda_{\min} = 0$, while black curves depict those obtained with the nonstationary version with $\lambda_{\min} > 0$ (see Section 8.4.1). Dashed lines display the evolution of parameters p_i and m (right axis, linear scale). (a) & (b) Monotonically decreasing and oscillating p_1. (c) & (d) Monotonically increasing and oscillating p_2. Reprinted with permission from [11].

We obtained similar results with the parameter p_2 (shown in Figure 8.14(c)–(d)), as well as p_3 and m (see [11]). Letting the reservoir computer adapt the readout weights by setting $\lambda_{\min} > 0$ produces notably lower error rates for a given channel, while stopping the training with $\lambda_{\min} = 0$ results in quickly growing SERs.

8.4.3.3 Equalization of a switching channel

In addition to slowly drifting parameters, the channel properties may be subject to abrupt variations due to sudden changes of the environment. For better practical equalization performance, it is crucial to be able to detect significant channel variations and adjust the RC readout weights in real time. We consider here the case of a "switching" channel, where the channel model switches instantaneously. The reservoir computer has to detect such changes and automatically trigger a new training phase, so that the readout weights get adapted for the equalization of the new channel. Specifically, we introduce three channels differing in nonlinearity (corresponding to 3 values of p_1):

$$u_1(n) = 1.00q(n) + 0.036q^2(n) - 0.011q^3(n), \tag{30a}$$

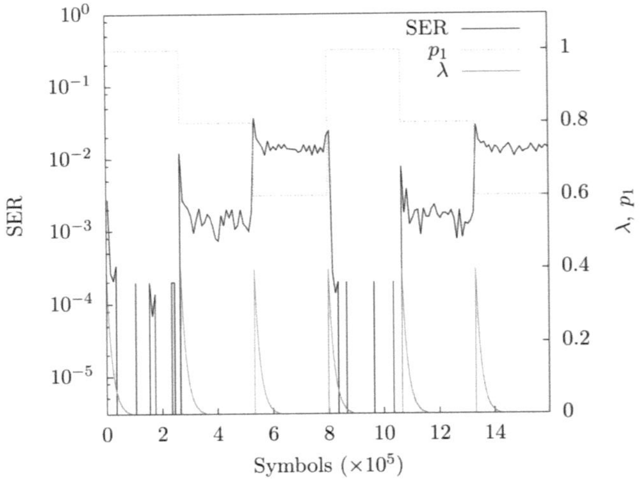

Figure 8.15: Symbol error rate (left axis, black curve), averaged over 10000 symbols, produced by the FPGA in case of a switching channel. The value of p_1 (right axis, dotted line) is modified every 266000 symbols. The change in channel is followed immediately by a steep increase of the SER. The λ parameter (right axis, grey curve) is automatically reset to $\lambda_0 = 0.4$ every time a performance degradation is detected, and then returns to its minimum value, as the equalizer adjusts to the new channel, bringing down the SER to its asymptotic value. Reprinted with permission from [11].

$$u_2(n) = 0.80q(n) + 0.036q^2(n) - 0.011q^3(n), \tag{30b}$$

$$u_3(n) = 0.60q(n) + 0.036q^2(n) - 0.011q^3(n), \tag{30c}$$

and switch regularly from one channel to another, keeping equation (21) unchanged.

Figure 8.15 shows the error rate produced by the experiment in case of a switching noiseless communication channel. The parameters of the channel are programmed to switch in cycle among equations (30) every 266000 symbols. Every switch is followed by a steep increase of the SER, as the reservoir computer is no longer optimized for the channel it is equalizing. The performance degradation is detected by the algorithm, causing the learning rate λ to be reset to the initial value λ_0, and the readout weights are retrained to new optimal values.

For each value of p_1, the reservoir computer is trained over 45000 symbols, then its performance is evaluated over the remaining 221000 symbols. In case of $p_1 = 1$, the average SER is 1×10^{-5}, which is the expected result. For $p_1 = 0.8$ and $p_1 = 0.6$, we compute average SERs of 7.1×10^{-4} and 1.3×10^{-2}, respectively, which are the best results achievable with such values of p_1 according to the previous experiment (see Figure 8.14 in Section 8.4.3.2). This shows that after each switch the readout weights are updated to new optimal values, producing the best error rate for the given channel.

Note that the current setup is rather slow for practical applications. With a round-trip time of $T = 7.94\,\mu\text{s}$, its bandwidth is limited to 126 kHz and training the reservoir over 45000 samples requires 0.36 s to complete. However, it demonstrates the potential of such systems in equalization of nonstationary channels. For real-life applications, such as for instance Wi-Fi 802.11g, a bandwidth of 20 MHz would be required. This could be realized with a 15 m fiber loop, thus resulting in a delay of $T = 50$ ns. This would also decrease the training time down to 2.2 ms and make the equalizer more suitable for realistic channel drifts. The speed limit of our setup is set by the bandwidth of the different components, and in particular of the ADC and DAC. For instance with $T = 50$ ns and keeping $N = 50$, reservoir states should have a duration of 1 ns, and hence the ADC and DAC should have bandwidths significantly above 1 GHz (such performance is readily available commercially). As an illustration of how a fast system would operate, we refer to the optical experiment [6] in which information was injected into a reservoir at rates beyond 1 GHz.

8.5 Output feedback

Forecasting is one of the primary problems in science: how can one predict the future from the past? Over the past few decades, artificial neural networks have gained a significant recognition in the time series forecasting community. Similar to previously employed statistics-based techniques, they are both data driven and nonlinear. Differently, they are more flexible and do not require an explicit model of the underlying process. A review of artificial neural networks models for time series forecasting can be found in [47]. Reservoir computing can be readily applied to short-term prediction tasks, that focus on generating a few future timesteps. As for long-horizon forecasting, that involves predicting the time series for as long as possible, it requires a small modification of the architecture, namely by feeding the RC output signal back into the reservoir. This additional feedback significantly enriches the internal dynamics of the reservoir, enabling it to generate time series autonomously, i. e., without receiving any input signal. With this modification, reservoir computing can be used for long-term prediction of chaotic series [14, 48–51]. In fact, to the best of our knowledge, this approach holds the record for such chaotic time series prediction [14, 51]. A reservoir computer with output feedback can also achieve the easier task of generating periodic signals [52–54], and sine waves with a tunable frequency [55–57].

The aim of the project described in this section and published in [19] is to explore these novel applications experimentally. This requires, in general, a readout layer fast enough to generate and feed the output signal back in real-time. Several analogue solutions have been investigated (see Section 8.3) but none are as yet capable of performing sufficiently well in this application. In fact, successful training of an analogue readout layer with offline learning methods, used in most experimental RC setups up to now, requires a very precise model of the readout setup, which is hardly achievable experimentally. As demonstrated in [10], it is virtually impossible to characterize each hardware component of the setup with sufficient level of accuracy. The reason for this difficulty is that the output is a weighted sum with positive and negative coefficients of the internal states of the reservoir. Therefore, errors in the coefficients can build up and become comparable to the value of the desired output. For this reason, we chose the approach of a real-time digital readout layer implemented on a FPGA chip. The use of high-speed dedicated electronics makes it possible to compute the output signal in real time, as has been demonstrated in Section 8.4, and feed it back into the reservoir. In order to keep the experiment simple, we used as reservoir the optoelectronic delay system presented in Section 8.2.

We show that an experimental reservoir computer can successfully solve two periodic time series generation tasks: frequency and random pattern generation, that have been previously investigated numerically [53, 55, 56]. The first task allows to reveal different timescales within the neural network, and the second can be used to quantify the memory of the reservoir. The photonic computer manages to generate both sine waves and random patterns with high stability (verified on the timescale of several days). Furthermore, we apply the RC to emulation of two chaotic attractors: Mackey–Glass [58] and Lorenz [59] systems. In the literature, the emulation performance on these tasks is quantified in terms of the prediction horizon, i. e., the duration for which the RC can accurately follow a given trajectory on the chaotic attractor [14]. However, this method is not applicable in the presence of a relatively high level of experimental noise, with a Signal-to-Noise Ratio (SNR) of roughly 40 dB, as will be discussed in Section 8.5.2.1. This noise was not problematic in previous experiments using the same optoelectronic reservoir [4, 11], but turns out to be intolerable for a system with output feedback. This raises the question of how to evaluate a system that emulates a known chaotic time series in the presence of noise. We introduce several new approaches, such as frequency spectrum comparison and randomness tests, based on well-known signal analysis techniques. Our results show that, although the RC struggles at following the target trajectory on the chaotic attractor, its output accurately reproduces the core characteristics of the target time series.

This section is structured as follows: Section 8.5.1 outlines the specific features of the experimental setup, and Section 8.5.2 contains the results of this investigation.

8.5.1 Experimental setup

The introduction of the output feedback requires a minor change of notation, used in Section 8.2. Since the RC can now receive two different signals as input, we shall denote $I(n)$ the input signal, which can be either the external input signal $I(n) = u(n)$, or its own output, delayed by one timestep $I(n) = y(n-1)$.

The reservoir computer is operated in two stages, depicted in Figure 8.16: a training phase and an autonomous run. During the training phase, the reservoir computer is driven by a time-multiplexed teacher signal $I(n) = u(n)$, and the resulting states of

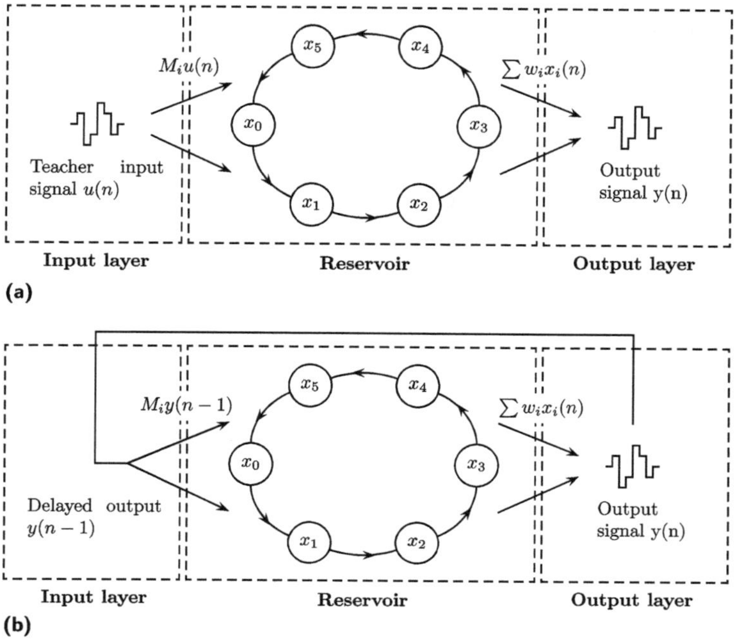

Figure 8.16: Schematic representation of (a) the training stage and (b) the autonomous run of a reservoir computer. For simplicity, a small network with $N = 6$ nodes is depicted. During the training phase, the reservoir is driven by a teacher input signal $u(n)$, and the readout weights w_i are optimized for the output $y(n)$ to match $u(n+1)$ as accurately as possible. During the autonomous run, the teacher signal $u(n)$ is switched off and the reservoir is driven by its own output signal $y(n)$. The readout weights w_i are fixed and the performance of the system is measured in terms of how long or how well it can generate the desired output. Reprinted with permission from [19].

the internal variables $x_i(n)$ are recorded. The teacher signal depends on the task under investigation. The system is trained to predict the next value of the teacher time series from the current one, i. e., the readout weights w_i are optimized so as to get as close as possible to $y(n) = u(n+1)$. The error is measured in terms of the mean-square error (MSE), defined as

$$\text{MSE} = \langle (y(n) - d(n))^2 \rangle. \tag{31}$$

Then the reservoir input is switched from the teacher sequence to the reservoir output signal $I(n) = y(n-1)$, and the system is left running autonomously. The reservoir output $y(n)$ is used to evaluate the performance of the experiment.

The experimental setup, schematized in Figure 8.17, consists of two main components: the optoelectronic reservoir and the FPGA board. The structure and operation of the optoelectronic reservoir have been discussed in Section 8.2. In this section, we will focus on a few particular aspects of this experiment.

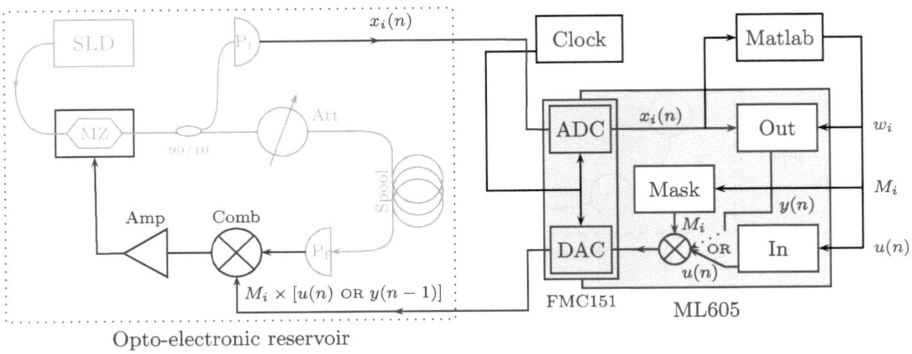

Figure 8.17: Schematic representation of the experimental setup used for output feedback. The optoelectronic reservoir has been introduced in Section 8.2. The FPGA board implements the readout layer and computes the output signal $y(n)$ in real time. It also generates the analogue input signal $I(n)$ and acquires the reservoir states $x_i(n)$. The computer, running Matlab, controls the devices, performs the offline training and uploads all the data ($u(n)$, w_i and M_i) on the FPGA. Reprinted with permission from [19].

With time-multiplexed neurons, the maximal reservoir size is imposed by the ratio between the delay from the fiber spool (Spool) and the sampling frequency of the ADC. While increasing the latter involves relatively high costs, one can lengthen the delay line fairly easily. In this work, we used two spools of single mode fiber with approximate lengths of 1.6 km and 10 km. The first spool produced a delay of 7.93 µs and could take in $N = 100$ neurons when sampled at 203.7831 MHz. The second spool was used to increase the delay up to 49.2 µs and the reservoir size up to $N = 600$, with a sampling frequency of 195.4472 MHz. In both cases, each state was averaged over 16 samples in order to decrease the noise and remove the transients induced by the finite bandwidth of the Digital-to-Analogue converter (DAC).

The experiment is operated as follows. First, the input mask M_i and the teacher signal $u(n)$ are generated in Matlab and uploaded on the FPGA board. The latter generates the masked input signal $M_i \times u(n)$ and sends it to the reservoir via the DAC. The resulting reservoir states $x_i(n)$ are sampled, averaged, and transferred to the computer in real time by the FPGA. That is, the FPGA design uses minimal memory (a small buffer for the Ethernet frames), and thus allows to capture the reservoir states without limitation of the time interval. After acquisition of the desired amount of data (reservoir states) in Matlab, the optimal readout weights w_i are computed and uploaded on the FPGA board. Because of the relatively long delay (compared to the microsecond timescale of the experiment) needed for the offline training, the reservoir needs to be reinitialized in order to restore the desired dynamics of the internal states prior to running it autonomously. To this end, we drive the system with an initialization sequence of 128 timesteps (as illustrated in Figure 8.23), before coupling the output signal with the input and letting the reservoir computer run autonomously. In this stage, the FPGA computes the output signal $y(n)$ in real time, then creates a masked version $M_i \times y(n)$ and sends it to the reservoir via the DAC.

As the neurons are processed sequentially, the output signal $y(n)$ can only be computed in time to update the 24th neuron $x_{23}(n+1)$. In other words, the first 23 neurons do not "see" the input signal $I(n)$ because it can not be computed and delivered in time. Therefore, we set the first 23 elements of the input mask M_i to zero. That way, all neurons can contribute to solving the task, despite the first 23 lacking the input information. Note that this reflects an aspect that is inherent to any experimental time-multiplexed reservoir computer with output feedback. Indeed, the output $y(n)$ has to be computed after the acquisition of the last neuron $x_{N-1}(n)$ at timestep n, but before the first neuron $x_0(n+1)$ of the following timestep. However, in time-multiplexed implementation of reservoir computing, these states are consecutive, and the experiment cannot be paused to let $y(n)$ be computed. Therefore, there may be a delay, whose duration depends on the hardware used, before $y(n)$ is computed and can be fed back into the reservoir. In the present experiment, this delay is approximately 115 ns, which corresponds to 23 neuron durations. This delay is mainly due to propagation times between the intensity modulator (MZ) and the ADC in one hand, and the DAC and the resistive combiner on the other. The FPGA computation time also plays a role here, but it does not exceed 20 ns with our design. This delay can have an impact on system performance [19].

8.5.2 Results

The output feedback allows the computer to generate time series autonomously, i. e., without any external input. We tested the capacity of the experiment to produce both periodic and chaotic signals, with two tasks in each category, that will be presented in Sections 8.5.2.2–8.5.2.5. The two periodic signal generation tasks were solved using a small reservoir with $N = 100$ and a fiber spool of approximately 1.6 km. The chaotic signal generation tasks, being more complex, required a large reservoir of $N = 600$ for sufficiently good results that we fit in a delay line of roughly 10 km. But first, we focus on the question of the experimental noise.

8.5.2.1 Experimental noise

For most tasks studied here, we found the experimental noise to be the major source of performance degradation in comparison to previously reported numerical investigations [49]. This disparity stems from an ideal noiseless reservoir considered in [49], while our experiment is noisy. This noise can come from the active and even passive components of the setup: the amplifier, which has a relatively high gain and is therefore very sensitive to small parasitic signals (e. g., from the power source), the DAC, the photodiodes, and the optical attenuator (shot noise). In-depth experimental investigations have shown that, in fact, each component contributes more or less equally to the overall noise level. Thus, it cannot be reduced by replacing a single component. Neither can it be averaged out, as the output value has to be computed at each timestep. This noise was found to have a marked impact on the results, as will be shown in the following sections. To further investigate this issue, we estimated the level of noise present in the experimental system and incorporated it to the numerical models. In particular, we developed three models that simulate the experiment to different degrees of accuracy.

Idealized model. It incorporates the core theoretical characteristics of our reservoir computer: the ring-like architecture, the sine nonlinearity, and the linear readout layer (as described by equations (1), (3), and (4) from (8.2)). All experimental considerations are disregarded. We use this model to define the maximal performance achievable in each configuration.

Noiseless experimental model. This model emulates the most influential features of the experimental setup: the high-pass filter of the amplifier, the finite resolution of the ADC and DAC, and precise input and feedback gains. This model allows to cross-check the experimental results and to identify the problematic points. That is, if the experiment performs much worse than the model, then most likely, something does not work as it should.

Noisy experimental model. To compare our experimental results with a more realistic model, we estimated the level of noise present in the experimental system (see below), and incorporated this noise into the noisy version of the experimental model.

Our custom Matlab scripts are based on [4, 49]. This modeling allows us to examine different levels of noise, and even switch it "off" completely, which is impossible experimentally.

Figure 8.18 shows numerical and experimental states of a reservoir with $N = 100$ neurons, as received by the readout photodiode. That is, the curves depict the time-multiplexed neurons: each point represents a reservoir state $x_{0,\dots,99}(n)$ at times $n = 1$ and $n = 2$. The system does not receive any input signal $I(n) = 0$. The experimental signal is plotted with a solid grey line. We use it to compute the experimental noise level by taking the standard deviation of the signal, which gives 2×10^{-3}. We then replicate this noise level in the noisy experimental model to compare experimental results to numerical simulations. The dotted black curve in Figure 8.18 shows the response of the noisy experimental model, with the same amount of Gaussian noise (standard deviation of 2.0×10^{-3}) as in the experiment. The choice of a Gaussian noise distribution was validated by experimental measurements.

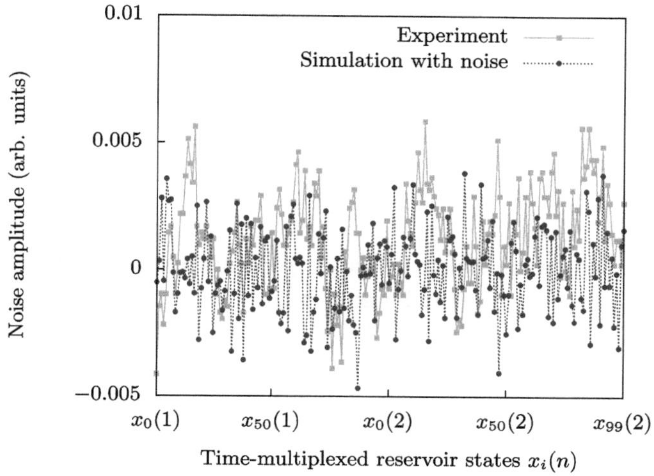

Figure 8.18: Illustration of the noisiness of the experimental reservoir. Experimental (solid grey line) and numerical (dotted black line) reservoir states $x_i(n)$ are scaled so that in normal experimental conditions (nonzero input) they would lie in a $[-1, 1]$ interval. Despite the null input signal $I(n) = 0$, the actual neurons are nonzero because of noise. Numerical noise was generated with a Gaussian random distribution with standard deviation of 1×10^{-3} so that to reproduce the noise level of the experiment. Reprinted with permission from [19].

The experimental noise level can also be characterized by the Signal-to-Noise Ratio (SNR), defined as [60]

$$\mathrm{SNR} = 10 \log_{10}\left(\frac{\mathrm{RMS}^2_{\mathrm{signal}}}{\mathrm{RMS}^2_{\mathrm{noise}}}\right),$$

where RMS is the Root Mean Square value, given by

$$\mathrm{RMS}(x_i) = \sqrt{\frac{1}{N}\sum_{i=1}^{N} x_i^2}.$$

We measured $\mathrm{RMS}_{\mathrm{signal}} = 0.2468$ and $\mathrm{RMS}_{\mathrm{noise}} = 0.0023$, so the SNR is equal to approximately 40 dB in this case. Note that this figure is given as an indicator of order of magnitude only as the RMS of the reservoir states depends on the gain parameters (α and β in equation (4) from (8.2)) and varies from one experiment to another.

8.5.2.2 Frequency generation

Frequency generation is the simplest time series generation task considered here. The system is trained to generate a sine wave given by

$$u(n) = \sin(\nu n), \tag{32}$$

where ν is a real-valued relative frequency. The physical frequency f of the sine wave depends on the experimental roundtrip time T (see Section 8.5.1) as follows:

$$f = \frac{\nu}{2\pi T}. \quad (33)$$

This task allows to measure the bandwidth of the system and investigate different timescales within the neural network.

We found the frequency generation task to be the only one not affected by noise: our experimental results matched accurately the numerical predictions reported in [56]. From this study, we expected a bandwidth of $\nu \in [0.06, \pi]$ with a 100-neuron reservoir. The upper limit is a signal oscillating between −1 and 1 and is given by half of the sampling rate of the system (the Nyquist frequency [61]). The lower limit is due to the reservoir memory. In fact, low-frequency oscillations correspond to longer periods, and the neural network can no longer "remember" a sufficiently long segment of the sine wave so as to keep generating a sinusoidal output. These numerical results are confirmed experimentally here.

We tested our setup on frequencies ν ranging from 0.01 to π. We found that frequencies within the $[0.1, \pi]$ interval are generate accurately with any random input mask. Lower frequencies between 0.01 and 0.1, on the other hand, were produced properly with some random masks, but not all. Since this is where the lower limit of the bandwidth lies, we investigated the [0.01, 0.1] interval more precisely. For each frequency, we ran the experiment 10 times for 10^4 timesteps with different random input masks and counted the number of times the reservoir produced a sine wave with the desired frequency (MSE $< 10^{-3}$) and amplitude of 1. The results are shown in Figure 8.19. Frequencies below 0.05 are not generated correctly with most input masks. At $\nu = 0.7$ the output is correct most of the times, and for $\nu = 0.08$ and above the output sine wave is correct with any input mask. Thus, we can conclude that the bandwidth of this experimental RC is $\nu \in [0.08, \pi]$. Considering the round-trip time $T = 7.93\,\mu\text{s}$, this results in a physical bandwidth of 1.5 – 63 kHz. Note that frequencies within this interval can be generated with any random input mask M_i. Lower frequencies, down to 0.02, could also be generated after choosing a suitable mask.

Figure 8.20 shows an example of the output signal during the autonomous run. The system was trained for 1000 timesteps to generate a frequency of $\nu = 0.1$, and successfully accomplished this task with a MSE of 5.6×10^{-9}.

These results were obtained by scanning the input gain β and the feedback gain α to obtain the best results. It was found that β has little impact on the system performance so long as it is chosen in the interval $\beta \in [0.02, 0.5]$, while the feedback gain α, on the contrary, has to lie within a narrow interval of $\alpha \in [4.25, 5.25]$ dB (this corresponds approximately to $\alpha \in [0.85, 0.95]$), otherwise the reservoir yields very poor results. The DC bias V_ϕ of the MZ modulator was set to 0.9 V to ensure a symmetric transfer function ($\phi = 0$). These parameters are summarized in Table 8.2.

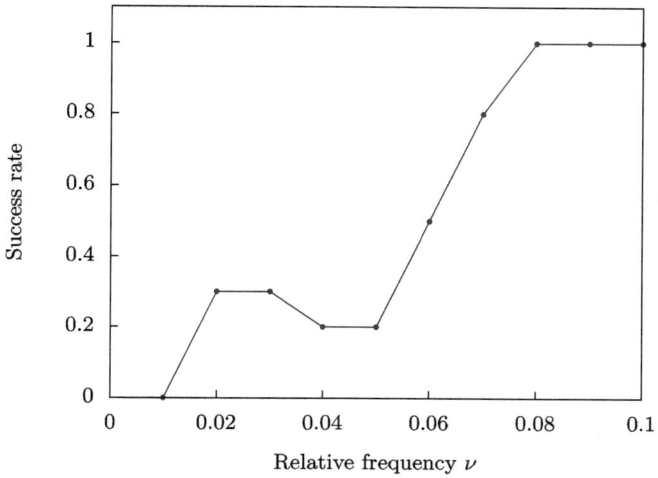

Figure 8.19: Verification of the lower limit of the reservoir computer bandwidth on the frequency generation task. Frequencies above 0.1 (not shown on the plot) are generated very well with any of the 10 random input mask. Frequencies below 0.05 fail with most input masks. We thus consider 0.08 as the lower limit of the bandwidth, but keep in mind that frequencies as low as 0.02 could also be generated, but only with a carefully picked input mask. Reprinted with permission from [19].

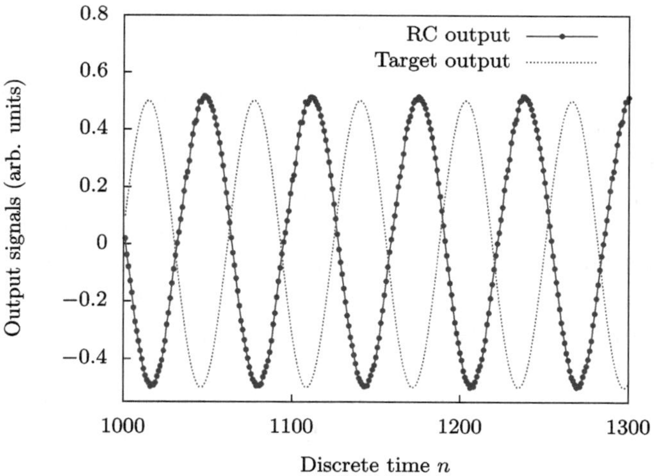

Figure 8.20: Example of an autonomous run output signal for frequency generation task with $\nu = 0.1$. The experiment continues beyond the range of the figure. Reprinted with permission from [19].

Table 8.2: Optimal experimental parameters for the benchmark tasks.

	α (dB)	β	V_ϕ (V)
Frequency generation	4.25–5.25	0.02–0.5	0.9
Pattern generation	4.25–5.25	0.1–1	0.9
Mackey–Glass series prediction	4.25–5.25	0.1–0.3	0.9
Lorenz series prediction	5.1	0.5	0.9

8.5.2.3 Random pattern generation

A natural step forward from the frequency generation task is random pattern generation. Instead of a regularly-shaped continuous function, the system is trained to generate an arbitrarily-shaped discontinuous periodic function. In this context, a pattern is a sequence of L randomly chosen real numbers (here within the interval $[-0.5, 0.5]$) that is repeated periodically to form an infinite periodic time series [49]. Similar to the physical frequency in Section 8.5.2.2, the physical period of the pattern is given by $\tau_{\text{pattern}} = L \cdot T$. The aim is to obtain a stable pattern generator, that reproduces precisely the pattern and does not deviate to another periodic behavior. To evaluate the performance of the RC, we compute the MSE (see equation (31)) between the reservoir output signal and the target pattern signal during both the training phase and the autonomous run, and arbitrarily set the maximal threshold to 10^{-3}. As will be illustrated in Figure 8.24, the 10^{-3} level corresponds to the point where the RC strongly deviates from the target signal. For consistency, we have used this threshold in all our experiments, for all tasks. If the error does not grow above the threshold during the autonomous run, the system is considered to accurately generate the target pattern. We also tested the long-term stability on several patterns by running the system for several hours.

The random pattern generation task is more complex than frequency generation and is slightly affected by the experimental noise. The goal of this task is two-fold: "remember" a pattern of a given length L and reproduce it for an unlimited duration. We have shown numerically that a noiseless reservoir with $N = 51$ neurons is capable of generating patterns up to 51-element long [49]. This is a logical result, as, intuitively, each neuron of the system is expected to "memorize" one value of the pattern. Simulations of a noisy reservoir with $N = 100$ neurons, similar to the experimental setup, show that the maximum pattern length is reduced down to $L = 13$. That is, the noise significantly reduces the effective memory of the system. In fact, the noisy neural network has to take into account the slight deviations of the output from the target pattern so as to be able to follow the pattern disregarding these imperfections. Figure 8.21 illustrates the manifestation of noise. Periodic oscillations of one neuron of the reservoir are shown, with intended focus on the upper values and an adequate magnification so as to see the small variations. The plot shows that the neuron oscillates between similar, but not identical values. This makes the generation task much more complex and requires more memory, which, in turn, shortens the maximal pattern length.

We obtained similar results in the experiments. Figure 8.22 shows the evolution of the MSE measured during the first 1000 timesteps of 10^4-timestep autonomous runs with different pattern lengths. Plotted curves are averaged over 100 runs of the experiment, with 5 random input masks and 20 random patterns for each length. The initial minimum (at $n = 128$) corresponds to the initialization of the reservoir (see Section 8.5.1), then the output is coupled back and the system runs autonomously. Patterns with $L = 12$ or less are generated very well and the error stays low. Patterns

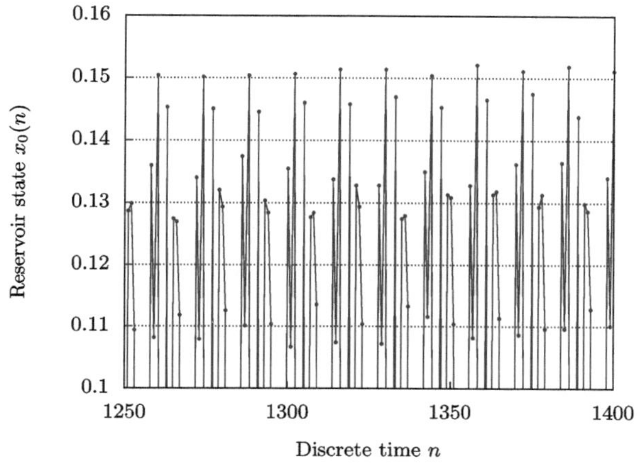

Figure 8.21: Example of behavior of one neuron in a noisy experimental reservoir. For clarity, the range of the Y axis is limited to the area of interest. Because of noise, despite a periodic input signal $u(n)$, the reservoir state takes similar, but not identical values. Reprinted with permission from [19].

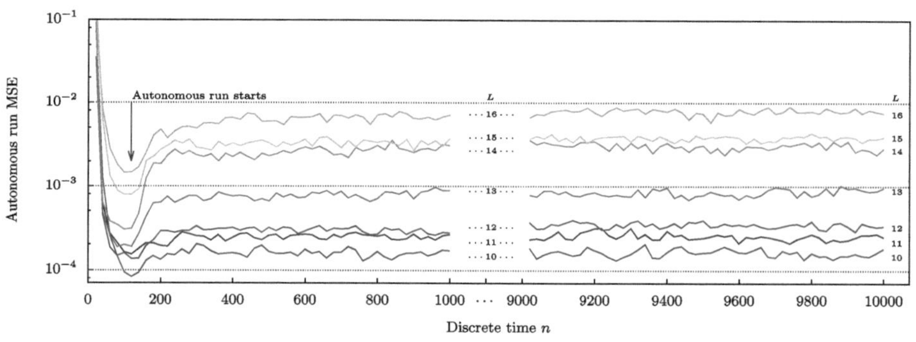

Figure 8.22: Evolution of MSE(n) during experimental autonomous generation of periodic random patterns of lengths $L = 10, \ldots, 16$. The autonomous run starts at $n = 128$, as indicated by the arrow. Patterns shorter than 13 are reproduced with low MSE $< 10^{-3}$. Patterns longer than 14 are not generated correctly with MSE $> 10^{-3}$. In the latter case, the reservoir dynamics remains stable and periodic, but the output only remotely resembles the target pattern (see Figure 8.23 and Figure 8.24 for illustration). Reprinted with permission from [19].

of length 13 show an increase in MSE, but they are still generated reasonably well. For longer patterns, the system deviates to a different periodic behavior, and the error grows above 10^{-3}.

Figure 8.23 shows an example of the output signal during the autonomous run. The system was trained for 1000 timesteps to generate a pattern of length 10. The reservoir computer successfully learned the desired pattern and the output accurately matches the target signal. Figure 8.24 illustrates a case with a longer pattern ($L = 14$), that could not be learned by the system. As can be seen from the plot, the RC captured the general shape of the pattern, but cannot accurately generate individual points. The MSE of this run is 5.2×10^{-3}, which is above the acceptable 10^{-3} threshold.

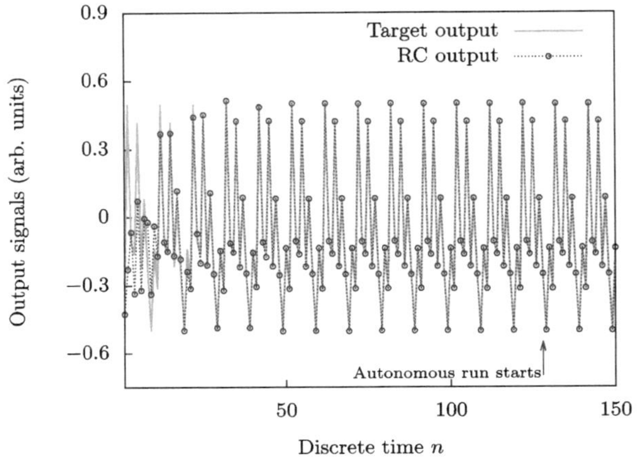

Figure 8.23: Example of an output signal for random pattern generation task, with a pattern of length 10. The reservoir is first driven by the desired signal for 128 timesteps (see Section 8.5.1), and then the input is connected to the output. Note that in this example the reservoir output requires about 50 timesteps to match the driver signal. The autonomous run continues beyond the range of the figure. Reprinted with permission from [19].

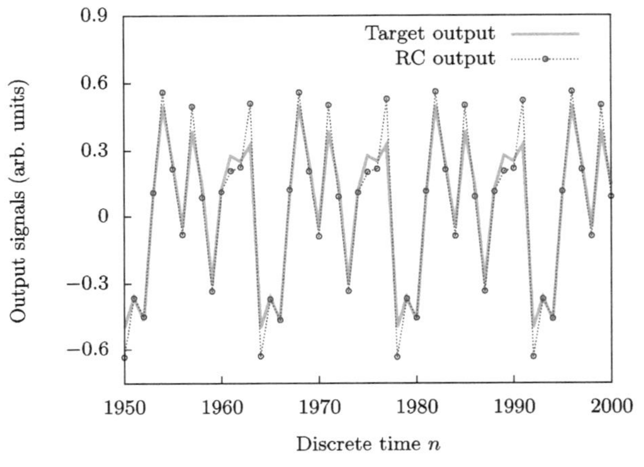

Figure 8.24: Example of an autonomous run output after 1950 timesteps, with a pattern of length $L = 14$. The RC outputs a periodic signal that clearly does not match the target pattern ($\text{MSE} = 5.2 \times 10^{-3}$). Reprinted with permission from [19].

We also tested the stability of the generator by running it for several hours (roughly 10^9 timesteps) with random patterns of lengths 10, 11, and 12. The output signal, observed on a scope, remained stable and accurate through the whole test.

The above results were obtained by scanning the input gain β and the feedback gain α to obtain the best results. As for frequency generation, it was found that β has little impact on the system performance so long as it is chosen in the interval $\beta \in [0.1, 1]$, while the feedback gain α, on the contrary, has to lie within a narrow interval of $\alpha \in [4.25, 5.25]$ dB (this corresponds approximately to $\alpha \in [0.85, 0.95]$). The DC bias of the MZ modulator was set to $V_\phi = 0.9$ V to ensure a symmetric transfer function ($\phi = 0$). These parameters are summarized in Table 8.2.

Since the noise plays such an important role, we performed a series of numerical experiments with different levels of noise to find out to what extent it affects the performance of the computer (these results have been published in [62]). We used the noisy

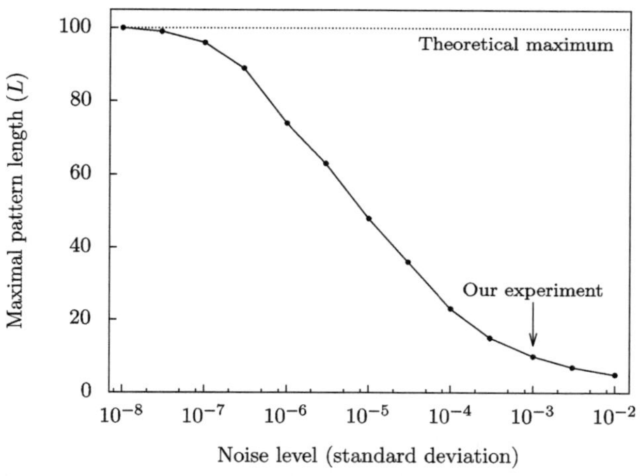

Figure 8.25: Impact of experimental noise on the performance of a reservoir computer with output feedback. The graph presents numerical results obtained with an accurate model of the experimental setup. Noise levels are shown as standard deviations of the Gaussian noise used in the simulations. The system was tested on the random pattern generation task and the performance metric is the maximal length L of a pattern that the reservoir could generate. The theoretical maximum is $L = 100$, since we used a reservoir with $N = 100$ neurons. Noise levels of 10^{-8} and below are equivalent to an ideal noiseless system. The arrow indicates the experimental results discussed here. Reprinted with permission from [62].

model of the experiment with Gaussian white noise with zero mean and standard deviations ranging from 10^{-2} to as low as 10^{-8}. These simulations allow to estimate the expected performance of the experiment for different levels of noise.

Figure 8.25 shows the maximum pattern length L that the reservoir computer is able to generate for different levels of noise. The maximal length is determined using the 10^{-3} autonomous error threshold. That is, if the NMSE does not grow above 10^{-3} during the autonomous run, the reservoir computer is considered to have successfully generated the given pattern. For statistical purposes, we used 10 different random patterns for each length L, and only counted the cases where the system have succeeded in all 10 trials. The results show that the noise level of 10^{-8} is equivalent to an ideal noiseless reservoir. As the noise level increases, the memory capacity of the reservoir deteriorates. At a level of 10^{-3}, the maximum pattern length is decreased down to 10, which matches the experimental results presented here. For higher noise levels, the results are, obviously, even worse.

Overall, these results show what level of noise one should aim for in order to obtain a certain performance from an experimental reservoir computer with output feedback. Our experiments have confirmed the numerical results for the noise level of 10^{-3}. In principle, one could double the maximal pattern length by carefully rebuilding the same experiment with low-noise components, namely a weaker amplifier and a low-V_π intensity modulator, which would lower the noise to 10^{-4}. Switching to a passive setup, such as the coherently driven cavity reported in [38], could potentially lower the noise down to 10^{-5} or even 10^{-6}, with performance approaching the maximum memory capacity.

8.5.2.4 Mackey–Glass series prediction

The Mackey–Glass delay differential equation

$$\frac{dx}{dt} = \beta \frac{x(t-\tau)}{1 + x^n(t-\tau)} - \gamma x, \tag{34}$$

with $\tau, \gamma, \beta, n > 0$ was introduced to illustrate the appearance of complex dynamics in physiological control systems [58]. To obtain chaotic dynamics, we set the parameters as in [14]: $\beta = 0.2$, $\gamma = 0.1$, $\tau = 17$, and $n = 10$. The equation was solved using the Runge–Kutta 4 method [63] with a stepsize of 1.0. To avoid repeating computations, we pre-generated a sequence of 10^6 samples that we used for all numerical and experimental investigations. The MSE is used to evaluate both the training phase and the autonomous run. During the latter, the system does not receive the correct teacher signal anymore, and thus slowly deviates from the desired trajectory. Therefore, we compute the number of correct prediction steps, i. e., steps for which the MSE stays below the 10^{-3} threshold (see Section 8.5.2.3), and use this figure to evaluate the performance of the system.

Chaotic time series generation tasks were the most affected by the experimental noise. This is not surprising, since, by definition, chaotic systems are very sensitive to initial conditions, which are affected by noise. Reservoir computing was first applied to this class of tasks in [14]. The authors numerically investigated the capacity of the computer to follow a given trajectory in the phase space of the chaotic attractor. We also followed this approach at first, but since our experimental system performs as a "noisy" emulator of the chaotic attractor, its trajectory deviates very quickly from the target one, especially with a SNR as low as 40 dB (see Section 8.5.2.1). For this reason, we considered alternative methods to evaluate the performance of the system, as will be described below.

The system was trained over 1500 input samples and was running autonomously for 600 timesteps. In particular, we prepared 2100 steps of the Mackey–Glass series for each run of the experiment and used the first 1500 as a teacher signal $u(n)$ to train the system and the last 600 both as an initialization sequence (see Section 8.5.1) and as a target signal $d(n)$ to compute the MSE of the output signal $y(n)$. These 2100 samples were taken from several starting points t (see equation (34)) in order to test the reservoir computer on different instances of the Mackey–Glass series. We scanned the input gain and the feedback attenuation (β and α in equation (4) from Section 8.2) to find optimal dynamics of the optoelectronic reservoir for this task. We used $\beta \in [0.1, 0.3]$

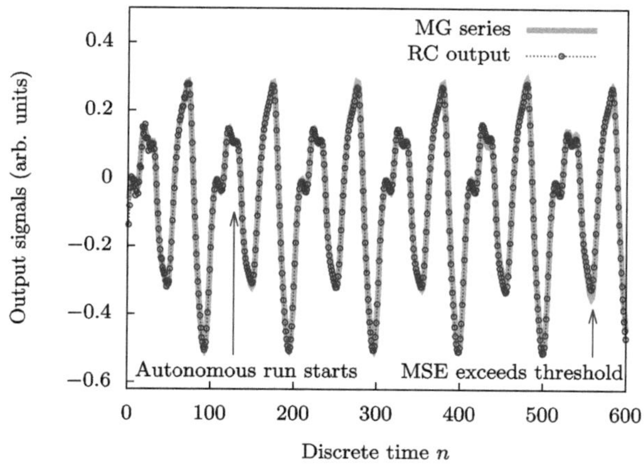

Figure 8.26: Example of reservoir computer output signal $y(n)$ (dotted black line) during autonomous run on the Mackey–Glass task. The system was driven by the target signal (solid grey line) for 128 timesteps and then left running autonomously, with $y(n)$ coupled to the input $I(n)$ (see Section 8.5.1). The MSE threshold was set to 10^{-3}. The photonic reservoir computer with $N = 600$ was able to generate up to 435 correct values. Reprinted with permission from [19].

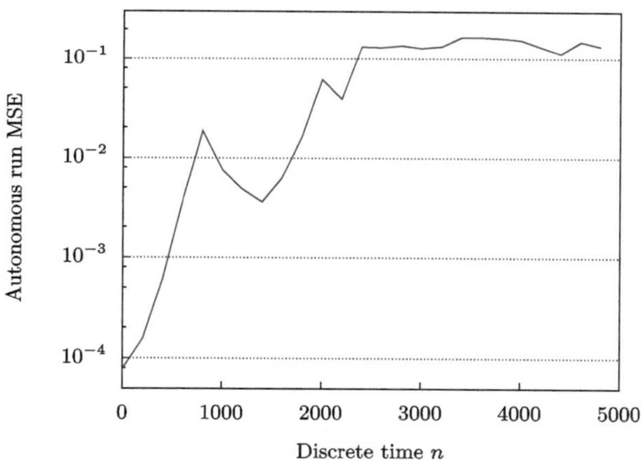

Figure 8.27: Evolution of MSE during experimental autonomous generation of the Mackey–Glass chaotic time series (same run as in Figure 8.26). The error curve, averaged over 200 timesteps, and crosses the 10^{-3} threshold approximately between $n = 500$ and $n = 600$. Reprinted with permission from [19].

and tuned the optical attenuator in the range [4.25, 5.25] dB, which corresponds approximately to $\alpha \in [0.85, 0.95]$, with slightly different values for different instances of the Mackey–Glass series. The DC bias of the MZ modulator was set to $V_\phi = 0.9\,\mathrm{V}$ to ensure a symmetric transfer function ($\phi = 0$). These parameters are summarized in Table 8.2.

Figure 8.26 shows an example of the reservoir output $y(n)$ (dotted black line) during the autonomous run. The target Mackey–Glass series is shown in grey. The MSE threshold was set to 10^{-3} and the reservoir computer predicted 435 correct values in this example. Figure 8.27 displays the evolution of the MSE recorded during the same autonomous run. The plotted error curve was averaged over 200-timestep intervals. It exceeds the 10^{-3} threshold within $n \in [500, 600]$ and reaches a constant value of approximately 1.1×10^{-1} after 2500 timesteps. At this point, the generated time series is completely off the target (see Figure 8.28 for illustration).

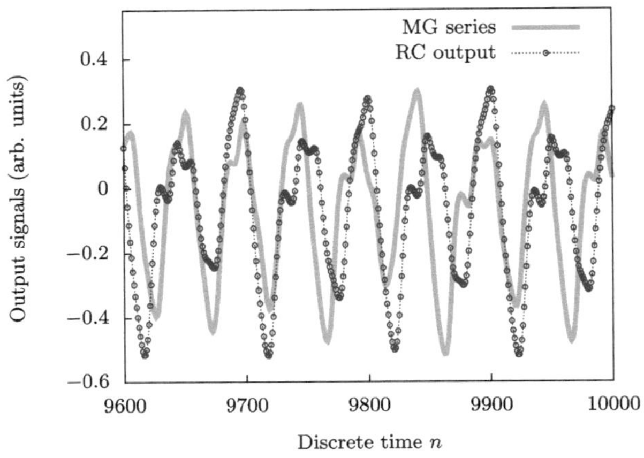

Figure 8.28: Output of the experimental reservoir computer (dotted black line) at the end of a long run of 10^4 timesteps. Although the system does not follow the starting trajectory (solid grey line), its output still resembles visually the target time series. Reprinted with permission from [19].

The noise inside the optoelectronic reservoir, discussed in Section 8.5.2.1, makes the outcome of an experiment inconsistent. That is, repeating the experiment with same parameters may result in significantly different prediction lengths. In fact, the impact of noise varies from one trial to another. In some cases it does not disturb the system much. But in most cases it induces a significant error on the output value $y(n)$, so that the neural network deviates from the target trajectory. To estimate the variability of the results, we performed 50 consecutive autonomous runs with the same readout weights and the same optimal experimental parameters. The system produced several very good predictions (of order of 400), but most of the outcomes were rather poor, with an average prediction length of 63.7 and a standard deviation of 65.2. We obtained similar behavior with the noisy experimental model, using the same level of noise as in the experiments. This suggests that the reservoir computer emulates a "noisy" Mackey–Glass system and, therefore, using it to follow a given trajectory does not make much sense with such a high noise level. Nevertheless, the noise does not prevent the system from emulating the Mackey–Glass system—even if the output quickly deviates from the target, it still resembles the original time series. Therefore, we tried a few distinct methods of comparing the output of the system with the target time series.

We performed a new set of experiments, where, after a training phase of 1500 timesteps, the system was running autonomously for 10^4 timesteps in order to collect enough points for data analysis. We then proceeded with a simple visual inspection of the generated time series, to check whether it still looks similar to the Mackey–Glass time series, and does not settle down to simple periodic oscillations. Figure 8.28 shows the output of the experimental reservoir computer at the end of the 10^4-timestep autonomous run. It shows that the reservoir output is still similar to the target time series, that is, irregular and consisting of the same kind of uneven oscillations.

A more thorough way of comparing two time series that "look similar" is to compare their frequency spectra. Figure 8.29 shows the Fast Fourier Transforms of the original Mackey–Glass series (solid grey line) and the output of the experiment af-

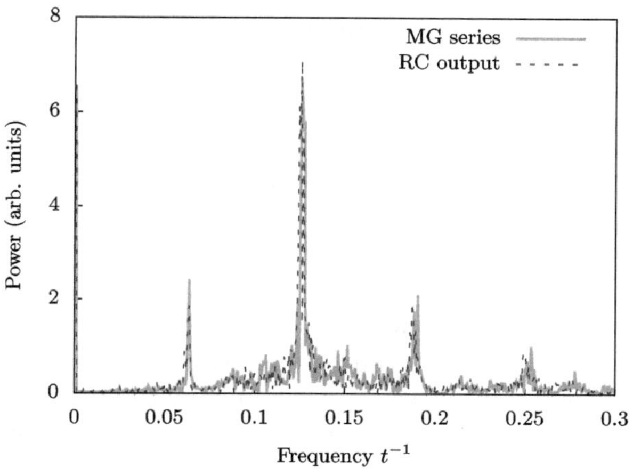

Figure 8.29: Comparison of Fast Fourier Transforms of the original Mackey–Glass series (solid grey line) and the time series generated by the photonic reservoir computer (dashed black line). The plot is limited to low frequencies as the power at higher frequencies is almost null. Dominant frequencies correspond to multiples of $1/\tau \approx 0.06$. The experiment reproduces the target spectrum notably well. Reprinted with permission from [19].

ter a long run (dotted black line). Remarkably, the reservoir computer reproduces very accurately the spectrum of the chaotic time series, with its main frequency and several secondary frequencies.

Finally, we estimated the Lyapunov exponent of the generated time series, using the method described in the Supplementary Material of [14]. We obtained 0.01 for our experimental implementation. For the Mackey–Glass system considered here, the value commonly found in the literature is 0.006. The slightly higher value of the Lyapunov exponent may simply reflect the presence of noise in the emulator.

8.5.2.5 Lorenz series prediction

The Lorenz equations, a system of three ordinary differential equations

$$\frac{dx}{dt} = \sigma(y - x), \tag{35a}$$

$$\frac{dy}{dt} = -xz + rx - y, \tag{35b}$$

$$\frac{dz}{dt} = xy - bz, \tag{35c}$$

with $\sigma, r, b > 0$, was introduced as a simple model for atmospheric convection [59]. The system exhibits chaotic behavior for $\sigma = 10$, $b = 8/3$, and $r = 28$ [64] that we used in this study. These parameters yield a chaotic attractor with the highest Lyapunov exponent of $\lambda = 0.906$ [14]. The system was solved using Matlab's `ode45` solver and a stepsize of 0.02, as in [14]. We used all computed points, meaning that one timestep of the reservoir computer corresponds to a step of 0.02 in the Lorenz time scale. To avoid unnecessary computations and save time, we pregenerated a sequence of 10^5 samples that we used for all numerical and experimental investigations. Following

[14], we used the x-coordinate trajectory for training and testing, that we scaled by a factor of 0.01.

This task was investigated in a similar way to the Mackey–Glass. The reservoir computer was trained over 3000 input samples and ran autonomously for 1000 timesteps. The 4000 samples were taken after discarding the initial transients of 1000 timesteps, as in [14]. For optimal performance of the optoelectronic reservoir, we set the input gain to $\beta = 0.5$ and the feedback attenuation to $\alpha = 5.1$ dB. The DC bias of the MZ modulator was set to $V_\phi = 0.9$ V to ensure a symmetric transfer function ($\phi = 0$). These parameters are summarized in Table 8.2.

Figure 8.30 shows an example of the reservoir output $y(n)$ (dotted black line) during the autonomous run. The target Lorenz series is shown in grey. With the MSE threshold set to 10^{-3}, the system predicted 122 correct steps, including two transitions between the wings of the attractor. As in the Mackey–Glass study, we performed 50 autonomous runs with identical parameters and readout weights and obtained an average prediction horizon of 46.0 timesteps with a standard deviation of 19.5. Taking into account the higher degree of chaos of the Lorenz attractor, and given the same problems related to noise, it is hard to expect a better performance of the reservoir computer at following the target trajectory. Figure 8.31 depicts the evolution of the MSE during the autonomous run. The error curve was averaged over 100-timestep intervals. The initial dip corresponds to the teacher-forcing of the reservoir computer with the target signal for 128 timesteps, as discussed in Section 8.5.1. The error exceeds

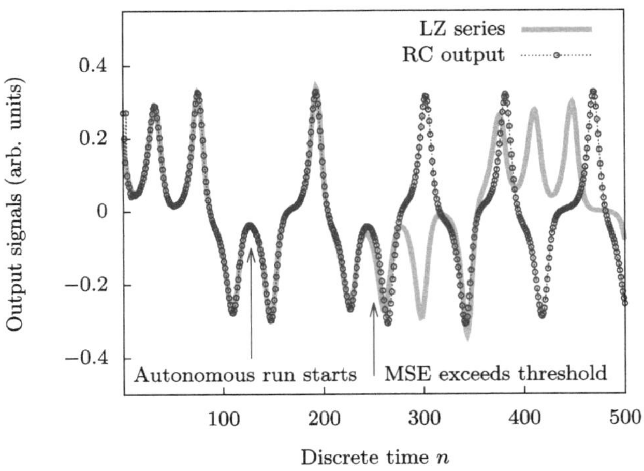

Figure 8.30: Example of reservoir computer output signal $y(n)$ (dotted black line) during autonomous runs on the Lorenz task. The system was driven by the target signal (solid grey line) for 128 timesteps before running autonomously (see Section 8.5.1). The MSE threshold was set to 10^{-3}. The photonic system with $N = 600$ generated 122 correct values in this example, and predicted two switches of the trajectory from one lobe of the attractor to the other. Reprinted with permission from [19].

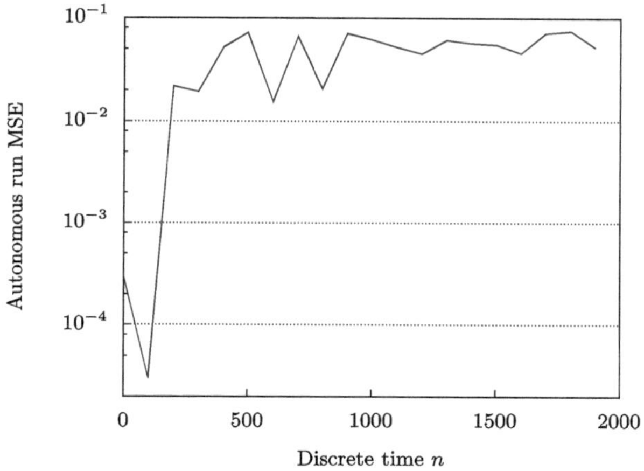

Figure 8.31: Evolution of MSE during experimental autonomous generation of the Lorenz chaotic time series (same run as in Figure 8.30). The error curve, averaged over 100 timesteps, crosses the 10^{-3} threshold near $n = 250$. The initial dip corresponds to the warm-up of the reservoir (see Section 8.5.1). Reprinted with permission from [19].

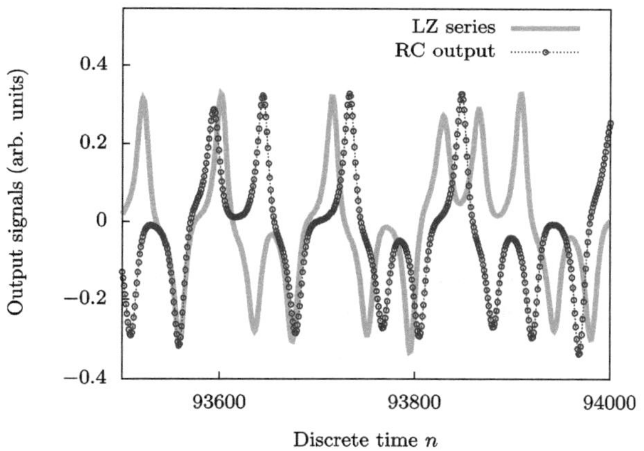

Figure 8.32: Output of the experiment (dotted black line) at the end of a long run of 95000 timesteps on the Lorenz task. Although the system does not follow the starting trajectory (solid grey line), it does a fairly good job at emulating the dynamics of the Lorenz system. Reprinted with permission from [19].

the 10^{-3} threshold around the $n = 250$ mark and reaches a constant value of approximately 1.5×10^{-2} after less than 1000 timesteps. At this point, the reservoir computer has lost the target trajectory, but keeps on generating a time series with properties similar to the Lorenz series (see Figure 8.32 for illustration).

Similar to the Mackey–Glass task, we performed a visual inspection of the generated Lorenz series after a long run, and compared the frequency spectra. Figure 8.32 shows the output of the experiment near the end of a 95000 autonomous run. Although the system is quite far from the target trajectory (plotted in grey) at this point, it is apparent that it has captured the dynamics of the Lorenz system very well. Figure 8.33 displays the Fast Fourier Transforms of the generated time series (dotted black line) and the computed Lorenz series (solid grey line). Unlike the Mackey–Glass system, these frequency spectra do not have any dominant frequencies. That is, the power

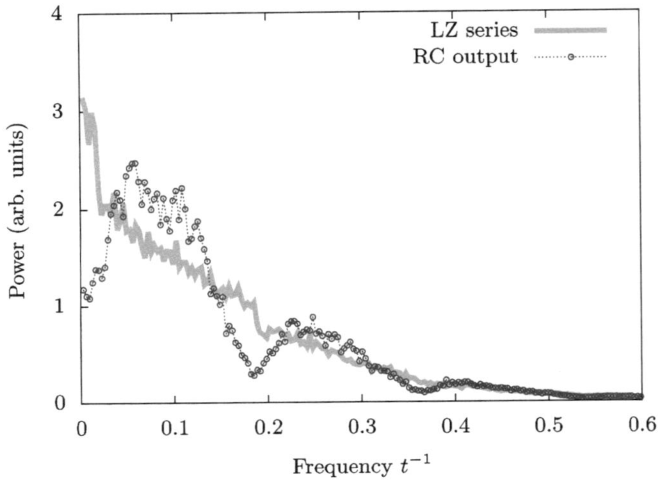

Figure 8.33: Comparison of Fast Fourier Transforms of the Lorenz series (solid grey line) and the time series generated by the photonic reservoir computer (dotted black line) during 95000 timesteps. Both spectra are normalized so as to have equal total power. The curves are smoothened by averaging over 50 samples and the plot is limited to lower frequencies (the higher ones being close to zero). Despite some mismatch, the shape of the dotted curve is roughly similar to the grey line. Reprinted with permission from [19].

distribution does not contain any strong peaks, that could have been used as reference points for comparison. Therefore, comparing the two spectra is much more subjective in this case. Although the curves do not match, one can still see a certain similarity between them.

In addition to those visual comparisons, we performed a specific randomness test of the generated series. We exploited an interesting property of the Lorenz dynamics. Since it basically switches between two regions (the wings of the butterfly-like Lorenz attractor), with random transitions from one to the other, one can assign binary "0" and "1" to these regions and thus transform the Lorenz series into a sequence of random bits. We used this technique to check the randomness of the generated series. To this end, we both solved the Lorenz equation and ran the experiment for 95000 timesteps, and converted the resulting time series into two sequences of approximately 2400 bits. The two were then analyzed with the ENT program [65]—a well-known software for testing random number sequences—with the results shown in Table 8.3. Their interpretation requires a few explanations.

- The first test computes the entropy per byte (8 bits). Since the entropy can be seen as a measure of disorder or randomness, a totally random sequence should have 8 bits of entropy per byte. Both sequences are close to the maximum value, with the Lorenz series being slightly more random.

Table 8.3: Results returned by the ENT program for the bit sequences generated by the experiment (RC) and the integrated Lorenz system. The Lorenz sequence shows better figures, but the RC output is not far behind. All these figures are poor compared to common random series, but this is due to the very short sequences used here (roughly 300 bytes).

	RC	Lorenz
Entropy (byte)	6.6	7.1
Compression (%)	17	10
Mean (byte)	134.3	125.8
π	2.88	3.00
Correlation	−0.08	−0.02

- The compression, i. e., how efficiently a sequence of bytes could be reduced in size by a compression algorithm, such as, e. g., the Lempel–Ziv–Renau algorithm, used by the Zip program, is a commonly used indirect method of estimating the randomness of bytes in a file. These algorithms basically look for large repeating blocks, that should not appear in a totally random sequence. Again, both sequences could only be slightly compressed.
- The mean value is the arithmetic mean of the data bytes. A random sequence should be evenly distributed around the mean value of 127.5. The Lorenz series is very close to this value, and the RC sequence is fairly close.
- The Monte Carlo method of computing the value of π randomly places points inside a square and computes the ratio of points located inside an inscribed circle, that is proportional to π. This complex test requires a long sequence of bytes to yield accurate results. We note that, nevertheless, both sequences produce a plausible estimation of π.
- Finally, the serial correlation, i. e., the degree of similarity between the sequence and a delayed copy of itself, in a totally random sequence is zero. Both series present a very low correlation, yet again the Lorenz series demonstrating a better score.

These results do not strictly prove that the generated sequence is random. One obviously has to use a much longer sequence of bits for that task, and should also consider more sophisticated and complete tests, such as Diehard [66] or NIST Statistical Test Suite [67]. The purpose of these tests was to show that the output of the RC generator does not consist of trivial oscillations, that only remotely resemble the Lorenz system. The figures in Table 8.3 show that the randomness of the RC output is similar to the Lorenz system, which gives reasons to believe in the similarity between the properties of the two time series. This, in turn, indicates that our photonic reservoir computer was capable of learning to effectively emulate the dynamics of the Lorenz chaotic system.

8.6 Conclusion

Most experimental works on reservoir computing so far have focused on demonstrating the basic concept. However, as we have discussed in this chapter, it is possible to go beyond this and start making a full information processing system. We have discussed three such advances. First, we showed that it is possible to build analogue input and output layers, with the aim of making the reservoir computer an autonomous system. Second, we discussed the possibility of modifying the output layer in real time, which allows the reservoir computer to cope with drifts in the experimental parameters and with tasks which change in time. Third, we presented a reservoir computer with output feedback which allowed it to solve new tasks such as generation of periodic patterns and emulation of chaotic systems.

One important direction which we did not have the time to discuss here concerns the optimization of the internal parameters of the reservoir. Indeed, in the traditional reservoir computing architecture the internal parameters are not modified: they are (often) simply chosen at random, except, possibly, for a few meta parameters. But it is known that if more parameters are optimized, the reservoir performance can significantly improve. This has been shown in [29, 68]. Remarkably, in certain cases the algorithm used to train the internal parameters can be implemented physically on the reservoir that is being optimized. This was demonstrated in a simple case in [69], and then on a more complex optoelectronic system in [34] where the same level of improvement as in [29, 68] was reached.

An exciting challenge for the future would be to combine the different advances discussed in this chapter. For instance, can one build a reservoir computer with analogue input and output layers which also allows for real time optimization of the output layer and/or output feedback? To this end, it would be useful to develop robust and easy to implement analogue input and output layers, which could then be used as plug-and-play building blocks in these complex systems. Another exciting challenge would be to demonstrate a reservoir computer that processes, in real time, real-world analogue data, such as the output signal of a telecommunication line, and produces a high-quality output ready for further use. Hopefully, in such a demonstration one would also exhibit a gain in energy consumption and/or footprint compared to traditional digital systems that solve the same task. Such a demonstration would pave the way toward concrete applications of reservoir computers.

Thinking of reservoir computers from a systems point of view raises many new questions and challenges. Whether or not significant progress can be made in this direction will determine if reservoir computing stays a laboratory curiosity or can move into the real world. The present chapter illustrates some of the first steps made in this direction.

Bibliography

[1] R. Martinenghi et al. "Photonic nonlinear transient computing with multiple-delay wavelength dynamics". In: *Physical Review Letters* 108 (2012), p. 244101.

[2] L. Appeltant et al. "Information processing using a single dynamical node as complex system." In: *Nature Communications* 2 (2011), p. 468.

[3] L. Larger et al. "Photonic information processing beyond Turing: an optoelectronic implementation of reservoir computing". In: *Optics Express* 20 (2012), p. 3241.

[4] Y. Paquot et al. "Optoelectronic reservoir computing". In: *Scientific Reports* 2 (2012), p. 287.

[5] F. Duport et al. "All-optical reservoir computing". In: *Optics Express* 20 (2012), p. 22783.

[6] D. Brunner et al. "Parallel photonic information processing at gigabyte per second data rates using transient states". In: *Nature Communications* 4 (2013), p. 1364.

[7] J. Torrejon et al. "Neuromorphic computing with nanoscale spintronic oscillators". In: *Nature* 547 (2017), p. 428.

[8] G. Dion, S. Mejaourdi, and J. Sylvestre. "Reservoir computing with a single delay-coupled non-linear mechanical oscillator". In: *Journal of Applied Physics* 124 (2018), p. 152132.

[9] K. Vandoorne et al. "Experimental demonstration of reservoir computing on a silicon photonics chip". In: *Nature Communications* 5 (2014), p. 3541.

[10] F. Duport et al. "Fully analogue photonic reservoir computer". In: *Scientific Reports* 6 (2016), p. 22381.

[11] P. Antonik et al. "Online training of an opto-electronic reservoir computer applied to real-time channel equalization". In: *IEEE Transactions on Neural Networks and Learning Systems* 28 (Nov. 2017), p. 2686.

[12] P. Antonik, M. Haelterman, and S. Massar. "Online training for high-performance analogue readout layers in photonic reservoir computers". In: *Cognitive Computation* 9 (2017), p. 297.

[13] J. Bueno et al. "Reinforcement learning in a large scale photonic recurrent neural network". In: *Optica* 5 (2018), p. 756.

[14] H. Jaeger and H. Haas. "Harnessing nonlinearity: predicting chaotic systems and saving energy in wireless communication". In: *Science* 304 (2004), p. 78.

[15] Z. Lu et al. "Reservoir observers: model-free inference of unmeasured variables in chaotic systems". In: *Chaos: An Interdisciplinary Journal of Nonlinear Science* 27 (2017), p. 41102.

[16] J. Pathak et al. "Using machine learning to replicate chaotic attractors and calculate Lyapunov exponents from data". In: *Chaos: An Interdisciplinary Journal of Nonlinear Science* 27 (2017), p. 121102.

[17] J. Pathak et al. "Model-free prediction of large spatiotemporally chaotic systems from data: a reservoir computing approach". In: *Physical Review Letters* 120 (2018), p. 24102.

[18] P. Antonik et al. "Using a reservoir computer to learn chaotic attractors, with applications to chaos synchronisation and cryptography". In: *Physical Review E* 98 (2018), p. 012215.

[19] P. Antonik, M. Haelterman, and S. Massar. "Brain-inspired photonic signal processor for generating periodic patterns and emulating chaotic systems". In: *Physical Review Applied* 7 (2017), p. 54014.

[20] H. Jaeger. "The "echo state" approach to analysing and training recurrent neural networks – with an erratum note". In: *GMD Report* 148 (2001).

[21] P. Antonik. *Application of FPGA to real-time machine learning: hardware reservoir computers and software image processing*. Springer, 2018.

[22] M. C. Soriano et al. "Minimal approach to neuro-inspired information processing". In: *Frontiers in Computational Neuroscience* 9 (2015), p. 68.

[23] L. Larger et al. "From flow to map in an experimental high-dimensional electrooptic nonlinear delay oscillator". In: *Physical Review Letters* 95 (2005), p. 43903.

[24] Y. C. Kouomou et al. "Chaotic breathers in delayed electro-optical systems". In: *Physical Review Letters* 95 (2005), p. 203903.

[25] M. Peil et al. "Routes to chaos and multiple time scale dynamics in broadband bandpass nonlinear delay electro-optic oscillators". In: *Physical Review E* 79 (2009), p. 026208.

[26] A. Smerieri et al. "Analog readout for optical reservoir computers". In: *Advances in Neural Information Processing Systems*. 2012, p. 944.

[27] M. C. Soriano et al. "Optoelectronic reservoir computing: tackling noise-induced performance degradation". In: *Optics Express* 21 (2013), p. 12.

[28] M. C. Soriano et al. "Delay-based reservoir computing: noise effects in a combined analog and digital implementation". In: *IEEE Transactions on Neural Networks and Learning Systems* 26 (2015), p. 388.

[29] M. Hermans et al. "Photonic delay systems as machine learning implementations". In: *Journal of Machine Learning Research* 16 (2015), p. 2081.

[30] F. Duport et al. "Analogue input layer for optical reservoir computers". In: *arXiv preprint* arXiv:1406.3238 (2014).

[31] D. Sussillo and L. F. Abbott. "Generating coherent patterns of activity from chaotic neural networks". In: *Neuron* 63 (2009), p. 544.

[32] H. Jaeger et al. "Optimization and applications of echo state networks with leaky-integrator neurons". In: *Neural Networks* 20 (2007), p. 335.

[33] L. Appeltant et al. "Constructing optimized binary masks for reservoir computing with delay systems". In: *Scientific Reports* 4 (2014), p. 3629.

[34] M. Hermans et al. "Embodiment of learning in electro-optical signal processors". In: *Physical Review Letters* 117 (2016), p. 128301.

[35] A. N. Tikhonov et al. *Numerical methods for the solution of ill-posed problems*. Vol. 328. Springer Netherlands, 1995.

[36] A. Rodan and P. Tino. "Minimum complexity echo state network". In: *IEEE Transactions on Neural Networks* 22 (2011), p. 131.

[37] H. Jaeger. "Adaptive nonlinear system identification with echo state networks". In: *Advances in neural information processing systems*. Vol. 15. MIT Press, 2003, p. 593.

[38] Q. Vinckier et al. "High-performance photonic reservoir computer based on a coherently driven passive cavity". In: *Optica* 2 (2015), p. 438.

[39] Y. Xue, L. Yang, and S. Haykin. "Decoupled echo state networks with lateral inhibition". In: *Neural Networks* 20 (2007), p. 365.

[40] K. Vandoorne et al. "Toward optical signal processing using photonic reservoir computing". In: *Optics Express* 16 (2008), p. 11182.

[41] A. Dejonckheere et al. "All-optical reservoir computer based on saturation of absorption". In: *Optics Express* 22 (2014), p. 10868.

[42] L. Bottou. "Online algorithms and stochastic approximations". In: *Online learning and neural networks*. Cambridge University Press, 1998.

[43] S. Benedetto and E. Biglieri. *Principles of digital transmission: with wireless applications*. Springer Science & Business Media, 1999.

[44] J. Singh, S. Ponnuru, and U. Madhow. "Multi-gigabit communication: the ADC bottleneck". In: *Ultra-wideband, 2009. ICUWB 2009. IEEE international conference on*. IEEE. 2009, p. 22.

[45] S. Ruder. "An overview of gradient descent optimization algorithms". In: *arXiv preprint* arXiv:1609.04747 (2016).

[46] P. Antonik et al. "Online training of an opto-electronic reservoir computer". In: *APNNA's 22th international conference on neural information processing*. Vol. 9490. LNCS. 2015, p. 233.

[47] G. P. Zhang. "Neural networks for time-series forecasting". In: *Handbook of natural computing*. Ed. by G. Rozenberg, T. Back, and J. N. Kok. Springer Berlin Heidelberg, Berlin, Heidelberg, 2012, p. 461.

[48] F. Wyffels and B. Schrauwen. "A comparative study of reservoir computing strategies for monthly time series prediction". In: *Neurocomputing* 73 (2010), p. 1958.

[49] P. Antonik et al. "Towards pattern generation and chaotic series prediction with photonic reservoir computers". In: *SPIE's 2016 laser technology and industrial laser conference*. Vol. 9732. 2016, 97320B.

[50] M. Xu, M. Han, and S. Kanae. "L1/2 norm regularized echo state network for chaotic time series prediction". In: *APNNS's 23th international conference on neural information processing (ICONIP)*. Vol. 9886. LNCS. 2016, p. 12.

[51] The 2006/07 forecasting competition for neural networks & computational intelligence. URL: http://www.neural-forecasting-competition.com/NN3/. 2006.

[52] F. Wyffels, B. Schrauwen, and D. Stroobandt. "Stable output feedback in reservoir computing using ridge regression". In: *International conference on artificial neural networks*. Springer, 2008, p. 808.

[53] K. Caluwaerts et al. "Locomotion without a brain: physical reservoir computing in tensegrity structures". In: *Artificial Life* 19 (2013), p. 35.

[54] R. F. Reinhart and J. J. Steil. "Regularization and stability in reservoir networks with output feedback". In: *Neurocomputing* 90 (2012), p. 96.

[55] F. Wyffels et al. "Frequency modulation of large oscillatory neural networks". In: *Biological Cybernetics* 108 (2014), p. 145.

[56] P. Antonik et al. "Towards adjustable signal generation with photonic reservoir computers". In: *25th international conference on artificial neural networks*. Vol. 9886. 2016.

[57] H. Jaeger. "Echo state network". In: *Scholarpedia* 2 (2007), p. 2330.

[58] M. C. Mackey and L. Glass. "Oscillation and chaos in physiological control systems". In: *Science* 197 (1977), p. 287.

[59] E. N. Lorenz. "Deterministic nonperiodic flow". In: *Journal of the Atmospheric Sciences* 20 (1963), p. 130.

[60] P. Horowitz and W. Hill. *The art of electronics*. Cambridge University Press, 1980.

[61] A. V. Oppenheim and R. W. Schafer. *Discrete-time signal processing*. Prentice-Hall signal processing series. Prentice Hall, 1989.

[62] P. Antonik et al. "Random pattern and frequency generation using a photonic reservoir computer with output feedback". In: *Neural Processing Letters* 47 (2018), p. 1041.

[63] K. E. Atkinson. *An introduction to numerical analysis*. John Wiley & Sons, 2008.

[64] M. W. Hirsch, S. Smale, and R. L. Devaney. *Differential equations, dynamical systems, and an introduction to chaos*. Academic Press, Boston, MA, 2003.

[65] J. Walker. *ENT program*. URL: http://www.fourmilab.ch/random/.

[66] G. Marsaglia. *The Marsaglia random number CDROM including the Diehard battery of tests of randomness*. URL: http://stat.fsu.edu/pub/diehard/.

[67] A. Rukhin et al. "A statistical test suite for random and pseudorandom number generators for cryptographic applications". In: Technical report, National Institute of Standards and Technology, 2001.

[68] M. Hermans, J. Dambre, and P. Bienstman. "Optoelectronic systems trained with backpropagation through time". In: *IEEE Transactions on Neural Networks and Learning Systems* 26 (2015), p. 1545.

[69] M. Hermans et al. "Trainable hardware for dynamical computing using error backpropagation through physical media". In: *Nature Communications* 6 (2015), p. 6729.

Outlook

After the first demonstrations of reservoir computing (RC) in nonlinear physical substrates, the field has grown at an astonishing rate. From a distance, one could consider that demonstrating the feasibility of neural network computing with high-performance photonic components is the greatest achievement of photonic RC. Using mostly off-the-shelf devices as nonlinear neurons, recent results unlocked the potential of optical communication and integration technology for the next generation neural network processors. This development has pushed processing speeds that were previously out of reach and, furthermore, the field is identifying unique applications at an increasing pace. The interest into optical computing and photonic neural networks has been reinvigorated, notably due to the work reported in the preceding chapters.

The current success of photonic RC, as well as the considerable general interest, certainly benefited from breakthroughs achieved in neural network computing. This creates an excellent situation, almost unique in the history of optical computing. Breakthroughs are achieved in parallel in almost all relevant aspects: computing concepts, hardware substrates, and commercial applications. The result is a surge of novel ideas spanning various disciplines. Along the recent years, a relatively skeptical view on optical computing has been transformed into a rather euphoric search for the future computing substrate. As discussed in earlier sections of the book, it remains essential that criteria like energy-efficiency, practicality and scalability in all aspects are continuously considered. Otherwise, the development in this exciting new field might be cut-short by too short-term focused strategies. Yet, future prospects are outstanding due to the recent advances, which indicate strategies to overcome bottlenecks encountered in the past.

Using photonics, a better power efficiency at very high bandwidths than with electronic substrates can in principle be achieved. Now that several technologies for creating photonic and other types of physical reservoirs are starting to mature, the next step is to leverage their potential for appropriate industrial applications. However, it is equally clear that computations which can be performed based on a single reservoir are limited. In addition, most work on reservoir computing addresses tasks with a single input signal. This is a side effect of the upper bound on total computing capacity: if information needs to be extracted from multiple input channels, the system's capacity for remembering past inputs is reduced. Overcoming these limitations will at least require larger systems, also consisting of multiple reservoir layers. Although first steps have been taken, an efficient implementation and training strategy for creating such cascaded reservoir layers, especially if combined with multiple input channels, is still lacking. Overcoming this architectural bottleneck will be one of the focus of future efforts.

https://doi.org/10.1515/9783110583496-009

With respect to integrated photonic reservoir chips, one of the next challenges is moving from lab-based prototypes and toy benchmarks to fully integrated systems that can be deployed for industrially relevant applications. In this context, full-scale benchmarking and comparison with more traditional approaches will need to be carried out in terms of speed, power consumption, latency, and footprint. We see photonic reservoir computing making the most impact in applications where input information already is encoded in the optical domain. This demands extremely high information throughput, beyond anything which is easily achievable with current electronic-based infrastructure.

As it is often the case in the development of a novel technological platform, a detailed analysis with respect to all relevant processes is required. This is particularly true for the implementation of complex neural networks, which can be considered as ambiguous-by-design. Understanding the propagation and scaling of noise in hardware neural networks will be one of the crucial future endeavors. For this, the large scale spatiotemporal neural networks based on diffractive coupling are exceptionally well suited. Achievable network sizes easily reach the dimensions required by standard neural network applications, and the modular architecture enables detailed characterization of all components involved. Essential for this platform's future development is the demonstration of diffractive networks based on all-optical nonlinear substrates. First steps have been made with small networks of semiconductor lasers, but optical quality of the available laser-arrays needs to be considerably improved before large systems become realistic. Those, however, will then immediately unlock large-scale photonic neural networks operating at the rate of multiple GHz. Combining all-optical nonlinear substrates and the excellent scalability of the diffractive coupling concept, one can start considering extremely large and multilayer photonic neural networks.

In turn, the consideration of delay-based RC represents a further simplification of the hardware complexity for photonic RC. Following this approach, only a single nonlinear physical node and a delay feedback line are needed for the corresponding hardware implementations. Unsurprisingly, the first demonstrators of photonic RC were based on the delay-based approach. Furthermore, delay RC can profit of a delay-system generic property: unless intentionally broken, the implemented networks posses perfect symmetry. This creates exceptional conditions for investigating aspects of nonlinear dynamic systems and their impact on computational performance. As the field is advancing, it is likely that hybrid approaches combining some sort of time, frequency and spatial multiplexing will become commonplace. It is clear that there will be no *one size fits all* photonic RC system and it is encouraging to see that the different approaches presented in the previous chapters nicely complement each other.

Photonic delay dynamics have demonstrated their capability to process information according to RC concepts; moreover, with very attractive performances in terms of computational power, speed, and energy efficiency. Among the various research directions one could propose the following issues, spanning from fundamental to very applied perspectives. Much work is still to be performed in order to better understand computing mechanisms that can be obtained from nonlinear delayed feedback systems. Concerning the learning and the read-out layer, important breakthroughs might consist in a proper translation into a temporal signal processing scheme and address the problem on stabilizing delayed feedback patterns. Instead of employing digital computer-aided readout techniques, an in-line learning mechanism is still needed. Here, chimera states might have a role to play. Finally, a dynamical, iterative signal processing approach could lead to the emergence of the same read-out pattern as the one algorithmically obtained.

Most of the experimental systems developed so far for the investigation of delay-based RC, have essentially assumed a delay feedback architecture comprising a single nonlinear node with a single delay feedback loop. It is then easy to imagine slightly more complex delay architectures, exploring for example cascaded as well as parallel delay-based RC. Such advanced systems offer the possibility to combine multiple different filtering functions, and a more powerful RC-architecture can therefore be expected. Finally, since delay-based reservoirs are time-domain processors, it is obvious, and already has been revealed by several experiments, that timing issues are of key technical challenge. Correct timing and synchronized signal are very important aspects when optimizing performance of a delay RC. The underlying reasons are not yet fully understood, and delay-based RC could significantly progress if signal synchronization could be more clearly connected to fundamental reservoir concepts.

A semiconductor laser subject to optical feedback fulfills the requirements for a high-speed implementation of the reservoir in an elegant way. This single component creates a single-device, fast, and energy efficient all-optical nonlinearity. At the moment, it still remains unclear what the precise influence of the laser nonlinearity is in the computational performance. In this context, a precise experimental characterization of the amplitude and phase response of the semiconductor laser may shed some light on the role of the nonlinearity for the case of the optically driven semiconductor laser. Current trends for the advancement of photonic RC include the possibility to integrate most of the photonic components and to develop an all-optical implementation of the full system. Recent works suggest certain modifications to the original scheme that could still improve the performance of laser-based RC. These modifications include the use of two delay loops in order to extend the fading memory of the system, or the combination of responses for different laser parameters in order to enhance the computational power of the system.

Full delay-system RC implementations must also include the input and output layers, which for now remains a challenge due to the intricate time-multiplexing techniques needed. Most photonic RC experiments so far have focused on demonstrating the basic concept. Often only the reservoir was experimentally implemented as an analog system, and the rest was based on digital hardware. However, it is possible to go beyond such systems, and preliminary progress in this direction has been reported here. Indeed, a physical reservoir computer with analog input and output layers has been demonstrated. This includes the optimization of the output layer in real time, allowing the reservoir computer to cope with drifts in either experimental parameters as well as to address tasks which change in time. In addition, several works report how performance can be improved by optimization of the internal reservoir parameters. Finally, output feedback allows reservoir computers to solve new tasks such as generation of periodic patterns and emulation of chaotic systems.

Still, building a reservoir computer with analogue input and output layers which allow for real time optimization of the output layer remains one of the most important objectives ahead. This would be facilitated by the development of robust and easy to implement analogue input and output layers, preferably creating plug-and-play like building blocks for these complex systems.

A key challenge is to fully exploit the full potential of optics. Future reservoirs should leverage the coherence of light and optical parallelism, profit from the potentially very high speeds of optics, and to process optical inputs to create optical outputs. Progress along these lines has been realized, but much remains to be done. Ultimately, one would like to demonstrate a reservoir computer that processes, in real time, real-world data, such as the output signal of a telecommunication line, and produces a high-quality output ready for further use.

Index

https://doi.org/10.1515/9783110583496-010

Bei Fragen zur Produktsicherheit wenden Sie sich bitte an:
If you have any questions regarding product safety,
please contact:

Walter de Gruyter GmbH
Genthiner Straße 13
10785 Berlin
productsafety@degruyterbrill.com